Multiscale Mathematical Modeling

Multiscale Mathematical Modeling

Guest Editors

Julius Kaplunov
Igor Andrianov

MDPI

Basel • Beijing • Wuhan • Barcelona • Belgrade • Novi Sad • Cluj • Manchester

Guest Editors
Julius Kaplunov
School of Computer Science and Mathematics
Keele University
Keele
UK

Igor Andrianov
Chair and Institute of General Mechanics
RWTH Aachen University
Aachen
Germany

Editorial Office
MDPI AG
Grosspeteranlage 5
4052 Basel, Switzerland

This is a reprint of the Special Issue, published open access by the journal *Mathematics* (ISSN 2227-7390), freely accessible at: https://www.mdpi.com/journal/mathematics/special_issues/BAX40K5788.

For citation purposes, cite each article independently as indicated on the article page online and as indicated below:

Lastname, A.A.; Lastname, B.B. Article Title. *Journal Name* **Year**, *Volume Number*, Page Range.

ISBN 978-3-7258-7280-0 (Hbk)
ISBN 978-3-7258-7281-7 (PDF)
https://doi.org/10.3390/books978-3-7258-7281-7

Contents

mathematics

MDPI

Editorial

Preface to the Special Issue on "Multiscale Mathematical Modeling"

Igor Andrianov [1,*] and Julius Kaplunov [2]

1 Institute of General Mechanics, Rheinisch-Westfälische Technische Hochschule (RWTH) Aachen University, Eilfschornsteinstrasse 18, 52062 Aachen, Germany
2 School of Computer Science and Mathematics, Keele University, Newcastle ST5 5BG, UK
* Correspondence: igor.andrianov@gmail.com

MSC: 34; 35; 37; 39; 40; 41; 68; 74; 76

It is not easy to give a precise definition of the concept of multiscale modelling (MM). As is customary nowadays, we posed our question to AI and received the following response: "MM is an approach to system simulation that integrates multiple levels of detail, often employing diverse models to improve sampling efficiency and capture phenomena across different scales".

Having accepted this definition as a working one, let us say a few words about the essence of the matter. Predictions of large-scale behaviour based on fine-scale theory are a key goal of modern natural sciences and of particular importance to numerous engineering applications. However, for a long time, this description was understood as the construction of a hierarchy of models describing processes of different scales according to phase space variables (temporal, spatial, etc.). In this case, for a more accurate description, one simply needs to change the scale. This reductionist approach has proven effective in physics and mechanics. However, with the increasing complexity of the phenomena under consideration, especially with the transition to nanostructures and metamaterials, it has become clear that describing complex systems based solely on knowledge of the dynamics or behaviour of their components is often insufficient. For example, in modern materials science, it is often necessary to consider several different and distinct scales of the object of investigation simultaneously. The set of concepts and algorithms that enable the analysis of objects and phenomena in various fields of science and engineering, taking into account the interaction of different scales, has been named MM. For example, "in physics and chemistry, multiscale modeling is aimed at the calculation of material properties or system behavior on one level using information or models from different levels" (https://en.wikipedia.org/wiki/Multiscale_modeling (accessed on 25 December 2025)).

Recent major progress in this area is due to the development of advanced computational and experimental methodologies, along with the establishment of powerful mathematical frameworks operating on multiple length and time scales. Computational MM can make accurate predictions about various phenomena, simultaneously considering several different and distinct scales of the phenomena. However, as stated by R. Thom, "Prediction is not explanation".

What is meant by explanation? Once again, we encounter an ambiguous concept, the detailed interpretation of which would lead us too far into philosophical territory. For a significant portion of the current generation of mathematicians and physicists, explanation means the existence of a model within the framework of continuous mathematics. We

 https://doi.org/10.3390/math14030434

will not deny ourselves the pleasure of quoting Kolmogorov, who wrote about the relationship between discrete and continuous mathematics: "It is quite probable that with the development of modern computing techniques, it will become understood that in many cases it is reasonable to study real phenomena without making use of the intermediate step of their stylization in the form of infinite and continuous mathematics, passing directly to discrete models. Pure mathematics was successful developed mainly as the science of the infinite. Apparently, this state of affairs is deeply rooted in our consciousness, which operates with great ease with an intuitively clear understanding of unbounded sequences, limiting processes, continuous and even smooth manifolds, etc."

Whether this situation is related to the peculiarities of our subconscious or is caused by the modern education system, which focuses primarily on continuous mathematics, is a matter of debate. However, the prevalence of continuous mathematics models in modern science and engineering is evident. At the same time, we are witnessing a rapidly growing trend towards discrete models. Of course, discrete and continuous approaches to modelling the real world are not competing but complementary. Modelling multiscale phenomena both mathematically and computationally allows us to describe the phenomenon under study in its entirety. Although this SI is mainly devoted to issues of mathematical MM, it also addresses computational aspects of MM.

The problems of MM are often interdisciplinary in nature, as confirmed by this SI, which includes articles devoted to the application of MM in the theories of discrete systems (contributions 1 and 2), elasticity (contributions 3 and 4) and hydroelasticity (contribution 5), composite materials, and metamaterials and quasicrystals (contributions 6, 7 and 8), as well as in hydromechanics (contribution 9), biomechanics (contribution 10), geomechanics (contribution 11), and tribology (contribution 12). Among the methods of mathematical MM, asymptotic approaches stand out. The articles in this SI used methods of continuation (contributions 1 and 2), homogenisation (contribution 3), the regular (contribution 9) and singular (contributions 4, 5 and 7), perturbation methods, and Padé approximants (contribution 1). The theory of self-adjoint operators (contribution 6) and the integral formalism based on two-dimensional Green tensors (contribution 8) were also applied. Multiscale numerical approaches (contributions 10 and 11) and multiscale statistical analysis (contribution 12) were also used. The results obtained confirm the importance of considering discreteness for constructing refined continuous models (contributions 1, 2 and 3). It has been shown that when describing composite materials, especially those with high-contrast components, it is not sufficient to determine only the effective characteristics; the microstructure must also be taken into account (contribution 6). In general, MM requires, along with the study of the global behaviour of systems, the analysis of local effects (contributions 4, 5, 6 and 7).

As guest editors, we would like to sincerely thank the authors who responded to our invitations and submitted their work to this SI. The constructive and insightful feedback from reviewers contributed to improving the quality of the accepted articles and enhancing their style of presentation. We would also like to note the professionalism of the editors of the MDPI journal *Mathematics*.

The aim of this SI was to demonstrate effective MM methods and their applicability for in-depth analysis of systems and processes in various fields of science and engineering. We hope that the published works will be of interest to readers, both for the specific results obtained and for the general ideology of the research and the methods and approaches used.

Conflicts of Interest: The author declares no conflicts of interest.

https://doi.org/10.3390/math14030434

List of Contributions:

1. Challamel, N.; Nguyen, H.P.; Wang, C.M.; Ruta, G. In-Plane Vibrations of Elastic Lattice Plates and Their Continuous Approximations. *Mathematics* **2024**, *12*, 2312. https://doi.org/10.3390/math12152312.
2. Andrianov, I.V.; Khajiyeva, L.A.; Kudaibergenov, A.K.; Starushenko, G.A. On Aspects of Continuous Approximation of Diatomic Lattice. *Mathematics* **2024**, *12*, 1456. https://doi.org/10.3390/math12101456.
3. Rallu, A.; Boutin, C. Modal Analysis of a Multi-Supported Beam: Macroscopic Models and Boundary Conditions. *Mathematics* **2024**, *12*, 1844. https://doi.org/10.3390/math12121844.
4. Aghalovyan, L.A.; Ghulghazaryan, L.G.; Kaplunov, J.; Prikazchikov, D. Degenerated Boundary Layers and Long-Wave Low-Frequency Motion in High-Contrast Elastic Laminates. *Mathematics* **2023**, *11*, 3905. https://doi.org/10.3390/math11183905.
5. Shamsi, S.; Prikazchikova, L. Asymptotic Analysis of an Elastic Layer under Light Fluid Loading. *Mathematics* **2024**, *12*, 1465. https://doi.org/10.3390/math12101465.
6. Zhunussova, Z.; Mityushev, V. Spectral ℝ-Linear Problems: Applications to Complex Permittivity of Coated Cylinders. *Mathematics* **2025**, *13*, 1862. https://doi.org/10.3390/math13111862.
7. Cherednichenko, K.D.; Ershova, Y.Y.; Kiselev, A.V. Norm-Resolvent Convergence for Neumann Laplacians on Manifold Thinning to Graphs. *Mathematics* **2024**, *12*, 1161. https://doi.org/10.3390/math12081161.
8. Lazar, M. Line Defects in One-Dimensional Hexagonal Quasicrystals. *Mathematics* **2025**, *13*, 1493. https://doi.org/10.3390/math13091493.
9. Kudish, I.I.; Volkov, S.S. A General Case of a Line Contact Lubricated by a Non-Newtonian Giesekus Fluid. *Mathematics* **2023**, *11*, 4679. https://doi.org/10.3390/math11224679.
10. Syomin, F.A.; Danilov, A.A.; Liogky, A.A. Numerical Study on Excitation–Contraction Waves in 3D Slab-Shaped Myocardium Sample with Heterogeneous Properties. *Mathematics* **2025**, *13*, 2606. https://doi.org/10.3390/math13162606.
11. Bratov, V.; Murachev, A.; Kuznetsov, S.V. Utilization of a Genetic Algorithm to Identify Optimal Geometric Shapes for a Seismic Protective Barrier. *Mathematics* **2024**, *12*, 492. https://doi.org/10.3390/math12030492.
12. Borodich, F.M.; Pepelyshev, A.; Jin, X. A Multiscale Statistical Analysis of Rough Surfaces and Applications to Tribology. *Mathematics* **2024**, *12*, 1804. https://doi.org/10.3390/math12121804.

https://doi.org/10.3390/math14030434

mathematics

Article

Numerical Study on Excitation–Contraction Waves in 3D Slab-Shaped Myocardium Sample with Heterogeneous Properties

Fyodor A. Syomin [1,*], Alexander A. Danilov [1,2,3,4] and Alexey A. Liogky [1,2]

[1] Institute of Mechanics, Lomonosov Moscow State University, 119192 Moscow, Russia; a.danilov@inm.ras.ru (A.A.D.); al.liogky@yandex.ru (A.A.L.)
[2] Marchuk Institute of Numerical Mathematics, Russian Academy of Sciences, 119333 Moscow, Russia
[3] Scientific Center for Information Technologies and Artificial Intelligence, Sirius University of Science and Technology, 354340 Sirius Federal Territory, Russia
[4] Institute for Computer Science and Mathematical Modeling, Sechenov First Moscow State Medical University (Sechenov University), 119048 Moscow, Russia
* Correspondence: f.syomin@imec.msu.ru

Abstract

In this study, we have performed 3D numerical simulations of the excitation and contraction of thin slab-like samples of myocardium tissue. The samples included a narrow region of almost non-excitable tissue simulating impaired myocardium. In the numerical experiments, we considered the heterogeneity of myocardium excitation and the Ca^{2+} activation of its contraction, as well as the orientation of the muscle fibers. Those characteristics varied throughout the thin wall of the sample. The simulations were performed in our numerical framework for the problems of cardiac electromechanics developed recently. The framework was previously tested for the benchmark problems in which formulations took into account only myocardium electrophysiology and passive mechanics. The study could be considered as an approbation of the framework performance with the fully coupled mathematical model of myocardium electromechanics. Here we dealt with the problems requiring a multiscale approach, taking into account cell-level electrophysiology, cell-level mechano-chemical processes, macromechanics (strain and stress) of the 3D sample, and interconnections between the levels. It was shown how the tissue heterogeneity and its strain affected the propagation of excitation–contraction waves in the sample, including, in particular, the formation of spiral waves.

Keywords: numerical simulation; cardiac modeling; cardiac electromechanics; computational framework; multiscale modeling

MSC: 92C10; 74L15; 92-10; 74-10; 74H15

1. Introduction

Mathematical modeling of heart excitation and contraction has been an actively developing tool for application in the treatment of heart disease. Among others, arrhythmia is a serious and frequently encountered problem, which could be caused by different factors, including dysfunctions of ionic currents and impaired tissue excitation or conductance, and even could be affected by myocardium deformation. The mathematical modeling of such complex cardiac dysfunctions always involves numerical simulations, as the models have to include descriptions for the complex cell-level (for a single cardiomyocyte) and tissue-level (for myocardium) processes leading to the functioning of heart chambers at the organ level. The computational implementation of the multiscale simulation of cardiac activity is not easy from the computational point of view, which encourages researchers to develop new numerical suits and optimize numerical schemes and methods.

The computation frameworks and libraries for the numerical simulations of myocardium excitation and contraction commonly have implementations of multiscale mathematical models at their core. For the correct simulation, one should take into account the mechanics of the myocardium sample undergoing large strain, cell-level models of cardiomyocyte excitation and contraction, and a macro-level model for the propagation of the excitation waves within the tissue. Heterogeneity and anisotropy of the material should also be accounted for at all levels of the modeling. The model blocks are typically interconnected with each other. In particular, the processes of excitation–contraction coupling and mechano-electric feedback (MEF) should be specified in the models. In addition, while considering a whole-organ mathematical formulation, one should couple the mechanics of the heart chambers with a model of blood circulation, specifying corresponding boundary conditions. Having a numerical framework for cardiac modeling in mind, one should note that the variety of the problem formulations and models requires the framework to allow its users to incorporate new models, specify the equations, along with their boundary and initial conditions, and operate with different geometry and meshes with ease.

A 0D model of cardiomyocyte electrophysiology commonly defines a set of ionic currents specified by a system of ordinary differential equations (ODEs). As most of the currents are very fast, the systems of ODEs are stiff and should be handled with the numerical methods applicable to that case. Excitation propagation through the tissue is commonly described by a reaction–diffusion equation for the transmembrane potential, or by a system of the equations if some other electric potentials, like the potential of external medium, are considered. The excitation models are required to reproduce the dynamics of the transmembrane potential (called action potential (AP) when it grows from its resting value) and the dependence of the action potential duration (APD) on the stimulation frequency. Complex electrophysiological simulations at the heart level are challenging from the computational point of view, even if one does not take cardiac mechanics into account [1]. Speaking of cardiac mechanics, the myocardium is generally treated as an incompressible or nearly incompressible anisotropic (orthotropic or transversely isotropic) nonlinear material with finite strains. Active stress in cardiac muscle during its contraction is much higher than its passive response to strain. Because the active forces are aligned with the fiber directions, the corresponding components of the stress tensor are especially high during the contraction. All of these factors lead to ill-conditioned stiffness matrices emerging from finite element methods. That means that the linear algebraic equations, which one has to deal with while applying solvers for the nonlinear equations, require efficient methods for the matrix preconditioning before applying a reliable iterative solver. Examples of the studies on numerical methods implemented for the simulation of cardiac electromechanics can be found in [2–5]. Moreover, the solver choice sometimes depends on the problem setup. Therefore, it would be convenient to be able to choose from several different solvers for systems of ODEs and nonlinear and linear algebraic equations. Recent studies on the MEF in myocardium, both experimental and involving numerical simulation, show that the tissue deformations could be crucial for the initiation of spiral waves or their breakup, as conductance velocity in the tissue decreases upon its stretch [6–10]. For example, experimental data show that the rotation dynamics of the excitation waves in the left ventricle occurring during ventricular fibrillation is associated with the propagation of deformation waves in myocardium, which was studied numerically [11]. Thus, in order to reproduce arrhythmogenic phenomena accurately, numerical studies on the problems of heart electrical activation should take cardiac mechanics into consideration, at least for the heart ventricles, as the myocardium in the ventricular walls undergoes large strains. In addition, the computational tools for cardiac simulation should provide ways to couple the macroscopic strains with cell-level models and allow their users to specify the dependencies required for the coupling without resorting to modification of the major code segments.

The computational packages, libraries, and frameworks developed by various groups for modeling in cardiology or for more general multiphysics problems include LifeX (https://gitlab.com/lifex/lifex (accessed on 30 May 2025)), OpenCARP (https://opencarp.org/ (accessed on 30 May 2025)), Cardioid (https://github.com/LLNL/cardioid (accessed on 30 May 2025)), and Chaste (https://github.com/Chaste (accessed on 30 May 2025)). Some of the libraries developed by scientific groups are based on general-purpose computing platforms, such as FEniCS (https://fenicsproject.org/ (accessed on 30 May 2025)). Some of those computational suits and libraries were used successfully to simulate both excitation and contraction of myocardium samples or heart chambers. However, there are only a few computational tools that provide the options to specify electrophysiology and mechanics model blocks and their two-way coupling in a user-friendly manner, while, in addition, providing the instruments for flexible discretization of equations in each model block.

Our study focuses on the numerical experiments with spiral waves in anisotropic myocardium samples undergoing myocardial contraction. The spiral waves can cause or accompany some types of arrhythmia, and thus their dynamics seems to be an important subject of numerical simulation. The application of a fully coupled electromechanical model in the computations could also be especially useful, as the strain and heterogeneity of the tissue structure both affect the waves' dynamics [12,13]. In this paper, we present several 3D problems formulated for a myocardium sample of simple geometrical shape, which had a region with impaired myocardium excitation. The sample was deformed, followed by a periodic stimulation of one of its sides. For several numeric experiments, we also considered the variation of the characteristics of the regulation of myocardial contraction and the orientation of muscle fibers through the tissue. The simulations were run with our new computational framework for mathematical modeling in cardiology, CarNum [14], to test its capabilities and observe the modeled effects of the sample strain on the propagation of the excitation waves. The test problems were also supposed to demonstrate what changes in the behavior of the excitation–contraction waves we could obtain by introducing the heterogeneity of the myocardium stated above.

We pursued two goals in this study. On the one hand, we expected that the simulation results would be able to reveal important changes in the dynamics of the excitation–contraction waves upon the introduction of the spatial heterogeneity of the contraction regulation, since we used the model accounting for strain dependence of both the regulation and the propagation of the electrical activation in the myocardium. The features of the myocardium model used here, in particular, the presence of mechano-electric feedback, in combination with the specifics of the modeled samples determined our interest in the study. On the other hand, we intended to test and debug our framework CarNum while working on the formulation and solution of the problems of cardiac electromechanics, in which material anisotropy and heterogeneity of different myocardium cell-level properties were assumed as being important to take into account. CarNum is still under development, and this paper demonstrates its applicability to research tasks rather than stating its superiority over other packages, which differ in software architecture, approaches to solving problems, user interfaces, etc. The numerical methods used in CarNum and the related issues have already been described in [15] and will only be briefly mentioned here.

2. Materials and Methods

2.1. Numerical Implementation and CarNum Framework

We performed the simulations in the framework for cardiac electromechanics, CarNum. It was used to build the coupled numerical scheme, assemble the linear systems, and customize the exchanges and interactions between the individual submodels. CarNum has been developed on the basis of several other frameworks and libraries with flexibility in mind, providing the ability for a user to use their own discretizers and several pre-built solvers, as well as to implement custom cell-level models of cardiac electrophysiology and mechanics. The platform core is built on the parallel computing framework INMOST [16] and extended by the FEM toolkit AniFem++ (https://github.com/INMOST-DEV/INMOST-FEM (accessed on 30 May 2025)). We used heavily modified OpenCARP's open-source library, Limpet (https://opencarp.org/doxygen/master/limpet.html (accessed on 30 May 2025)), to set the variables and equations of cell-level models in EasyML language. We used the SUNDIALS (v7.0.0) computation suit for the solvers of nonlinear algebraic and ordinary differential equations [17]. We have verified [14] the platform and the implemented solvers with electrophysiological and mechanical benchmarks. Our results of the test problems showed good agreement with the results obtained by other research groups. Parallel computations demonstrated the good performance and scalability of our software. The framework allows one to specify several parameters to set up the mathematical problem and numerical methods using the command line.

The algorithm for the problem solution and the numerical methods were as follows. The equations of continuous mechanics (the motion and incompressibility equations for the sample) and the partial differential equation for the transmembrane potential (monodomain model) were solved by the finite element method, while the cellular ODEs (for the cell-level contraction model, ionic balance for cell compartments, ionic currents, or for the phenomenological variables of simplified models) were solved independently at the points of the quadratures on the corresponding meshes. The values of the cell-level variables computed at the current time step in the elements' points of quadrature integration were used to compute the "ionic current" terms of electrophysiological equations. Those "micro-scale" variables also included active stress developed by myocardium contractile proteins, and that stress term was included in the motion equation. To solve the ODEs of the cellular activation–contraction model, we used the CVODE solver from the SUNDIALS suit using a backward differentiation formula (BDF) of variable order. The nonlinear system of equations obtained after discretization of the continuous mechanics equations was solved by the inexact Newton method implemented in the KinSol package (numerical solvers included in the SUNDIALS suit). We solved linear systems arising at the Newton steps of the KinSol solvers, as well as the equations obtained after discretization of the monodomain equation, by the iterative BiCGStab solver with the parallelized second-order Crout-ILU preconditioner implemented in the INMOST framework [16].

The following features of CarNum that we find important and convenient for the cardiac modeling can be highlighted.

- The flexible processing of parallel computational grid meshes in common formats (inherited from the INMOST computational suit).
- The cellular block of the model (ODEs for cardiomyocyte mechanics, ionic currents or their phenomenological approximations, supplementary equations of electromechanical coupling) is specified in a human-readable language. One is able to link the block's parameters with spatial coordinates and strains, as well as import cell-level variables into tensors of the macromechanical formulation. Macro- and micro-level blocks are interconnected by the cell-level variables.

- The framework utilizes the processing of symbolic algebraic expressions and their automatic differentiation, simplifying the mathematical formulation of the cardiology related problems, including the specification of myocardium properties and boundary conditions.
- Separate time steps can be used to solve equations of mechanics, electrophysiology, and cell-level activation–contraction models.
- Users can choose from a variety of numerical methods. Generally, methods based on backward differential formula are used to solve stiff ODEs, and inexact Newton methods are used to solve nonlinear algebraic equations that arise from PDEs for the electrical propagation and continual mechanics. Those methods utilize a custom implementation of adaptive time-step and allow one to choose one of several preconditioners used in linear iterative solvers. Parallel computation using OpenMP and MPI are supported.

2.2. Excitation–Contraction Model of Cardiomyocyte

We used the myocardium model by Syomin et al. [18] in the simulations. The model combines the phenomenological model of excitation by Aliev and Panfilov [19] with the model of cardiac mechanics developed by our group earlier. The Aliev–Panfilov model at the cell level is set by a system of two ODEs for non-dimensional transmembrane potential U and a phenomenological variable v simulating, to some extent, the ionic currents that affect the AP repolarization. The equations in non-dimensional form were as follows:

$$\begin{aligned} \tau\frac{\partial C_m U}{\partial t} + k_{AP}U(U-a_{AP})(U-1) + Uv &= I_{stim}, \\ \tau\frac{\partial v}{\partial t} &= -\left(\varepsilon + \frac{\mu_1 v}{\mu_2 + U}\right)(v + k_{AP}U(U - b_{AP} - 1)). \end{aligned} \tag{1}$$

Here, C_m is the membrane capacitance, a_{AP} sets the excitation threshold, and all other parameters do not have any physical meaning, while their values were taken from [19]. In this study, we used the model modification taking into account mechano-electric feedback (MEF) implemented through the non-instant dependence of the membrane capacitance on strain upon muscle stretching. Assuming the rate of the capacitance variation to be much slower than the rate of change in the potential, the first equation was transformed into the following equations,

$$\begin{aligned} \tau C_m\frac{\partial U}{\partial t} + k_{AP}U(U-k_{exc}a_{AP})(U-1) + k_{rep}Uv &= I_{stim}, \\ \frac{\partial C_m}{\partial t} &= k_m\left(K_m\left(\lambda_f - 1\right)_+ - (C_m - 1)\right), \end{aligned} \tag{2}$$

where $(x)_+ = (|x|+x)/2$, $\lambda_f = \sqrt{(\mathbb{F}\mathbf{f}, \mathbb{F}\mathbf{f})}$ is the strain of muscle units, sarcomeres, with deformation gradient $\mathbb{F} = \mathbb{I} + \nabla\mathbf{u}$ and fiber direction given by the vector $\mathbf{f}$, $||\mathbf{f}|| = 1$. k_{rep} is equal to 1 in the Aliev–Panfilov model, while in this study we varied this "repolarization rate" parameter to change the APD through the sample tissue (see Section 2.3). We also varied this parameter and an additional multiplier k_{exc} to make the AP very short and the excitation threshold higher in the tissue regions with simulated impaired excitability.

The mechanical block was described by a system of ODEs for a set of state variables for contractile and regulatory proteins in sarcomeres, an ensemble-average distortion (muscle microscopic strain) of the formed contractile structures, cross-bridges, per sarcomere, the concentration of Ca^{2+} ions in different cell compartments, and a few other variables, which affected the calcium currents and provided correct functioning of excitation–contraction coupling. The general forms of the equations for cell-level mechanics and electromechanical coupling were

$$
\begin{aligned}
\dot{n} &= f_{cb1}(n, A_1, \delta), \\
\dot{\theta} &= f_{cb2}(n, \theta, A_1, \delta), \\
\dot{\delta} &= f_{\delta}\left(\dot{\lambda}_f, n, A_i\right), \\
\dot{A}_i &= f_A\left(n\theta, A_i, \delta, c_{cyt}, \lambda_f\right), \\
\dot{c}_{cyt} &= f_{c1}\left(U, c_{cyt}, c_{oth}, \lambda_f\right). \\
\dot{\mathbf{c}}_{\mathbf{oth}} &= \mathbf{f}_{\mathbf{c2}}(U, c_{cyt}, \mathbf{c}_{\mathbf{oth}}).
\end{aligned}
\tag{3}
$$

Here, n is the non-dimensional concentration of the contractile proteins interacting through the formation of cross-bridges, θ is a fraction of the cross-bridges being in the additionally deformed strongly bound state, and δ is the cross-bridges' average distortion. A_1 and A_2 are the non-dimensional concentrations of complexes formed by muscle regulatory proteins binding Ca^{2+} ions, and c_{cyt} is the concentration of free Ca^{2+} in the cytoplasm. The equations for A_1 and A_2 take into account the length-dependence and so called cooperativity of the muscle activation. $\mathbf{c}_{\mathbf{oth}}$ notes a vector of variables for Ca^{2+} concentration in other cell compartments and additional variables regulating the calcium currents. The equations for $\mathbf{c}_{\mathbf{oth}}$ describe calcium-induced calcium release and were sufficient to reproduce the dependencies of the peak force and c_{cyt} on the stimulation frequency, as well as the speed-up in muscle relaxation at increased stimulation frequency. Active stress T_{act} generated by the cross-bridges depended on their number (n), strain (the average strain δ and additional constant strain h of the strongly bound cross-bridges, whose fraction among all attached cross-bridges equals θ), and the sarcomere length (can be substituted by the fiber strain λ_f).

$$
T_{act} = k_{act} n(\delta + \theta h) W_{ov}\left(\lambda_f\right). \tag{4}
$$

The right-hand side of (4) includes the coefficient k_{act} defined from the stiffness of a cross bridge, the maximal number of cross bridges per sarcomere, and the number of contractile filaments per unit area. W_{ov} is a function of sarcomere length corresponding to the normalized length of the overlap zone of thick and thin contractile filaments, which affects the maximal number of binding sites for the cross-bridges. Details of the model of cell-level mechanics and electromechanical coupling can be found in Appendix A.1.

2.3. Mathematical Formulation for the Simulation of a Slab-like Myocardial Sample

The mathematical problem was specified by the following partial differential equations. Electrical propagation in the tissue domain Ω was described by the monodomain electrophysiology model, which was the first equation of (2) expanded to the tissue scale.

$$
C_m \frac{\partial U}{\partial t} - \nabla \cdot (\mathbf{D} \nabla U) + I_{ion}(U, v) = I_{stim}(t), \ \mathbf{X} \in \Omega \tag{5}
$$

with an insulating boundary condition on the surface with normal vector $\mathbf{N}$,

$$
(\mathbf{D} \nabla U) \cdot \mathbf{N} = 0, \mathbf{X} \in \partial\Omega, \tag{6}
$$

where the diffusion term emerges with the conductivity tensor $\mathbf{D}$ given by

$$
\mathbf{D} = D_{iso}\, \mathbb{I} + D_{aniso}\, \mathbf{f} \otimes \mathbf{f},
$$

which contains the terms for isotropic D_{iso} and anisotropic (aligned with the muscle fibers) D_{aniso} conductivity. The continuum mechanics were described by the equilibrium equation for the first Piola–Kirchhoff stress tensor **P** with the assumption of quasi-stationary contraction.

$$\sum_{j=1}^{3} \nabla_j P^{ij} = 0,\ i \in \overline{1,3},\ \mathbf{X} \in \Omega, \tag{7}$$

with the general form of boundary conditions specified as

$$\mathbf{PN} + (k_{\parallel}\mathbf{N} \otimes \mathbf{N} + k_{\perp}(\mathbb{I} - \mathbf{N} \otimes \mathbf{N}))(\mathbf{u} - \mathbf{u_{spr}}) = -p_{ext} J \mathbb{F}^{-T}\mathbf{N},\ \mathbf{X} \in \partial\Omega. \tag{8}$$

Here, $k_{\perp} \geq 0$, $k_{\parallel} \geq 0$ are stiffness coefficients for normal and tangential mounting springs, and displacements $\mathbf{u_{spr}}$ were either zero or equal to the displacements of the initially free boundaries of the sample after its preliminary stretch or inflation, preceding their fixation. p_{ext} is an external pressure at the boundary, and $\mathbb{F} = \mathbb{I} + \nabla \mathbf{u}$ and $J = \det \mathbb{F}$ are the deformation gradient and its third invariant. The stress tensor included the isotropic stress defined by a strain-energy function for the hyperelastic isotropic material, the stress caused by incompressibility (optional), and the anisotropic term for the stress caused by the forces applied along the muscle fibers. That last term was the sum of the nonlinear response of the elastic protein titin and the active tension developed by the contractile proteins in muscle (T_{act} in the Equation (4)). The expression for the titin stress was modified compared to the one from [18] and was described in [15]. The stress–strain curve was fitted by the exponential polynomial function of λ_f to avoid values close to zero emerging in the denominator of the original expression during the numerical simulations and to make the myocardium stiffer at sarcomeres shortening. The isotropic part of the stress–strain relationship was specified by a modified Fung model for hyperelastic materials [20]. The details can be found in Appendix A.2.

The initial conditions for the cell-level ODEs were chosen from the solution of the 0D problem of the isometric contraction at the sarcomere length of 2.25 µm with the transitions to the isotonic contraction at fixed load. Under the load, the sarcomeres shortened to approximately 1.8 µm and then stretched to their initial length; after that, the contraction mode switched back to the isometric one. Dimensional minimal and maximal values for AP were −20 and 80 mV.

The interconnections between the model blocks are shown in Figure 1, and a summary of the model equations and associated processes is presented in Table 1.

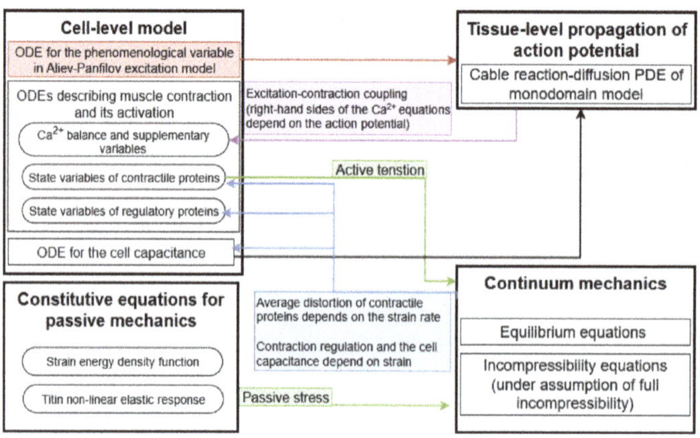

Figure 1. Simplified diagram showing the model blocks and their interconnections.

Table 1. Mathematical formulation of the model blocks.

Equations (Reference Numbers)	Physical Processes	Mesh Associated Variables	Numerical Solvers
PDEs: (5), (6)	AP propagation in the tissue	Nodal (P1, linear basis function) values of the action potential	Solution of the corresponding system of linear equations by iterative BiCGStab solver with the parallelized second-order Crout-ILU preconditioner
PDEs: (7), (8)	Continuum mechanics: equilibrium equations	Nodal (P2, quadratic basis functions) displacements	Solution of the corresponding system of nonlinear equations by the inexact Newton method, solution of linear equations at the Newton method's iterations by iterative BiCGStab solver with the parallelized second-order Crout-ILU preconditioner
ODEs: the second eq. in (1), the second eq. in (2)	Phenomenological description of ionic currents through the cell membrane and strain-dependence of the membrane capacitance	Values of v and C_m in integration points of the mesh cells (elements)	Solution of ODEs system by BDF of variable order.
ODEs: (3)	Interaction of muscle contractile and regulatory proteins, Ca^{2+} currents	Values of the state variables describing muscle contraction and its regulation, calcium concentrations. The variables are solved for integration points of the mesh cells (elements)	Solution of ODEs system by BDF of variable order.

The sample of myocardial tissue studied in the numerical experiments had a simple shape of a thin parallelepiped slab with a size of 9 × 9 × 1 cm. The material was considered to be transversally isotropic, and the anisotropy was defined by the orientation of the cardiac fibers, whose distribution differed between the numerical experiments. The left edge plane of the slab ($X = 0$) and one of the wide planes ($Z = 0$) were fixed by stiff linear elastic springs with a stiffness of 10 kPa for the left side and 50 kPa for the right one (X, Y, Z are Lagrangian reference coordinates). For the first test series of numerical experiments, the pressure was applied to the right plane ($X = L = 9$ cm) of the sample. The pressure was grown linearly in time up to 1 kPa during 100 ms in order to stretch the sample to approximately the length of $1.1L$. After that, the pressure was fixed and the right side was fastened to its current position by springs, more flexible than the ones used at the left side (2 kPa). All other planes were free of load. After the stretching, an external stimulus of 3 ms duration was applied with a period of 0.5 s, initiating the sample contraction. A non-excitable region, surrounded by the gray zone, was simulated by the parameters k_{exc} and k_{rep}, which provided high excitation threshold and fast repolarization. The details on the parameters and equations for their spatial variation are presented in Appendix A.3. We performed the computations of the myocardial sample contraction with variation of the following factors. The first one was the fiber orientation, which was horizontal and linearly changing through the Z-axis from 80° at $Z = 0$ to $-60°$ at $Z = 1$ cm; the second one was the MEF, which was taken as the strain-dependent membrane capacitance (the second equation in Equation (2)) or turned off (C_m being constant).

Another series of numerical experiments was set with the sample more similar to a part of the ventricular wall. For that purpose, the initial slab was inflated by incrementally increasing pressure applied to one of its wide surfaces. The pressure was increased linearly up to 0.05 kPa. Two edges ($X = 0$ and $X = L$) of the pressured surface were fixed by springs with a stiffness of 10 kPa. The orientation of the muscle fibers was specified in the initial non-inflated slab: the fibers were located in the XY plane and had their angle with the X-axis changing from 80° in the pressured inner, "endo", surface to $-60°$ in the unloaded "epi" surface. We introduced a heterogeneity of the cell-level properties of the electromechanical model to reproduce two effects observed in the myocardium of the left ventricle: the variations of the APD and length-dependence of the contraction activation from the endocardial layer to the epicardial layer. The activation length-dependence was specified by the model parameters, which defined the dependence of the kinetics of calcium-troponin complexes on $lambda_f$ (Equation (A3) in Appendix A.1) to reproduce experimental data [21]. The authors of that paper showed how the Ca^{2+} sensitivity differed through the ventricular wall. The sensitivity was measured by a difference in levels of calcium concentrations corresponding to half-maximal force (EC_{50}) or by a difference of $pCa_{50} = -\lg EC_{50}$ in steady-state isometric contraction at two different sarcomere lengths. Another effect taken into account was the APD variation: shorter AP in epicardium and longer AP in endocardium. In the case of the detailed electrophysiological model, that effect would be partially provided by the length-dependent change in the Ca^{2+} regulation. However, the Aliev–Panfilov model used here does not account for this type of calcium-excitation feedback. Therefore, we varied the APD through the repolazitation rate k_{rep}. The modified parameters and the introduced spatial dependencies are presented in Appendix A.3. To observe the heterogeneity effect more clearly without making the sample thicker (larger in the Z-direction), which would require more cells of the computational mesh, we decreased the isotropic conductance in some of the simulations with the sample.

Previously, we conducted a comprehensive study on the convergence of the computations of cardiac electromechanics problems, similar to those presented here, published in [15]. Here, for the sake of faster computations, we considered larger spatial steps than the optimal ones found in the convergence study. The chosen mesh sizes sufficed to show the qualitative and relative quantitative effects of various factors on the propagation of excitation–contraction waves. Our computational mesh for the numerical simulations shown below in Section 3 had the following specifications. The time steps for the cell-level ODEs, equations of the monodomain model, and equilibrium equations were 0.01 ms, 0.01 ms, and 1 ms, respectively. The mesh element size was equal to 2.5 mm. Linear discretization was used for the AP variables, and quadratic discretization was used for the displacements, together with fifth-order quadrature integration formulas. Relative tolerances of 10^{-8} and 10^{-12} were used for the nonlinear and linear solvers. Overall, 1 s of simulated contraction took approximately 7 h to compute on the computing cluster with 16 MPI tasks and 4 OpenMP processes per task.

3. Results

3.1. Excitation–Contraction of Uniaxially Stretched Slab-like Myocardial Sample

First, we ran the computations of the sample contraction with fiber orientation aligned with the X-axis. The sample was stretched along the fiber direction prior to the first stimulus. We analyzed the waves' behavior in the simulated excitation–contraction cycles, initiated by 2 Hz stimulation of the left side ($X = 0$) of the sample. This problem setup resulted in relatively large strain of the initially elongated fibers, as well as high strain gradients arising during the sample contraction. The results also showed the concentrations of stress and strain in the boundary areas of the "scar". So, the computations demonstrated both characteristic propagation of the excitation waves passing around the non-excitable region and mechanical stress and strain accompanying the excitation. Thus, we believe that this numerical experiment, which involved the model of myocardial electromechanics, was illustrative as a test problem. The computations were also useful for the debugging process of the framework implementation of the numerical methods. Analyzing the results, we focused on the waves' detachment from the non-excitable region and checked how the MEF affected the contraction slowing the conduction speed. Figures 2 and 3 show the AP maps and the maps of relative sarcomere strain at 150 ms and 300 ms after a periodic external stimulus.

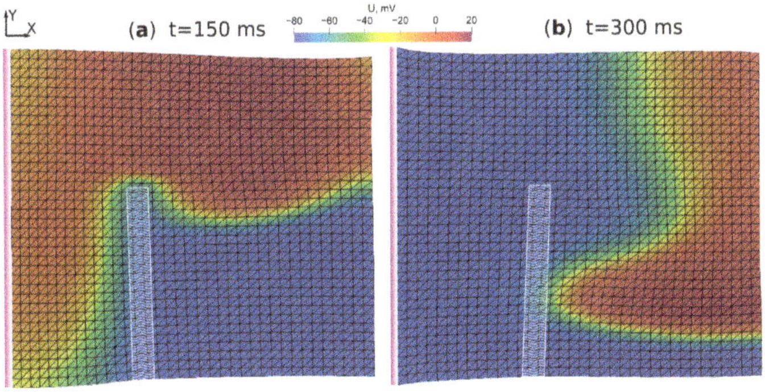

Figure 2. The excitation and contraction of the myocardial slab-like sample with narrow tissue region with impaired excitability and homogeneous properties of myocardium outside the region. The sample was stretched along the muscle fibers, aligned horizontally, before the first stimulation. Panels (**a**,**b**) correspond to 150 ms and 300 ms after the external stimulation. The color map is for the values of action potential in mV. The stimulated edge is highlighted with a magenta gradient.

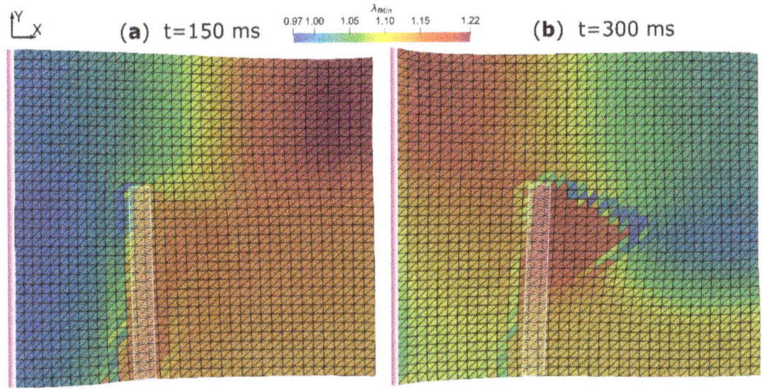

Figure 3. Strains in the periodically stimulated myocardial slab-like sample with narrow tissue region with impaired excitability and homogeneous properties of myocardium outside the region. The sample was stretched along the muscle fibers, aligned horizontally, before the first stimulation. Panels (**a**,**b**) correspond to 150 ms and 300 ms after the external stimulation. The color map is for the values of relative minimal strain per element. The stimulated edge is highlighted with a magenta gradient.

At the time instant of 150 ms after stimulation, the fiber elongations (minimal per element and relative to initial relaxed fiber length) above the "scar" area varied from 3% to 9% over 2 cm (along the X-axis), while the fiber lengthened by a maximum of 22% in the right side of the sample and shortened by 3% in the left side. It should be recalled that the sample was stretched by 10% before the stimulation, and thus the area behind the wave front shortened, while the relaxed fibers located in the right inactivated area elongated. After the excitation–contraction wave rotated around the "scar", a region of high strains of 20% was formed near the non-excitable zone with a boundary of shortened (by 10%) fibers, which can be seen at the time instant of 300 ms after the stimulus.

While the first wave of electrical activation ran through the "scar" gray zone without any visible disturbance and almost did not change its horizontal direction, the following waves detached from the "scar" and tended to rotate around its boundary as soon as a stable periodic mode for the variable v, which controls the repolarization in Aliev–Panfilov model, was set up. However, when the MEF was turned off in the cell model through the usage of a constant cell capacitance instead of a non-instant length-dependent one, no wave rotation was observed (Figure 4).

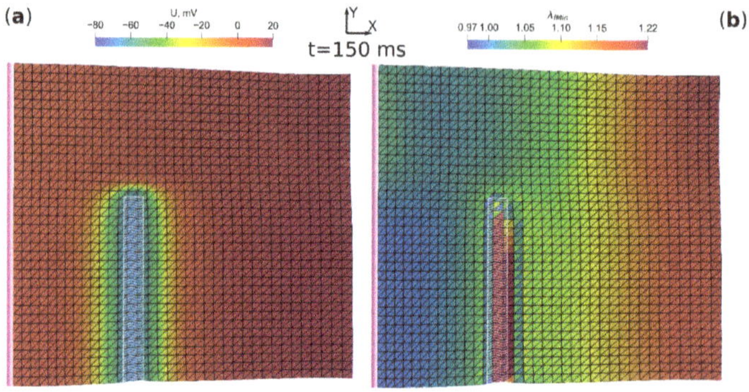

Figure 4. The simulations of the excitation–contraction waves in the pre-stretched myocardial slab-like sample with constant cardiomyocyte capacitance (no MEF taken into account). Panels (**a**,**b**) show action potential in mV and the relative muscle strain, minimal per element, 150 ms after the external stimulation, correspondingly. The color map is for the values of action potential in mV. The stimulated edge is highlighted with a magenta gradient.

The results demonstrated that the MEF influence taken into account in the simulations shown in Figures 2 and 3 led to an increase in the membrane capacitance ahead of the wavefront by 22–25% due to the initial stretching. That resulted in a slowdown in the excitation propagation. In particular, in the computations with the MEF, the wavefront reached the right edge of the sample in 150 ms after the stimulation compared to 110 ms in the simulations without the MEF taken into account.

We also performed similar numerical experiments with the horizontally pre-stretched myocardium sample, but varying the orientation of the muscle fibers linearly through the coordinate Z as described in Section 2.3. In those simulations, the fiber orientation affected the wave propagation, changing its direction rather predictably. Figure 5 shows the snapshot taken 220 ms after the stimulation, which was more suitable to compare to the simulation results for the slab with horizontally oriented fibers (Figure 2b), 300 ms after the stimulus, because of the difference in wave velocity. The variation in fiber orientation resulted in a slightly slower wave propagation in the horizontal direction (155 ms to reach the right edge) compared to the simulation shown in Figure 2, despite the fact that the MEF effect was weaker. Although the cell capacitance ahead of the wavefront was high (increased by 35%) in the middle (over Z-axis) of the slab, it increased by only 5% in the "epi" side and by less than 1% in the "endo" side. However, due to the non-horizontal fiber orientation, the vertical propagation of the front was much faster than in the case of horizontally oriented fibers, as the wave reached the bottom side of the sample in 155 ms instead of 240 ms after the moment of passing through the top side of the "scar". The detachment of the excitation wave was less clear, noticeably tending to a rotation mode only in the "epi" side, at the $Z = L, \alpha_f = -60°$ surface. The wave propagated along the boundary of the "scar" nearly without a detachment in the "endo" surface, where the fibers were oriented almost vertically. The wavefront dynamics are shown by the small magenta colored squares approximately representing the trajectory of the point with local maximum curvature on the wavefront.

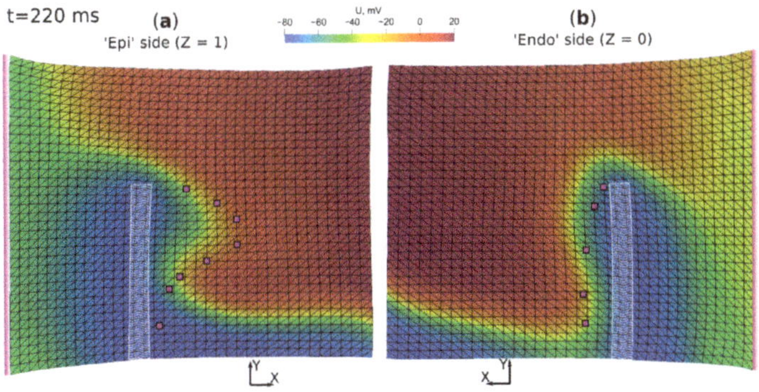

Figure 5. The excitation and contraction of the myocardial slab-like sample with narrow tissue region with impaired excitability and the distribution of the fibers' orientation through the Z-axis. The fiber's angle to the X-axis varied from 80° at $Z = 0$ to $-60°$ at $Z = 1$ cm. The sample was stretched along the X-axis before the first stimulation. Panels (**a**,**b**) correspond to the views of "epi" and "endo" (X-axis direction is from right to left) surfaces, correspondingly, 220 ms after the external stimulation. The color is for the values of action potential in mV. Magenta colored points refer to approximate positions of the point with maximal curvature on the wavefront. The stimulated edge is highlighted with a magenta gradient.

3.2. Excitation–Contraction of Convex Slab-like Myocardial Sample

We then proceeded to the series of simulations of the excitation and contraction of the convex slab, embowed by pressure applied to the $Z = 0$ surface, whose value was growing linearly up to 0.05 kPa for the duration of 50 ms before the first external stimulus. Myocardial regulatory and excitation properties and the orientation of muscle fibers were changing through the thin wall of the sample. Below, we would like to present the results demonstrating noticeable effects, caused by the variations of the following factors: the fibers' orientation, excitation threshold, and mechano-electric feedback. We also ran the simulations without the heterogeneity of APD and activation length-dependency to see its effect on the propagation of the excitation–contraction waves specifically. Hereafter, we refer to the simulation of 2 Hz stimulation of the convex slab with "normal" excitation threshold, the heterogeneity of fibers' orientation, heterogeneity of cell-level properties, and with the MEF taken into account as a "default" one. The excitation waves obtained in the default simulation are shown in Figure 6.

The results were similar to those obtained in the simulations of the horizontally pre-stretched sample with linearly varying fiber orientation. However, the wave detachment and its consequent rotation, observed in the "epi" surface of the sample, were even less prominent, presumably because of faster propagation speed caused by lower elongation of the majority of the fibers during the slab inflation, in comparison with the case of its uniaxial stretch. Although the fibers with zero spiral angle, located in the mid-wall of the slab, were stretched by higher values than in the "endo" and "epi" layers, the strains in that area seemed to have a negligible effect on the waves' dynamics. The increase in the capacitance of cardiomyocyte membrane reached 15% in the middle layers of the sample, decreasing towards its "endo" and "epi" sides to values of about 3% and less than 1%, respectively. Excitation waves reached the right edge of the slab in 135 ms, and the trajectory of the local curvature maximum of the wavefront was observed to be located even closer to the "scar" boundary than in the previous simulation, which is presumably a result of the very weak MEF influence.

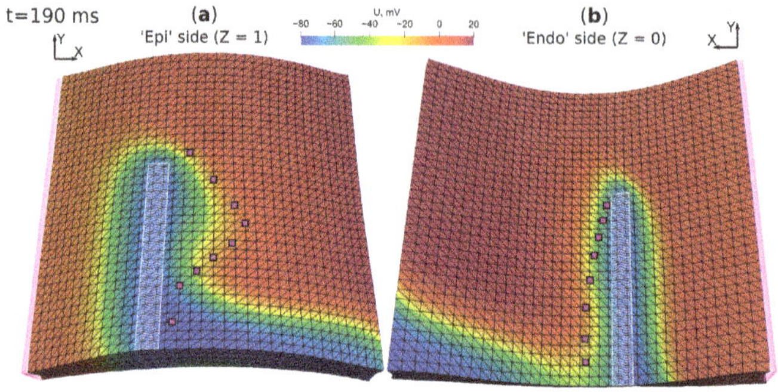

Figure 6. The excitation and contraction of the myocardial slab-like sample with narrow tissue region with impaired excitability and heterogeneous properties of myocardium outside the region. The sample was embowed to the convex form before the stimulation of its left surface. Results of the simulations with the "default" myocardium properties. Panels (**a**,**b**) correspond to the views of "epi" and "endo" (X-axis direction is from right to left) surfaces, correspondingly, 190 ms after the external stimulation. The color is for the values of action potential in mV. Magenta colored points refer to approximate positions of the point with maximal curvature on the wavefront. The stimulated edge is highlighted with a magenta gradient.

To look further into the heterogeneity, we focused on examining the effects of the distribution of the muscle regulatory properties and APD in the sample myocardium. To observe the effects, we compared the results of the "default" simulation with the results of computations for the sample, in which only the fiber orientation was changing through the sample's shortest dimension. The comparison demonstrated almost no difference in the propagation of the action potential and only a slight difference in contractility, which could be estimated by the concentration of activated regulatory complexes in the overlap region of thin and thick muscle filaments, A_1 (Equation (3)). The difference in speed of the action potential propagation was less than 0.5%, and the difference in the peak activation level locally reached 9%, while being less than 3% on average. In an attempt to reveal more significant effects of heterogeneity, we also ran the computations with five times slower isotropic conductance ($D_{iso} = 0.1\ \mathrm{mm}^2/\mathrm{ms}$ instead of $D_{iso} = 0.5\ \mathrm{mm}^2/\mathrm{ms}$). The results of those simulations demonstrated more noticeable difference between the samples with constant and transmurally varying properties, although the difference was still very small, except for some boundary areas. While the cell-level values in the "epi" surface differ by less than 3%, the shortened AP propagating in the "endo" surface, as well as higher A_1 values near the right ($X = L$) boundary of the heterogeneous sample, compared to the homogeneous sample, were observed. Higher levels of contractility are presumably caused by the increased length-dependence of Ca^{2+}-activation. The values of A_1 250 ms after the stimulation are shown in Figure 7, where the maximal difference in peak activation levels was equal to 15%.

The time courses of Ca^{2+} concentration and A_1 in an element in the upper right corner of the sample are presented in Figure 8. Although the solution had not reached a periodic mode during 3 s of 2 Hz stimulation, the result plots could still be used to compare the sample contractility showing the small difference of the variables between the "default" simulation and the simulation of the homogeneous sample. The values for the homogeneous sample were slightly lower than for the heterogeneous one, which was presumably caused by a longer APD in the "endo" layers of the heterogeneous sample. The maximal difference between those cell-level variables seemed to increase along with their peak values, while the simulation slowly approached the periodic mode. The differences in peak calcium concentrations and activation levels measured in the point were 7% and 6%, respectively.

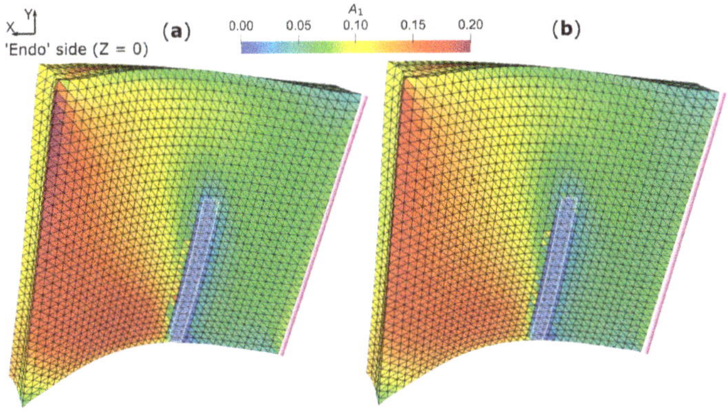

Figure 7. The excitation and contraction of the myocardial slab-like sample with narrow tissue region with impaired excitability and heterogeneous properties of myocardium outside the region. The sample was embowed to the convex form before the stimulation of its left surface. Myocardium isotropic conductance was set five times lower than its default value. The color map is for the concentration of activated regulatory complexes in the filaments overlap zone (A_1 value, average per each element's integration points). The "endo" sample surface 250 ms after the external stimulus is shown. Panel (**a**,**b**) present the results for the samples with heterogeneous and homogeneous distribution of the activation length-dependence and APD, correspondingly. The stimulated edge is highlighted with a magenta gradient.

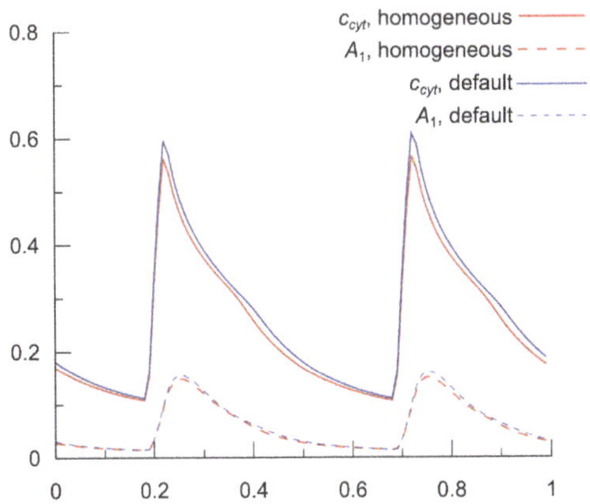

Figure 8. Computed time courses of cell-level variables in the upper right corner of "endo" surface of the convex-shaped myocardial slab with a narrows "scar" and the myocardium isotropic conductance five times lower than its default value. The curves for element-average values of non-dimensional free intracellular Ca^{2+} in µM (c_{cyt}, solid lines) and non-dimensional concentration of activated regulatory complexes in the filament overlap zone of a sarcomere (A_1, dashed lines) are plotted. The blue color corresponds to the "default" simulation and the red one is for the simulation without Z-dependent heterogeneity, except for the fiber orientation.

We then proceeded to the simulations in which we could obtain the results featuring general effects of the sample-wide increased excitation threshold and our implementation of MEF. Both modifications result in expected changes in the propagation speed of excitation–contraction waves, leading to different behaviors of the waves upon passing through the "scar" area. Figure 9 shows the results of the simulation with uniformly increased excitation threshold k_{exc} in Equation (2) (the parameter $k_{exc0} = 0.1$ was changed to $k_{exc0} = 0.15$). Action potential values 250 ms after another stimulus are presented, as the horizontal region of the wavefront at that time point was almost at the same location as the wavefront for the "default" simulation 190 ms after the stimulus. The results are shown in Figure 9.

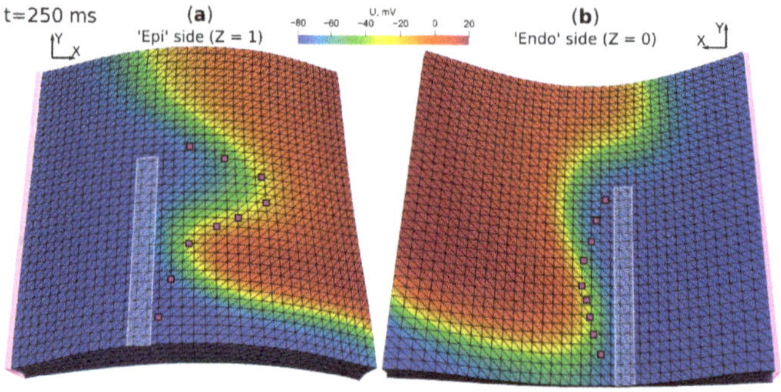

Figure 9. The excitation and contraction of the myocardial slab-like sample with narrow tissue region with impaired excitability and heterogeneous properties of myocardium outside the region; the tissue excitation threshold was increased by half uniformly. The sample was embowed to the convex form before the stimulation of its left surface. Panels (**a**,**b**) correspond to the views of "epi" and "endo" (X-axis direction is from right to left) surfaces, correspondingly, 250 ms after the external stimulation. The color is for the values of action potential in mV. Magenta colored points refer to approximate positions of the point with maximal curvature on the wavefront. The stimulated edge is highlighted with a magenta gradient.

The propagation speed of the waves was slower compared to the "default" simulation, as it took about 165 ms after the stimulus for the wave to reach the right edge of the sample. The waves' detachment in the "epi" surface of the sample was more pronounced than in all previous simulations, which is shown by the trajectory points located much further from the "scar" boundary in its upper half. In the "endo" side, the excitation waves only tended slightly to the transition to the rotation mode, though that transition was more distinctive than in the "default" numerical experiments. Proceeding to the discussion of the computations with the MEF "turned off", the absence of the strain-dependence of the membrane capacitance resulted only in mild speed-up of the excitation waves (135 ms from the stimulus to reach the right side). However, the consequences of the "turned off" MEF were different in the numerical experiments with either decreased isotropic conductance or horizontal orientation of the muscle fibers. In the absence of the MEF, those factors resulted in an almost uniform propagation of the waves after they passed the impaired region instead of the wave detachment from the "scar", as the wave was propagating through the gray zone of the impaired region near its bottom border. The effect is shown in Figure 10. We suppose that the reason for that wave passing through the "scar" area might be the non-impaired conductivity of the modeled scar. Apparently, both horizontal orientation of the fibers and reduced isotropic conductivity resulted in the diffusion of the excitation across the "scar" under our model assumptions, limited to high excitation threshold and repolarization rates of the scar tissue.

A summary on the performed simulations and their results is presented in Table 2.

Table 2. Summary on the numerical simulations performed during the study.

Preliminary Deformation	Heterogeneity over Z-Axis	Additional Conditions	Dynamics of Excitation–Contraction Waves
Horizontally stretched slab	Uniform myocardium properties over Z-axis; horizontal fiber orientation	MEF is taken into account	Apparent detachment of the wave from the "scar" region and its further rotation. Moderate slowdown in the wave propagation because of the MEF.
	Uniform myocardium properties over Z-axis; horizontal fiber orientation	No MEF taken into account	The wave propagation along the scar border without detachment, faster propagation speed.
	Uniform myocardium properties over Z-axis; linear variation of the fiber orientation	MEF is taken into account	Apparent detachment of the wave from the "scar" region and its further rotation, which was much less pronounced in the "endo" side. The wave propagation in horizontal direction was slower because of the varying fiber orientation, in spite of the weaker MEF effect (fibers' preliminary strains and cell capacitance were lower in average than at horizontal fibers).
Convex embowed slab	Linear variation of APD and the length-dependence of the contraction activation over Z-axis; linear variation of the fiber orientation	MEF is taken into account; simulations at normal and reduced isotropic conductance	The wave detachment was less pronounced and was almost absent at the "endo" side, where the waves propagated along the "scar" boundary. The wave propagation was faster than in the simulations presented above in rows 1 and 3 due to very weak MEF effect (the fiber strains and cell capacitance were much lower on average).
	Uniform myocardium properties over Z-axis; linear variation of the fiber orientation	MEF is taken into account; simulations at normal and reduced isotropic conductance	There is no difference in propagation of the excitation waves comparing to the previous case, except for the APD. While comparing the simulations under the condition of reduced isotropic conductance, a small difference in myocardial contractile properties was observed, which did not lead to the difference in excitation because of the weak MEF effect.
	Linear variation of APD and the length-dependence of the contraction activation over Z-axis; linear variation of the fiber orientation	MEF is taken into account; increased excitation threshold	The wave detachment and rotation were more pronounced than in the other simulations, while the horizontal wave propagation was much slower.
	Linear variation of APD and the length-dependence of the contraction activation over Z-axis; linear variation of the fiber orientation	No MEF taken into account; simulations at normal and reduced isotropic conductance	Only a slight increase in propagation speed at normal isotropic conductance was observed. However, at reduced isotropic conductance the wave detachment was barely distinguished, and the waves passed through the bottom part of the "scar" region.
	Linear variation of APD and the length-dependence of the contraction activation over Z-axis; horizontal fiber orientation	No MEF taken into account	A significant increase in propagation speed was observed. The waves passed through the "scar" region without rotation.

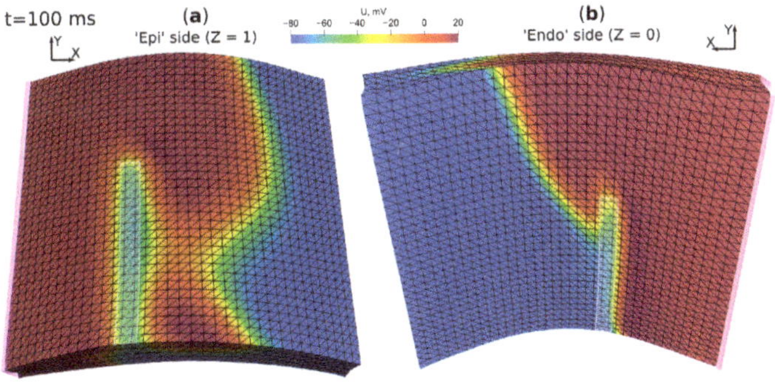

Figure 10. The excitation and contraction of the myocardial slab-like sample with narrow tissue region with impaired excitability and heterogeneous properties of myocardium outside the region. Myocardium isotropic conductance was set five times lower than its default value, and cardiomyocytes' membranes had constant capacitance (no MEF taken into account). The sample was embowed to the convex form before the stimulation of its left surface. Panels (**a**,**b**) correspond to the views of "epi" and "endo" (X-axis direction is from right to left) surfaces, correspondingly, 100 ms after the external stimulation. The color is for the values of action potential in mV. The stimulated edge is highlighted with a magenta gradient.

4. Discussion

4.1. Major Results and Limitations

The problems' formulations and their solutions presented here should be considered not as the computations that reproduce some important phenomena of myocardial excitation and contraction but rather as the numerical test experiments demonstrating how our computational framework deals with the modeling of coupled electromechanics in an anisotropic and heterogeneous myocardium sample. The main effects shown in the simulations are the tendency of the excitation–contraction waves to transform into spiral waves as they pass around a narrow obstacle. We have also shown how the wave propagation depended on some conditions varied by the model parameters. The demonstrated effects are well known and have been studied thoroughly in experiments and numerical simulations [22–24]. It is also known that the effects grow stronger when the excitation takes place in the tissue with an obstacle with impaired excitability or under localized or global slowed conductance [13,25]. We observed that the waves' behavior at the boundary of non-excitable regions in the "epi" and "endo" sides of the deformed slab-formed samples differed from each other, and the difference was larger at slower "transmural" (Z-axis oriented) propagation, when the isotropic term of conduction was decreased. Still, it could be said that the influence of fiber orientation and APD on the excitation was also rather predictable, as those factors changed the conduction velocity, including its direction near the "scar" boundary. That resulted in alternations of the wavefront's curvature, which is known to affect the initiation of spiral waves at fast-paced stimulation [26]. Even considering those phenomena as being well-studied, the application of the fully coupled electromechanical model allowed us to specify the heterogeneity in myocardium contraction and to make an attempt to examine how the variation of the properties of the contraction and electromechanical coupling in the sample would affect the sample's excitation and strains. Although we were not able to show any clear cases of re-entry formation caused by the contraction heterogeneity, the simulations seem to be useful as example problems. The successful simulations demonstrated that we should be able to set up and solve similar problems for myocardium in the ventricular wall using our computational framework. We would also like to note that the heterogeneity of the cell-level properties did affect the distribution of the activated regulatory units. That resulted in different fields of strain and stress, especially at the higher relation of anisotropic and isotropic conductance. With mechano-electric feedback of any kind being taken into account, this difference in strain might lead to disturbances in the excitation under some conditions, as it did, to some extent, in our numerical experiments with strain-dependent cardiomyocyte capacitance.

We would also like to comment on the "scar" tissue in our simulations. Although a scar region is usually considered as an insulated area, we used a less strict formulation here. Only the coefficients of excitation threshold and repolarization rate were increased in the scar zone, varying continuously from the values in the normal myocardium through the gray zone tissue. The equations of the cell-level model were solved for the scar region as if it was a part of the myocardial tissue and not a separate region of connective tissue. However, in the simulations, due to high values of the excitation threshold and fast repolarization, the action potential was never high enough to provide the calcium influx into the cell cytoplasm, and thus no active tension was generated. The transmembrane potential was also allowed to change due to the diffusion across the scar boundary, which could affect the behavior of the waves near the scar border differently, compared to the scar with no-flow boundary conditions. Normal conductivity of the scar region might be a reason for the absence of the waves' rotation in the simulation with the heterogeneous convex slab and the MEF being "turned off".

Another remark should be made about the contraction mode obtained in the performed simulations. While the simulations were run at the stimulation frequency of 2 Hz, the initial conditions for the cell-level equations were set to the diastolic values obtained for the 1 Hz periodic uniaxial contraction. Those conditions, including, in particular, low (for the 2 Hz contraction) diastolic Ca^{2+} concentration in sarcoplasmic reticulum, together with simplifications of the model (absence of the balance equations for sodium and potassium concentrations) resulted in low values of free cytosolic Ca^{2+} during the first several seconds of the simulation, while the peak Ca^{2+} values were slowly increasing to much higher values of the periodic solution. Thus, active tension and myocardium strain were rather low in the case of the convex slab, in which sarcomere shortening was less than 10%. Presumably, the effect of non-uniform strain-dependence of the contraction activation would be more significant at higher strain values. However, we believe that the qualitative effects shown in the results would be similar even at higher strains.

Discussing the choice of spatial grid size in our study, we would like to make the following remark. Several studies on the modeling of the propagation of excitation waves in myocardium have shown that a computation grid size of less than 500 μm should be used to obtain an accurate solution for the reaction–diffusion equations, while the motion equations have more relaxed requirements. In this study, we performed the simulations using a coarse grid with a spatial size of 2.5 mm, which can presumably lead to a notable difference in the results compared to the computations on a finer mesh. This difference can also involve the different effects of the mechano-electric feedback at the same model parameters because of the error in the conduction velocity. However, our study did not aim to estimate the precise values of characteristics, such as fibers' strains or excitation threshold, at which the wave rotation becomes apparent. Moreover, we suppose that the simplified phenomenological excitation model used here would not allow us to obtain accurate results even on a fine grid. To investigate possible numerical errors, we performed a test simulation on a homogeneous, except for the "scar" region, uniaxially pre-stretched sample on a mesh that was twice as fine as the one used in our study, with an element size of 1.25 mm. Some results are presented in the Supplementary Materials File S1.

4.2. CarNum Performance and Perspectives

In the study, we used the new cardiac modeling framework CarNum for the numerical simulations. The framework and its dependencies were entirely implemented in C/C++. The core methods of our computational platform were based on their implementations in the massively parallel framework INMOST and inherited its MPI/OpenMP parallelization properties with the computations running on high-performance computing systems. The CarNum platform was designed to solve the problems of coupled cardiac mechanics by dividing them into different physical processes (blocks), where continuous mechanics, electrical activation propagation, and cellular electrophysiology and mechanics were the major ones. This structure allows one to organize the computations of the individual model blocks with different space and time steps, as well as to use different discretization strategies for the different model blocks [15]. The linking of the blocks to each other and the organization of their interactions is managed by the software's user, which provides great flexibility. When setting up the model by the platform's methods, the user should provide descriptions of functional dependencies and parameters included in the model blocks. For interaction with user-provided functions and equations, the platform offers tools for the processing of symbolic expressions and methods of automatic differentiation. To make rapid implementation of individual submodels possible, some special capabilities have been implemented, which include support for a human-readable language for the descriptions of cellular models, expressions of the strain-energy functions used in passive continual mechanics, and configuration files. The interaction with the fully specified model is implemented with the use of configuration files and command line arguments.

The structure and features of the CarNum framework allowed us to switch quickly between different mathematical formulations, changing the boundary conditions and stimulation strategies, and to introduce tissue regions with special properties of the myocardium. We also used the framework's flexibility to modify the equations of the cell model, continuous mechanics, and the variables shared between the blocks of micromechanics and macromechanics with relatively small effort, while debugging and correcting the mathematical setup and its implementation. We believe that our platform proved to be both effective and efficient while dealing with the electromechanical problems studied here, and it can be modified further to simulate more complex objects, such as heart chambers.

4.3. Peculiarities of Numerical Implementation

Our initial attempts to run the computations for the electromechanical problems stated here were unsuccessful because of a divergence of the nonlinear solver. We found out that the divergence was caused by a very fast variation of the microscopic strain δ. The change rate of δ depended on the sarcomere strain rate and the fast cross-bridge kinetics. The first problem was that the active stress in myocardium depended on δ explicitly, leading to the dependence of the time derivative of the stress term on the strain rate. It was shown earlier [4] that the usage of the microscopic strain from a previous time step in a similar problem resulted in cycling through repeated consequential increments and decrements in strain, as the solution was unable to satisfy the equilibrium equation. This problem could be solved by the usage of the unknown (at the current time step) strain rate, expressed through unknown nodal displacements, instead of the one from the previous step. However, here we encountered another problem. The active stress depended also on the number of the cross-bridges and the fraction of strongly bound cross-bridges, while the variation of both of those variables depended on δ. Moreover, the δ variation depended on the cross-bridge kinetics, and these kinetics, together with the muscle active stress, reacted significantly to such low changes in δ as 2 nm, while the rate of sarcomeric strain could reach 1 μm/s and higher. This caused the numerical methods to perform poorly at the reasonable time-steps of 1 ms. We tried several approximation approaches, and our final workaround was to replace δ in the equilibrium equations by the value of δ obtained in the steady-state solution of the system (3). The constant strain rate was expressed in terms of the unknown nodal displacements. Then δ in the cell-level model block was calculated after the step of solution of the balance equations for macromechanics in accordance with its ODE. The time-derivative of λ_f in the right-hand part of the ODE for δ was expressed as $\left(\lambda_f^t - \lambda_f^{t-1}\right)/\Delta_t$, where the time indexes and time increments corresponded to the macromechanics time-mesh. We had also found that solving the ODE for θ in (3) (see also (A1)) instead of its quasi-steady-state approximation, which was used earlier in [18] and other studies on that model, resulted in better convergence due to the not instantaneous dependence of θ on δ in that case.

4.4. Novelty

The mathematical formulations of the problems of cardiac excitation and contraction presented here are rather simplistic, and similar numerical studies with 2D patches or thin 3D slabs, including ones with more sophisticated structure and conditions than in our paper, have been performed by different scientific groups [7,27,28]. However, we would like to highlight the following aspects of the study.

The problems examined in the study were formulated and solved to initially test our new computational framework. The results demonstrated that the framework performed well for 3D computations of coupled electromechanical problems. When performing a numerical analysis of myocardium excitation, researchers generally do not take into account muscle active contraction, and only a few research groups consider more or less detailed models of cardiac mechanics in the simulations. In addition, only a few groups presented the simulations of fully coupled cardiac electromechanics (see, for example, [29,30]). The effects of the MEF type considered here were studied in 3D simulations, but without active muscle contraction in consideration [10,31]. One can notice that the qualitative effects observed here in the sample of the simplistic geometrical form could be examined in 2D simulations, as our group did earlier. However, it was important to examine if the framework could deal with mechanics-related 3D problems, in particular, when the material anisotropy that was heterogeneous over the sample depth (Z-dimension) was taken into account.

As we have plans to apply our software to the case of the heart ventricles contracting within the circulatory system, we needed first to demonstrate that the simulations operated properly for a sample with more simple geometry under different conditions. We have performed the important numerical tests showing that we were able to consider the heterogeneity of the sample's material and introduce customizable parameters of the model's blocks without significant changes of the program code. Even though the heterogeneity taken into account for the cubic sample did not lead to results that would be as meaningful as the ones that could be obtained in the simulations of heart chamber contraction, this study's results are still useful and significant. The investigation of the test problems, which were numerically studied here, has provided insight into the possible difficulties that could arise while proceeding to the planned 3D simulation of the ventricles' function. The study also helped us to optimize the introduction of space-dependent and strain-dependent parameters of the cell-level models. The results demonstrated that the strain-dependence of the cardiomyocyte capacitance affects the conditions under which arrythmogenic phenomena occur within the myocardium. The variation of some parameters of the muscle activation combined with the variation in the fibers' orientation was also shown to have an effect on the dynamics of excitation–contraction waves.

5. Conclusions

In this study, we focused on a series of spatially simple problems related to myocardium excitation and contraction, which could potentially reveal the influence of strain and distribution of some myocardial properties in tissue on the propagation of excitation–contraction waves. At the same time, the results could help us to improve our computational framework and identify its strengths and weaknesses when applied to the problems requiring a two-way coupled model of cardiomyocyte electromechanics. We have presented the results of numerical simulations of the excitation–contraction waves in a thin-walled slab-shaped myocardial sample with heterogeneity of some of tissue properties and a narrow region imitating a non-excitable scar. Two different initial deformations were applied to the samples. The uniaxial stretch was useful to examine the effects of elongation of the muscle fibers, and the inflation of the sample imitated the region of the ventricular wall to some extent. The results of the simulations helped us to debug and modify the modules of our computational framework to deal with the problems of cardiac electromechanics more effectively, coupling macro- and microscopic deformations of a simulated sample and to set any changes of the tissue properties in specified locations. In addition, we could examine if the tissue heterogeneity, the fibers' orientation, muscle strain, and the increased excitation threshold affect propagation of the excitation–contraction waves through the sample. While the fibers' orientation strongly affected the waves' behavior, the effects of non-constant APD and length-dependence of contraction activation were almost negligible. However, the effects were slightly stronger in the case of slower conductivity across the muscle fibers.

At the thicker myocardial tissue and larger strains, those factors could be rather important for correct simulations of myocardium electromechanics. The clear effects of the increased excitation threshold and the MEF were also shown in the numerical studies. We plan to obtain more strongly-pronounced results by applying our computational framework to simulations of ventricular contraction within the cardiovascular system.

Supplementary Materials: The following supporting information can be downloaded at: https://www.mdpi.com/article/10.3390/math13162606/s1, File S1: Comparison of the test numerical simulations performed on spatial grids of 2.5 mm and 1.25 mm size. The simulations were performed using a homogeneous slab-like sample with a non-excitable region and horizontally oriented muscle fibers. The sample was preliminary stretched along the fibers' direction.

Author Contributions: Conceptualization, F.A.S. and A.A.L.; methodology, F.A.S.; software, A.A.L. and A.A.D.; validation, A.A.L. and A.A.D.; analysis, F.A.S.; writing—original draft preparation, F.A.S.; writing—review and editing, A.A.L., A.A.D. and F.A.S.; visualization, F.A.S.; supervision, F.A.S.; project administration, F.A.S.; funding acquisition, F.A.S. All authors have read and agreed to the published version of the manuscript.

Funding: This research was funded by Russian Science Foundation grant #22-71-10007 to F. Syomin (CarNum development, the numerical experiments with slab-shaped samples, analysis of their results shown in Figures 4–10). The part of the study was conducted under the state assignment of Lomonosov Moscow State university (fitting of model parameters for heterogeneity in the simulations of 0D uniaxial contraction, debugging the computations with pre-stretched homogeneous samples, whose results are shown in Figures 2 and 3).

Institutional Review Board Statement: Not applicable.

Informed Consent Statement: Not applicable.

Data Availability Statement: The datasets presented in this article are not readily available. The computational framework used here is a work in progress, and we would like to keep the source code private for now. Requests to access the datasets resulted from the numerical simulations shown in Figures 1–9 should be directed to F.A. Syomin (f.syomin@imec.msu.ru).

Conflicts of Interest: The authors declare no conflicts of interest. The funders had no role in the design of the study; in the collection, analysis, or interpretation of data; nor in the writing of the manuscript.

Abbreviations

The following abbreviations are used in this manuscript:

MEF	Mechano-electric feedback
AP	Action potential
APD	Action potential duration
ODE	Ordinary differential equation
BDF	Backward differentiation formula

Appendix A

Appendix A.1. Equations of the Cell-Level Model

Muscle contraction occurs due to the interaction of the contractile proteins, actin and myosin, which form filaments in muscle units called sarcomeres. Parts of myosin molecules, the so-called myosin heads, can bind to actin, forming cross-bridges. Main cell-level variables for contractile mechanics were the number of the cross-bridges bound to actin n, their ensemble-average non-dimensional distortion δ, and their fraction being in the strongly bound state, θ, in which the cross-bridges were additionally deformed by the value of h. The kinetic equations had the following form.

$$\begin{aligned}
\frac{dn}{dt} &= k_{cb}\left(k_{01}\left(A_1^2 - n\right) - k_{10}n\right),\\
\frac{d\theta}{dt} &= k_{cb}\left(k_{12}(1-\theta) - k_{21}\theta - \left(k_{01}\left(A_1^2 - n\right) - k_{10}n\right)/n\right),\\
\frac{d\delta}{dt} &= \begin{cases} 0, & \delta_{rhs} > 0 \text{ and } \delta > \delta_{max},\\ \delta_{rhs}, & \text{otherwise,}\end{cases} \qquad \delta_{rhs} = \frac{L_{s0}}{2h}\frac{d\lambda_f}{dt} - \delta k_{cb}k_{01}\left(A_1^2 - n\right)/n.
\end{aligned} \tag{A1}$$

The kinetic rates k_{ij} depended only on δ and were given by expressions

$$\begin{aligned}
&k_{21} = 20, k_{12} = k_{21}\cdot e^{-c_{cb}h\delta},\\
&k_{01} = \begin{cases}\left(\frac{\delta_*}{\delta_*-\delta}\right)^2, & \delta > 0,\\ 1, & \text{otherwise,}\end{cases} \qquad k_{10} = \begin{cases} a_{cb} + \frac{\delta}{\delta_*-\delta}, & \delta > 0,\\ a_{cb} + b_{cb}\delta^2, & \text{otherwise.}\end{cases}
\end{aligned} \tag{A2}$$

The kinetic equations for the activation of regulatory units in the overlap (A_1) and non-overlap (A_2) zones of the thick and thin filaments were based on the binding of Ca^{2+} ions from the cell cytoplasm (with the free calcium concentration c_{cyt}) by troponin. The length-dependence of the activation and the cooperativity effects were provided by the functions of λ_f, A_i, and $n\theta$ in the right-hand sides of the equations.

$$\begin{aligned}
\frac{dA_1}{dt} &= k_A\Bigg(\left(1 + k_l\cdot\left(\lambda_f - 1\right)\right)\frac{c_{cyt}}{c_{cyt}+K_B} -\\
&- K_A * \left(1 + k_{\theta 1}\left(e^{-k_{\theta 2}n\theta} - 1\right)e^{k_{ACoop}(A_1-0.5)}\right)\frac{A_1}{c_{cyt}+K_A}\Bigg),\\
\frac{dA_2}{dt} &= k_A\left(\left(1 + k_l\cdot\left(\lambda_f - 1\right)\right)\frac{c_{cyt}}{c_{cyt}+K_B} - K_A * \left(e^{k_{ACoop}(A_2-0.5)}\right)\frac{A_2}{c_{cyt}+K_A}\right).
\end{aligned} \tag{A3}$$

k_l was substituted by $k_l = k_{lENDO} + (k_{lEPI} - k_{lENDO})Z$ to simulate the heterogeneous length-dependency of the contraction activation.

The equations for Ca^{2+} concentrations in different cell compartments, free intracellular calcium c_{cyt}, calcium in the sarcoplasmic reticulum c_{sr}, and calcium in subspace between the cell membrane and the gates of the reticulum c_{ss}, all of which could bind to buffer proteins, were as follows.

$$\begin{aligned}
\alpha_{cyt}\frac{d}{dt}\left(c_{cyt} + \frac{B_{cyt}c_{cyt}}{c_{cyt}+K_{cyt}}\right) &= I_{NCX} - I_{up} + G_{xfer}(c_{ss} - c_{cyt}) + G_{leak}(c_{sr} - c_{cyt}) - I_{Tn},\\
\alpha_{sr}\frac{d}{dt}\left(c_{sr} + \frac{B_{sr}c}{c_{sr}+K_{sr}}\right) &= I_{up} - I_{rel} - G_{leak}(c_{sr} - c_{cyt}),\\
\alpha_{ss}\frac{d}{dt}\left(c_{ss} + \frac{B_{ss}c}{c_{ss}+K_{ss}}\right) &= I_{CaL} + I_{rel} - G_{xfer}(c_{ss} - c_{cyt}).
\end{aligned} \tag{A4}$$

Here, I_* are the calcium currents, G_* are the maximal conductance values for the corresponding currents, B_* are concentrations of the buffer proteins, and K_* are the equilibrium constants.

The currents between the compartments were set by the following equations. Ca^{2+} current through voltage-dependent L-channels into the cell was specified as

$$\begin{aligned}
I_{CaL} &= G_{CaL}\phi(U),\\
\phi(U) &= \frac{p_0 + p_1U + p_2U^2}{1 + e^{k_U(U_0-U)}}.
\end{aligned} \tag{A5}$$

Na-Ca exchange through the cell membrane depended only on the transmembrane potential U and free intracellular Ca^{2+} c_{cyt}, but we also took into account the slow length-dependence of the contraction by means of intracellular Na^{+} concentration depending on λ_f. The equation had the form

$$
\begin{aligned}
I_{NCX} &= G_{NCX}\frac{Na_i^3 Ca_o \psi_1(U) - \alpha_{NCX} c_{cyt} Na_o^3 \psi_2(U)}{(1+k_{sat}\psi_2(U))(Na_o^3 + K_{Na}^3)(Ca_o + K_{Ca})},\\
\psi_1 &= e^{\eta_{NCX}\frac{(-80+100U)eq}{k_B T}}, \psi_2 = e^{(\eta_{NCX}-1)\frac{(-80+100U)eq}{k_B T}},\\
Na_i &= Na_{i0}\left(1 + k_{NCXL}\left(\lambda_f - 1\right)\right).
\end{aligned}
\tag{A6}
$$

Uptake into the sarcoplasmic reticulum, which depended on the concentration of phosphorylated phospholamban p, was set as

$$
I_{up} = G_{up}\frac{c_{cyt}^2 p}{K_{up}^2 + c_{cyt}^2},
\tag{A7}
$$

and p kinetics was specified by equation

$$
\frac{dp}{dt} = k_p\left(c_{cyt}^2(1-p) - K_p^2 p\right).
\tag{A8}
$$

Calcium-induced calcium release into the subspace between the cell membrane and the gates of sarcoplasmic reticulum, whose activation was specified by variable R, was set by the equations

$$
\begin{aligned}
I_{rel} &= G_{rel}R(c_{sr} - c_{ss})\frac{c_{ss}^2}{c_{ss}^2 + K_{rel}^2},\\
\frac{dR}{dt} &= k_R\left(1 - R - K_R\frac{c_{ss}R}{c_{ss} + K_{ssR}}\right).
\end{aligned}
\tag{A9}
$$

I_{Tn} denotes the Ca^{2+} flow caused by the formation and dissociation of CaTnC complexes (the complexes formed by Ca^{2+} binding to the regulatory proteins).

$$
\begin{aligned}
I_{Tn} &= \frac{dC_{TnC}}{dt},\\
C_{TnC} &= \frac{C_{Tn}}{L_a}\left(\frac{c_{cyt}}{c_{cyt}+K_B}\left(W_{ovA}\left(\lambda_f\right)(1-A_1) + \left(L_a - W_{ovA}\left(\lambda_f\right)\right)(1-A_2)\right) + \right.\\
&\left. + \frac{c_{cyt}}{c_{cyt}+K_A}\left(W_{ovA}\left(\lambda_f\right)A_1 + \left(L_a - W_{ovA}\left(\lambda_f\right)\right)A_2\right)\right),
\end{aligned}
\tag{A10}
$$

where the normalized length of actin filaments overlap with myosin filaments W_{ovA}, and the length of actin filament L_a appears from our differentiation between the activated regulatory complexes in overlap zone and beyond it.

The last equation in Equation (A4) was solved in quasi-static approximation and, taking in account the expressions for Ca^{2+} currents, was transformed into a cubic equation.

$$
\begin{aligned}
&\left(G_{rel}R + G_{xfer}\right)c_{ss}^3 - \left(G_{rel}Rc_{sr} + G_{xfer}c_{cyt} + G_{CaL}\phi(U)\right)c_{ss}^2 + G_{xfer}K_{rel}^2 c_{ss} - \\
&- K_{rel}^2\left(G_{xfer}c_{cyt} + G_{CaL}\phi(U)\right) = 0.
\end{aligned}
\tag{A11}
$$

Appendix A.2. Mechanics Constitutive Equations

We used the following decomposition for the stress tensor:

$$\mathbb{P} = \mathbb{P}_{pas} + \mathbb{P}_{vol} + \mathbb{P}_{cell},$$

where the first two terms represent isotropic hyperelastic passive stress with volumetric penalty, introduced because of near-incompressibility being taken into account. The last term is the stress provided by the cell-level processes, which also includes the active stress. The isotropic passive terms were obtained from an elastic potential.

$$\mathbb{P}_{type} = \frac{\partial W_{type}(\mathbb{F})}{\partial \mathbb{F}}, \; type \in \{pas,\; vol\}.$$

To describe the passive properties of the myocardium, the isochoric variant of the Fung-type strain energy function for isotropic material [20] was considered,

$$W_{pas} = \frac{\mu_{pas}}{2}(\exp(b_{pas}\,\mathrm{tr}(\hat{\mathbb{E}}^2)) - 1),$$

and the volumetric term was given by

$$W_{vol} = \frac{K_{pas}}{4}(J^2 - 1 - 2\ln J)),$$

where $\hat{\mathbb{E}} = \frac{1}{2}(J^{-\frac{2}{3}}\mathbb{F}^T\mathbb{F} - \mathbb{I})$ is the isochoric variant of the Green–Lagrange strain tensor, $J = \det \mathbb{F}$. The stress caused by the cell-level contraction had the form

$$\mathbb{P}_{cell} = T_{cell}(\lambda_f,\; \partial_t\lambda_f,\; \mathbf{v_c})\,\frac{\mathbb{F}\mathbf{f} \otimes \mathbb{F}\mathbf{f}}{\lambda_f},$$

where $\mathbf{v_c}$ is the vector of the cell-level variables, $\mathbf{f}$ is the non-deformed unit vector aligned with the fiber orientation, $\lambda_f = \sqrt{(\mathbb{F}\mathbf{f}, \mathbb{F}\mathbf{f})}$ is the relative fiber strain, and $\partial_t\lambda_f = \partial\lambda_f/\partial t$ is the fiber stretch-rate. Scalar tension T_{cell} is the sum of two terms: $T_{cell} = T_{act} + T_{tit}$, which are active muscle tension (see Equation (4)) and the passive tension of titin protein. The titin stress was described by an exponential polynomial of λ_f approximating more complex expression for the worm-like chain model that was used in the previous studies on the myocardium model.

$$T_{tit} = t_{tit}\varepsilon_\lambda \cdot \begin{cases} \exp(q_1\varepsilon_\lambda + q_2\varepsilon_\lambda^2 + q_3\varepsilon_\lambda^3), & \varepsilon_\lambda > 0, \\ \exp(q_1\varepsilon_\lambda + q_2^*\varepsilon_\lambda^2), & \text{otherwise}, \end{cases} \qquad \varepsilon_\lambda = \lambda_f - 1.$$

The expanded expression for T_{act} was defined as

$$T_{act} = E_{cb} \cdot N_m \cdot N_{cb} \cdot n \cdot h(\delta + \theta) W_{ov}\left(\lambda_f\right).$$

It differs from Equation (4) because here δ is a non-dimensional variable. E_{cb}, N_m, and N_{cb} are the cross-bridge's stiffness, the number of myosin filaments per cross-section area of muscle fibers, and the number of cross-bridges per half of a myosin filament, respectively. The expression for the overlap zone W_{ov} was as follows.

$$W_{ov}(\lambda_f) = \begin{cases} 0, & L_s \le L_{min}, \\ \frac{L_s - L_{min}}{L_m + 2L_z - L_{min}} W_0, & L_{min} < L_s \le L_m + 2L_z, \\ W_0 + (1 - W_0)\frac{L_s - L_m - 2L_z}{2L_a - L_b - L_m - 2L_z}, & L_m + 2L_z < L_s \le 2L_a - L_b, \\ 1, & 2L_a - L_b < L_s \le 2L_a + L_b, \\ 1 - \frac{L_s - 2L_a - L_b}{L_m - L_b}, & 2L_a + L_b < L_s \le L_m + 2L_a, \\ 0, & L_s > L_m + 2L_a \end{cases},$$

$$W_0 = 2\frac{L_m + L_z - L_a}{L_m - L_b}, \; L_s = \lambda_f L_{s0},$$

$$W_{ovA} = (L_m - L_b) \cdot W_{ov} \cdot 0.5,$$

where $2L_a$ and L_m are the lengths of actin and myosin filaments, L_b is the length of the bare zone (free of myosin heads) of the myosin filament, $2L_z$ is the length of special structures, z-disks, acting as sarcomeres' borders. L_s and L_{s0} are actual and initial sarcomere lengths.

Appendix A.3. Specifications of "Scar" Area and the Heterogeneity

The non-excitable region in the simulations was specified by the parameters k_{exc} and k_{rep}, which were defined as follows:

$$k_{exc}(\mathbf{x}) = k_{exc0} + (k_{excSc} - k_{exc0}) f_{scar}(x, y),$$
$$k_{rep}(\mathbf{x}) = k_{rep0} + (k_{repSc} - k_{rep0}) f_{scar}(x, y),$$

$$f_{scar}(x,y) = \begin{cases} 1, \; (x,y) \in [x_1, \; x_2] \times [y_1, \; y_2], \\ 0, \; (x,y) \notin [x_1 - d_x, \; x_2 + d_x] \times [y_1 - d_y, \; y_2 + d_y], \\ 1 - \frac{\rho^2(x,[x_1,x_2]) + \rho^2(y,[y_1,y_2])}{I(x \notin [x_1,x_2]) d_x^2 + I(y \notin [y_1,y_2]) d_y^2}, \; \text{otherwise.} \end{cases}$$

Here, $\rho(x, [x_1, x_2]) = \min\limits_{y \in [x_1, x_2]} |y - x|$, $I(\cdot)$ is an indicator function, while x_i, y_i specify the "scar" position, and d_x, d_y specify the gray zone size. The coefficients k_{exc} and k_{rep} changed from their values in tissue, k_{exc0} and k_{rep0}, to their values in the "scar" region, k_{excSc} and k_{repSc}, setting the tissue excitability and APD. The parameters are presented in Table A1.

Table A1. The model parameters defining the "scar" area and the heterogeneity.

Parameter	Value
k_{exc0}	1—normal conditions 1.5—impaired excitation
k_{rep0}	1 (in the simulations with homogeneous APD properties)
$k_{repENDO}$	0.85
k_{repEPI}	1
k_{excSc}	4.5
k_{repSc}	6
k_l	4 (in the simulations with homogeneous APD properties)
k_{lENDO}	6.5
k_{lEPI}	2.5
x_1, x_2	$0.35 \cdot L$, $0.4 \cdot L$ ($L = 9$ cm)
y_1, y_2	$0, 0.6 \cdot L$
d_x, d_y	$0.005 \cdot L, 0.03 \cdot L$

Introducing the "endo"-to-"epi" heterogeneity, we varied the model parameter of linear dependence on λ_f in the equations for calcium-troponin complexes A_i ($k_l = 4$ in [18], see also Equation (A3) in Appendix A.1). We varied the length-dependence parameter linearly from 6.5 at the "endo" surface to 2.5 at the "epi" surface: $k_l = k_{lENDO} + (k_{lEPI} - k_{lENDO})Z$. We chose the values comparing the steady-state solution, in particular, the length-dependence shift of Ca^{2+}-force curves, with experimental data [21]. The APD was varied through the k_{rep0} parameter, which affected the decrease rate of AP, changing its value linearly from 0.85 in the "endo" surface to 1 in the "epi" surface: $k_{rep0} = k_{repENDO} + (k_{repEPI} - k_{repENDO})Z$. That corresponded to the variation of APD from 340 ms to 280 ms at the same external initiating stimulus in the 0D simulations of uniaxial contraction.

Other model parameters are listed in Table A2.

Table A2. Parameters for all model blocks.

	Cross-bridge kinetics	
k_{cb}	75	s^{-1}
a_{cb}	1.5	-
b_{cb}	8.5	-
c_{cb}	0.4	nm^{-1}
k_{21}	20	-
h	10	nm
δ_*	0.4	-
	Regulation kinetics	
k_A	50	s^{-1}
K_A	0.2	-
K_B	2.5	-
k_{ACoop}	3	-
$k_{\theta 1}$	0.35	-
$k_{\theta 2}$	40	-
	Monodomain model	
D_{iso}	5	cm^2/s
D_{aniso}	45	cm^2/s
	Aliev–Panfilov model and MEF	
τ	0.0129	s
k_{AP}	8	-
a_{AP}	0.1	-
b_{AP}	0.1	-
μ_1	0.2	-
μ_2	0.3	-
ε	0.01	-
k_m	1	s^{-1}
K_m	2.5	-
	Calcium currents	
C_{Tn}	70	µM
α_{cyt}	0.75	-
α_{sr}	0.05	-
B_{cyt}	130	µM
K_{cyt}	1	µM
B_{sr}	10	mM
K_{sr}	1	mM
G_{xfer}	3.8	s^{-1}
G_{rel}	150	s^{-1}
K_{rel}	400	µM
G_{leak}	0.03	s^{-1}
G_{CaL}	400	µM/s
p_0	6.211	-
p_1	−7.233	-
p_2	1.648	-

Table A2. *Cont.*

	Calcium currents	
k_U	13.33	-
U_0	0.75	-
G_{NCX}	52.5	mM/s
Na_{i0}	10.5	µM
Ca_o	2	mM
Na_o	140	µM
α_{NCX}	2.5	-
η_{NCX}	0.35	-
k_{sat}	0.1	-
k_{NCXL}	0.8	-
K_{Na}	87.5	µM
K_{Ca}	1300	µM
G_{up}	500	µM/s
K_{up}	0.4	µM
k_p	1	$\mu M^{-2}/s$
K_p	0.325	µM
k_R	0.25	s^{-1}
K_R	125	-
K_{ssR}	200	µM
	Passive myocardium stress	
μ_{pas}	0.55	kPa
b_{pas}	2.85	-
K_{pas}	50	kPa
	Titin tension	
t_{tit}	1.03	kPa
q_1	37.02	-
q_2	-257.47	-
q_2^*	772.41	-
q_3	556.32	-
	Active tension	
E_{cb}	2.5	pN/nm
N_{cb}	150	-
N_m	283	μm^{-2}
	Sarcomere dimensions	
L_{s0}	1.9	µm
L_{min}	1.3	µm
L_a	1.12	µm
L_m	1.63	µm
L_b	0.15	µm
L_z	0.035	µm

References

1. Plank, G.; Zhou, L.; Greenstein, J.L.; Cortassa, S.; Winslow, R.L.; O'Rourke, B.; Trayanova, N.A. From mitochondrial ion channels to arrhythmias in the heart: computational techniques to bridge the spatio-temporal scales. *Philos. Trans. R. Soc. A* **2008**, *366*, 3381–3409. [CrossRef]
2. Gurev, V.; Pathmanathan, P.; Fattebert, J.L.; Wen, H.F.; Magerlein, J.; Gray, R.A.; Richards, D.F.; Rice, J.J. A high-resolution computational model of the deforming human heart. *Biomech. Model. Mechanobiol.* **2015**, *14*, 829–849. [CrossRef]
3. Augustin, C.M.; Neic, A.; Liebmann, M.; Prassl, A.J.; Niederer, S.A.; Haase, G.; Plank, G. Anatomically accurate high resolution modeling of human whole heart electromechanics: A strongly scalable algebraic multigrid solver method for nonlinear deformation. *J. Comput. Phys.* **2016**, *305*, 622–646. [CrossRef] [PubMed]
4. Regazzoni, F.; Quarteroni, A. An oscillation-free fully staggered algorithm for velocity-dependent active models of cardiac mechanics. *Comput. Methods Appl. Mech. Eng.* **2021**, *373*, 113506. [CrossRef]
5. Cansız, B.; Kaliske, M. A comparative study of fully implicit staggered and monolithic solution methods. Part II: Coupled excitation–contraction equations of cardiac electromechanics. *J. Comput. Appl. Math.* **2024**, *442*, 115704. [CrossRef]
6. Nash, M.P.; Panfilov, A.V. Electromechanical model of excitable tissue to study reentrant cardiac arrhythmias. *Prog. Biophys. Mol. Biol.* **2004**, *85*, 501–522. [CrossRef]
7. Costabal, F.S.; Concha, F.A.; Hurtado, D.E.; Kuhl, E. The importance of mechano-electrical feedback and inertia in cardiac electromechanics. *Comput. Methods Appl. Mech. Eng.* **2017**, *320*, 352–368. [CrossRef]

8. Sung, D.; Mills, R.W.; Schettler, J.; Narayan, S.M.; Omens, J.H.; McCulloch, A.D. Ventricular filling slows epicardial conduction and increases action potential duration in an optical mapping study of the isolated rabbit heart. *J. Cardiovasc. Electrophysiol.* **2003**, *14*, 739–749. [CrossRef]
9. Pfeiffer, E.R.; Tangney, J.R.; Omens, J.H.; McCulloch, A.D. Biomechanics of cardiac electromechanical coupling and mechanoelectric feedback. *J. Biomech. Eng.* **2014**, *136*, 021007. [CrossRef]
10. Willems, E.; Janssens, K.L.P.M.; Dekker, L.R.C.; van de Vosse, F.N.; Cluitmans, M.J.M.; Bovendeerd, P.H.M. Strain-controlled electrophysiological wave propagation alters in silico scar-based substrate for ventricular tachycardia. *Front. Physiol.* **2024**, *15*. [CrossRef]
11. Molavi Tabrizi, A.; Mesgarnejad, A.; Bazzi, M.; Luther, S.; Christoph, J.; Karma, A. Spatiotemporal Organization of Electromechanical Phase Singularities during High-Frequency Cardiac Arrhythmias. *Phys. Rev. X* **2022**, *12*, 021052. [CrossRef]
12. Fenton, F.H.; Cherry, E.M.; Hastings, H.M.; Evans, S.J. Multiple mechanisms of spiral wave breakup in a model of cardiac electrical activity. *Chaos Interdiscip. J. Nonlinear Sci.* **2002**, *12*, 852–892. [CrossRef] [PubMed]
13. Kadota, S.; Kay, M.W.; Magome, N.; Agladze, K. Curvature-dependent excitation propagation in cultured cardiac tissue. *JETP Lett.* **2012**, *94*, 824–830. [CrossRef]
14. Liogky, A.A.; Chernyshenko, A.Y.; Danilov, A.A.; Syomin, F.A. CarNum: Parallel numerical framework for computational cardiac electromechanics. *Russ. J. Numer. Anal. Math. Model.* **2023**, *38*, 127–144. [CrossRef]
15. Danilov, A.A.; Liogky, A.A.; Syomin, F.A. Temporally and spatially segregated discretization for a coupled electromechanical myocardium model. *Russ. J. Numer. Anal. Math. Model.* **2024**, *39*, 243–258. [CrossRef]
16. Vassilevski, Y.; Terechov, K.; Nikitin, K.; Kapyrin, I., INMOST Platform Technologies for Numerical Model Development. In *Parallel Finite Volume Computation on General Meshes*; Springer International Publishing: Cham, Switzerland, 2020; pp. 127–177. [CrossRef]
17. Hindmarsh, A.C.; Brown, P.N.; Grant, K.E.; Lee, S.L.; Serban, R.; Shumaker, D.E.; Woodward, C.S. SUNDIALS: Suite of nonlinear and differential/algebraic equation solvers. *ACM Trans. Math. Softw. (TOMS)* **2005**, *31*, 363–396. [CrossRef]
18. Syomin, F.; Osepyan, A.; Tsaturyan, A. Computationally efficient model of myocardial electromechanics for multiscale simulations. *PLoS ONE* **2021**, *16*, e0255027. [CrossRef]
19. Aliev, R.R.; Panfilov, A.V. A simple two-variable model of cardiac excitation. *Chaos Solitons Fractals* **1996**, *7*, 293–301. [CrossRef]
20. Fung, Y.C.B. Biorheology of soft tissues. *Biorheology* **1973**, *10*, 139–155. [CrossRef]
21. Cazorla, O.; Szilagyi, S.; Le Guennec, J.Y.; Vassort, G.; Lacampagne, A. Transmural stretch-dependent regulation of contractile properties in rat heart and its alteration after myocardial infarction. *FASEB J.* **2005**, *19*, 88–90. [CrossRef]
22. Cluitmans, M.J.M.; Bayer, J.; Bear, L.R.; ter Bekke, R.M.A.; Heijman, J.; Coronel, R.; Volders, P.G.A. The circle of reentry: Characteristics of trigger-substrate interaction leading to sudden cardiac arrest. *Front. Cardiovasc. Med.* **2023**, *10*, 1121517. [CrossRef]
23. Pandit, S.V.; Jalife, J. Rotors and the Dynamics of Cardiac Fibrillation. *Circ. Res.* **2013**, *112*, 849–862. [CrossRef]
24. Pravdin, S.F.; Nezlobinsky, T.V.; Panfilov, A.V.; Dierckx, H. High-frequency pacing of scroll waves in a three-dimensional slab model of cardiac tissue. *Phys. Rev. E* **2021**, *103*, 042420. [CrossRef]
25. Nizamieva, A.A.; Kalita, I.Y.; Slotvitsky, M.M.; Berezhnoy, A.K.; Shubina, N.S.; Frolova, S.R.; Tsvelaya, V.A.; Agladze, K.I. Conduction of excitation waves and reentry drift on cardiac tissue with simulated photocontrol-varied excitability. *Chaos Interdiscip. J. Nonlinear Sci.* **2023**, *33*, 023112. [CrossRef] [PubMed]
26. Fast, V.G.; Kléber, A.G. Role of wavefront curvature in propagation of cardiac impulse. *Cardiovasc. Res.* **1997**, *33*, 258–271. [CrossRef] [PubMed]
27. Göktepe, S.; Kuhl, E. Electromechanics of the heart: A unified approach to the strongly coupled excitation–contraction problem. *Comput. Mech.* **2010**, *45*, 227–243. [CrossRef]
28. Alonso, S.; Santos, R.; Bär, M. Reentry and Ectopic Pacemakers Emerge in a Three-Dimensional Model for a Slab of Cardiac Tissue with Diffuse Microfibrosis near the Percolation Threshold. *PLoS ONE* **2016**, *11*, e0166972. [CrossRef]
29. Regazzoni, F.; Dedè, L.; Quarteroni, A. Biophysically detailed mathematical models of multiscale cardiac active mechanics. *PLoS Comput. Biol.* **2020**, *16*, e1008294. [CrossRef]
30. Gerach, T.; Schuler, S.; Fröhlich, J.; Lindner, L.; Kovacheva, E.; Moss, R.; Wülfers, E.M.; Seemann, G.; Wieners, C.; Loewe, A. Electro-Mechanical Whole-Heart Digital Twins: A Fully Coupled Multi-Physics Approach. *Mathematics* **2021**, *9*, 1247. [CrossRef]
31. de Oliveira, B.L.; Pfeiffer, E.R.; Sundnes, J.; Wall, S.T.; McCulloch, A.D. Increased cell membrane capacitance is the dominant mechanism of stretch-dependent conduction slowing in the rabbit heart: A computational study. *Cell. Mol. Bioeng.* **2015**, *8*, 237–246. [CrossRef]

mathematics

MDPI

Article

Spectral $\mathbb{R}$-Linear Problems: Applications to Complex Permittivity of Coated Cylinders

Zhanat Zhunussova [1,2,3,†] and Vladimir Mityushev [1,4,*,†]

1 Institute of Mathematics and Mathematical Modeling, Pushkina St., 125, Almaty 050010, Kazakhstan; zhunusova@math.kz
2 Faculty of Mechanics and Mathematics, Al-Farabi Kazakh National University, Al-Farabi Av., 71, Almaty 050040, Kazakhstan
3 Faculty of Computer Technology and Cybersecurity, International Information Technology University, Manasa St., 34/1, Almaty 050040, Kazakhstan
4 Faculty of Computer Science and Telecommunications, Cracow University of Technology, Warszawska St., 24, 31-155 Krakow, Poland
* Correspondence: wladimir.mitiuszew@pk.edu.pl
† These authors contributed equally to this work.

Abstract: A composite-coated inclusion is embedded in a matrix, where the conductivity (permittivity) of the phases is assumed to be complex-valued. The purpose of this paper is to demonstrate that a non-zero flux can arise under specific conditions related to the conductivities of the components in the absence of external sources. These conditions are unattainable with conventional positive conductivities but can be satisfied when the conductivities are negative or complex—a scenario achievable in the context of metamaterials. The problem is formulated as a spectral boundary value problem for the Laplace equation, featuring a linear conjugation condition defined on a smooth curve L. This curve divides the plane $\mathbb{R}^2$ into two regions, D^+ and $D^- \ni \infty$. The spectral parameter appears in the boundary condition, drawing parallels with the Steklov eigenvalue problem. The case of a circular annulus is analyzed using the method of functional equations. The complete set of eigenvalues is derived by applying the classical theory of self-adjoint operators in Hilbert space.

Keywords: complex-valued permittivity; functional equation; coated inclusion; spectral boundary value problem

MSC: 74Q15; 35P15

1. Introduction

One of the most prominent models in the theory of composites, the coated disk (or sphere) assemblage, represents a composite structure comprising circular inclusions of one phase embedded within a matrix of another phase. This model was first introduced by Hashin [1] and developed in [2] (Chapter 7) and [3–5]. The problem can be approached as a neutral inclusion problem, wherein researchers sought a specific non-zero solution to illustrate the cloaking phenomenon, the process of concealing an object from detection. This perspective played a pivotal role in the advancement of the metamaterial theory [6,7], a rapidly evolving domain in modern physics. In the quasistatic limit, the electromagnetic wave propagation equations are simplified to standard conductivity equations, albeit with complex local fields and complex-valued conductivity (or permittivity). Veselago first observed the negative permittivity values [8]. Complex permittivity values emerge due to energy loss in certain materials. The real part of permittivity represents the traditional dielectric constant, while the imaginary part characterizes energy dissipation, also known as loss permittivity [2,9]. Although the number of pure substances in nature is finite, their diverse compositions give rise to an infinite variety of materials at the macroscale. Consequently, the classical theory of composites covers a broad range of material property constants, including complex-valued parameters for metamaterials that adhere to some bounds [2]. In previous studies, analytical formulas and simulations of local fields were employed to demonstrate the physical attributes of the structures examined. The current work extends this exploration by linking such problems to a classical spectral framework.

Spectral boundary value problems for PDEs are classic analysis topics [10–13]. We omit an extended study on spectral problems for ODEs and instead refer to one of the latest papers [14]. To present the significance of the new result obtained in this paper, first, we outline three types of spectral problems and describe the third type related to a new physical trend concerning metamaterials.

Spectral problems involving a spectral parameter λ that is a control parameter of a partial differential equation have been sufficiently well studied. Equation $\Delta u + \lambda u = 0$ in a domain D with the Dirichlet boundary condition $u = 0$ on ∂D is one of the fundamental spectral problems. The main keys of the theory for this problem in a Riemannian manifold D are presented in [15]. A recent review of this problem can be found in [10–13], and the works cited therein.

The study of spectral boundary value problems with the spectral parameter in the boundary condition begins with the Steklov eigenvalue problem for the Laplace equation $\Delta u = 0$ in D with the boundary condition $\frac{\partial u}{\partial n} u = \lambda u$ on ∂D, where $\frac{\partial}{\partial n}$ denotes the normal derivative. Recent results on this problem can be found in [16–19], and the works cited therein. The present paper is devoted to the third type of spectral problem when, instead of a boundary condition, the following linear conjugation condition is given on a smooth curve L, dividing the plane $\mathbb{R}^2 \equiv \mathbb{C}$ into two domains D^+ and D^- [20–23]:

$$u^+ = u^-, \quad \frac{\partial u^+}{\partial n} = \lambda \frac{\partial u^-}{\partial n} \quad \text{on } L = \partial D^+, \tag{1}$$

where the signs $\pm$ correspond to the limit values of the function u harmonic in $D^\pm$ and continuously differentiable in the closures of $D^\pm$. Here, it is convenient to consider the extended complex plane $\widehat{\mathbb{C}} = \mathbb{C} \cup \{\infty\}$ and assume that $\infty \in D^-$. It should be noted that the domain D^+ may not be connected, and the constants λ may be different in the different components of L. The conjugation condition is called the perfect contact or transmission condition in the theory of composites.

Schiffer [24] considered an eigenvalue problem for a double-layer potential on a set of curves and showed that this problem can be reduced to the spectral problem (1). The spectral parameter λ was supposed to be the same for all components of the curve.

The present paper is devoted to studying a coated composite with complex-valued conductivity (permittivity) phases. This physical problem is stated as a spectral $\mathbb{R}$-linear problem. The problem is reduced to an iterative functional equation [25]. It is established that the theory of neutral inclusion related to metamaterials may be considered a new type of spectral problem with the spectral parameter λ in the conjugation condition (1).

2. Functional Spaces

Let $\mathcal{H}^2(\mathbb{U})$ denote the classic Hardy space of analytic functions in the unit disk $\mathbb{U} = \{\mathbb{C} : |z| < 1\}$ [26], and let $\mathcal{H}^2(\mathbb{U}_r)$ be the corresponding space on the disk $\mathbb{U}_r = \{\mathbb{C} : |z| < r\}$. The Hilbert space $\mathcal{H}^2(\mathbb{U}_r)$ is endowed with the norm

$$\|f\| = \sup_{0 \le r_0 < r} \frac{1}{2\pi} \int_0^{2\pi} |f(r_0 \exp(i\theta)|^2 d\theta. \tag{2}$$

Introduce a sequence space h_r^2 isomorphic to $\mathcal{H}^2(\mathbb{U}_r)$ as follows. Let a function

$$F(z) = \sum_{k=1}^{\infty} \alpha_k z^k \tag{3}$$

belong to $\mathcal{H}^2(\mathbb{U}_r)$. Then, the coefficients $\{\alpha_k\}_{k=0}^{\infty}$ of the function (3) can be considered as elements of a discrete space. The sets of these coefficients $\{\alpha_k\}_{k=0}^{\infty}$ form the space h_r^2 isomorphic to $\mathcal{H}^2(\mathbb{U}_r)$.

The isomorphism $\mathcal{H}^2(\mathbb{U}_r) \equiv h_r^2$ has to be checked on the level of the scalar product since the considered spaces are Hilbert. Consider the radial limit of $F(z)$ on the circle $z = r\exp\theta$ existing for almost every argument $\theta \in [0, 2\pi)$:

$$f(\theta) := F\left(re^{i\theta}\right), \quad 0 \le \theta < 2\pi. \tag{4}$$

The scalar product (F, G) in the Hardy space $\mathcal{H}^2(\mathbb{U}_r)$ coincides with the scalar product (f, g) in the space L_2 given by the integral

$$(f, g) = \frac{1}{2\pi} \int_0^{2\pi} f(\theta)\overline{g(\theta)} d\theta. \tag{5}$$

Using (4), find

$$(F, G) = (f, g) = \frac{1}{2\pi i} \int_{|z|=r} F(z)\overline{G(z)} \frac{dz}{z}. \tag{6}$$

Transform the integral (6)

$$(F, G) = \frac{1}{2\pi i} \int_{|z|=r} F(z)\overline{G\left(\frac{r^2}{\overline{z}}\right)} \frac{dz}{z}. \tag{7}$$

By substituting the Taylor series of functions $F(z)$ and $G(z)$, we find that

$$F(z) = \sum_{k=1}^{\infty} \alpha_k z^k, \quad G(z) = \sum_{k=1}^{\infty} \beta_k z^k, \quad |z| < r. \tag{8}$$

Find

$$\overline{G\left(\frac{r^2}{\overline{z}}\right)} = \sum_{k=1}^{\infty} \overline{\beta_k} \frac{r^{2k}}{z^k}, \quad |z| = r \tag{9}$$

and substitute the first expressions of (8) and (9) into (7). The obtained integral can be calculated by the residue theorem, by the coefficient in $\frac{1}{z}$ of the integrand

$$(F, G) = \sum_{k=1}^{\infty} r^{2k} \alpha_k \overline{\beta_k}. \tag{10}$$

Consequently, the norms in the considered spaces can be written in the form

$$\|f\| = \|F\| = \sum_{k=0}^{\infty} r^{2k} |\alpha_k|^2. \tag{11}$$

3. Spectral Problem

Let $D_1 = \{z \in \mathbb{C} : |z| < r\}$, $D_2 = \{z \in \mathbb{C} : |z| > 1\}$ and $D = \{z \in \mathbb{C} : r < |z| < 1\}$ be domains in the extended complex plane $\widehat{\mathbb{C}} = \mathbb{C} \cup \{\infty\}$ (see Figure 1). We are looking for functions $u_1(x_1, x_2)$, $u_2(x_1, x_2)$ and $u(x_1, x_2)$ that are harmonic in D_1, D_2, and D, respectively, and belonging to the Sobolev spaces $W^{(1,2)}$ [27] in the corresponding domains.

Consider the following conductivity problem. Let the domains D_1, D_2, and D be occupied by the materials of conductivities σ_1 and σ_2, and the normalized conductivity $\sigma = 1$, respectively. The perfect contact between the components of the considered composite has the form [21]

$$u_1 = u, \quad \sigma_1 \frac{\partial u_1}{\partial n} = \frac{\partial u}{\partial n}, \quad |z| = r, \tag{12}$$

$$u_2 = u, \quad \sigma_2 \frac{\partial u_2}{\partial n} = \frac{\partial u}{\partial n}, \quad |z| = 1, \tag{13}$$

where $\frac{\partial}{\partial n}$ are the outward normal derivatives to the circles. It is supposed that

$$u_2(\infty) = 0. \tag{14}$$

The boundary value problem in Equations (12)–(14) can be stated for the spectral parameter σ_1 or σ_2. We will demonstrate below its equivalence for a complex boundary value problem in relation to Equations (12)–(14).

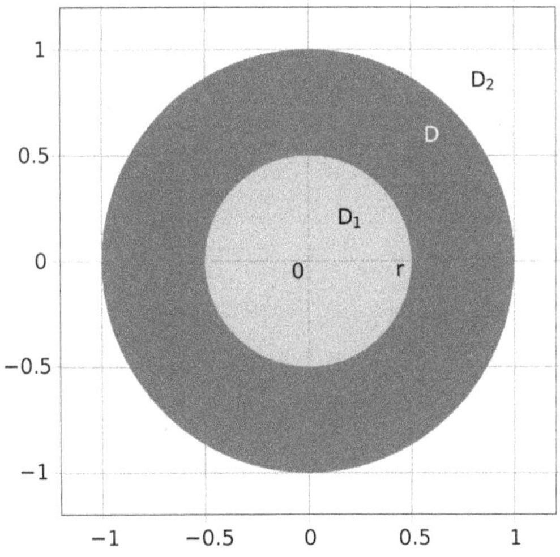

Figure 1. A coated composite on the complex plane. The domains D_1 and D_2 are occupied by materials of conductivities σ_1 and σ_2, respectively. The conductivity of the annulus $r < |z| < 1$ is normalized to unity.

Following [21], we introduce complex potentials analytically in the considered domain:

$$u(z) + iv(z) = \varphi(z), z \in D, \tag{15}$$

$$u_k(z) + iv_k(z) = \frac{2}{\sigma_k + 1} \varphi_k(z), \ z \in D_k \ (k = 1, 2), \tag{16}$$

where v_k are the imaginary parts of the considered analytic functions. Let $\frac{\partial}{\partial s}$ denote a tangent derivative, where s is a natural length parameter of arcs. The Riemann–Cauchy conditions on the circles $L_1 = \{z \in \mathbb{C} : |z| = r\}$ and $L_2 = \{z \in \mathbb{C} : |z| = 1\}$ have the form

$$\frac{\partial u}{\partial n} = \frac{\partial v}{\partial s}, \quad \frac{\partial u}{\partial s} = -\frac{\partial v}{\partial n}. \tag{17}$$

Then, the second Equations (12) and (13) become

$$\frac{\partial v}{\partial s}(t) = \sigma_k \frac{\partial v_k}{\partial s}(t), \; t \in L_k \; (k = 1, 2), \tag{18}$$

By integrating (18), we get

$$v(t) = \sigma_k v_k(t) + d_k, \; t \in L_k, \tag{19}$$

where d_k are arbitrary constants.

We then introduce the contrast parameters

$$\varrho_k = \frac{\sigma_k - 1}{\sigma_k + 1}, \text{ on } L_k \; (k = 1, 2). \tag{20}$$

The first real Equation (12) and the real Equation (19) for $k = 1$ can be written in the first of one complex equation. The same concerns the second pair of real equations for $k = 2$. The obtained complex equations can be written in the form of the $\mathbb{R}-$linear problem:

$$\varphi(t) = \varphi_k(t) - \varrho_k \overline{\varphi_k(t)} + C_k, \; t \in L_k \; (k = 1, 2), \tag{21}$$

where C_k are constants. The function $\varphi_2(z)$ is bounded at infinity.

We introduce the complex velocity

$$\psi(z) = \varphi'(z) = \frac{\partial u}{\partial x_1}(z) - i\frac{\partial u}{\partial x_2}(z), \quad z \in D, \tag{22}$$

and

$$\psi_k(z) = \varphi_k'(z), \quad z \in D_k \; (k = 1, 2). \tag{23}$$

The boundary conditions (21) can be differentiated along the curves L_k. It was shown in [20,21] that

$$\psi(t) = \psi_k(t) + \varrho_k \overline{n(t)^2} \; \overline{\psi_k(t)}, \quad t \in L_k \; (k = 1, 2), \tag{24}$$

where the unit outward normal vectors $n(t)$ are expressed in terms of complex values.

4. Functional Equation

In the present section, the boundary value problem is reduced to a functional equation that will be solved.

The normal vector can be written as follows:

$$n(t) = \begin{cases} \frac{t}{r}, & |t| = r, \\ t, & |t| = 1. \end{cases} \tag{25}$$

Substitute (25) into (24) and write the resulting equation in the extended form:

$$\begin{aligned} \psi(t) &= \psi_1(t) + \varrho_1 \frac{r^2}{t^2}\overline{\psi_1(t)}, \quad |t| = r, \\ \psi(t) &= \psi_2(t) + \varrho_2 \frac{1}{t^2}\overline{\psi_2(t)}, \quad |t| = 1. \end{aligned} \tag{26}$$

The function $\varphi_2(z)$ near infinity is represented by the series

$$\varphi_2(z) = constant + \frac{c_1}{z} + \frac{c_2}{z^2} + \dots. \tag{27}$$

Hence,

$$\psi_2(z) = -\frac{c_1}{z^2} - \frac{2c_2}{z^3} + \dots, \tag{28}$$

in particular,

$$\psi_2(\infty) = 0. \tag{29}$$

The spectral problem for the complex velocity can be stated as follows. It is required to find ψ_1, ψ_2 and constants such as ϱ_1 and ϱ_2 that the boundary value problem given by Equations (26) and (29) has a non-zero solution. One of the parameters ϱ_1 and ϱ_2 can be fixed. For example, ϱ_1 is fixed, and ϱ_2 has to be found. In the case of the unknown $\varrho_1 = \varrho_2 = \varrho$, we arrive at the Schiffer spectral problem [24].

Introduce the analytic function separately in $|z| < r$, $r < |z| < 1$, and $|z| > 1$, and ensure that it is separately continuous in the closures of the considered domains:

$$\Phi(z) = \begin{cases} \psi_1(z) - \varrho_2 \frac{1}{z^2}\overline{\psi_2\left(\frac{1}{\overline{z}}\right)}, & |z| \leqslant r, \\ \psi_2(z) - \varrho_1 \frac{r^2}{z^2}\overline{\psi_1\left(\frac{r^2}{\overline{z}}\right)}, & |z| \geqslant 1, \\ \psi(z) - \varrho_1 \frac{r^2}{z^2}\overline{\psi_1\left(\frac{r^2}{\overline{z}}\right)} - \varrho_2 \frac{1}{z}\overline{\psi_2\left(\frac{1}{\overline{z}}\right)}, & r < |z| < 1. \end{cases} \tag{30}$$

Calculate the jump of $\Phi(z)$ across $|t| = r$ using the relation $t = \frac{r^2}{\overline{t}}$,

$$\begin{aligned} \Phi^+(t) - \Phi^-(t) &= \psi_1(t) - \psi(t) + \varrho_1 \frac{r^2}{t^2}\overline{\psi_1\left(\frac{r^2}{\overline{t}}\right)} = \\ &= \psi_1(t) + \varrho_1 \frac{r^2}{t^2}\overline{\psi_1(t)} - \psi(t) = 0, \end{aligned} \tag{31}$$

according the first relation (26). Therefore, $\Phi(z)$ is analytically continued through the circle $|z| = r$ by the principle of analytic continuation. In the same way, one can check that $\Phi(z)$ is analytically continued through the circle $|z| = 1$. Find $\Phi(\infty) = 0$ using Equation (29). By Liouville's theorem, $\Phi(z) = 0$ everywhere in the extended complex plane. Writing $\Phi(z) = 0$ in the extended form (30) in $|z| \leqslant r$ and $|z| \geqslant 1$, we obtain the system of functional equations

$$\begin{aligned} \psi_1(z) &= \varrho_1 \frac{1}{z^2}\overline{\psi_2\left(\frac{1}{\overline{z}}\right)}, \quad |z| \leqslant r, \\ \psi_2(z) &= \varrho_2 \frac{r^2}{z^2}\overline{\psi_1\left(\frac{r^2}{\overline{z}}\right)}, \quad |z| \geqslant 1. \end{aligned} \tag{32}$$

Substituting $\frac{1}{\bar{z}}$ instead of z in the second Equation (32), we obtain

$$\overline{\psi_2\left(\frac{1}{\bar{z}}\right)} = \overline{\varrho_2} r^2 z^2 \psi_1\left(r^2 z\right), \quad |z| \leqslant 1. \tag{33}$$

Substituting Equation (33) into the first Equation (32) yields the functional equation

$$\psi_1(z) = \varrho_1 \overline{\varrho_2} r^2 \psi_1\left(r^2 z\right), \quad |z| \leqslant r. \tag{34}$$

The unknown function $\psi_1(z)$ is analytic in the disk $|z| < r$ and continuous in $|z| \leqslant r$. One can consider the functional equation in the wider Hardy space $\mathcal{H}^2(\mathbb{U}_r)$. The result does not depend on the space since the function $\psi_1(z)$ is analytically continued to the complex plane $\mathbb{C}$ by Equation (34).

4.1. Case of Equal Contrast Parameters

In the present section, consider the case $\varrho := \varrho_1 = \varrho_2$. Introduce the operator

$$(A\psi_1)(z) = r^2 \psi_1\left(r^2 z\right) \tag{35}$$

acting on $\psi_1(z)$ in the written part of Equation (34). This is a shift operator inside the domain of analyticity. According to Shapiro [28], the shift operator is a compact operator in $\mathcal{H}^2(\mathbb{U}_r)$. Thus, we arrive at the eigenvalue problem for the operator A:

$$\lambda\psi_1 = A\psi_1 \quad \Longleftrightarrow \quad \lambda\psi_1(z) = r^2\psi_1\left(r^2 z\right),\ |z| \leqslant r \tag{36}$$

with the spectral parameter $\lambda = |\varrho|^{-2}$.

Represent the unknown function in the form of its Taylor series as follows:

$$\psi_1(z) = \sum_{k=0}^{\infty} \alpha_k z^k,\ |z| < r, \tag{37}$$

Substitute it into Equation (36) and take the coefficients in the same power of z:

$$\lambda\alpha_k = r^2 \alpha_k r^{2k}, \quad k = 0, 1, \ldots. \tag{38}$$

Equation (38) has a countable number of non-zero solutions:

$$\lambda_k = r^{2(k+1)}, \quad k = 0, 1, \ldots. \tag{39}$$

The corresponding solutions of the functional Equation (36) have the form

$$\psi_1^{(k)}(z) = z^k, \quad k = 0, 1, \ldots. \tag{40}$$

Theorem 1. *The compact operator A defined by Equation* (35) *is self-adjoint in $\mathcal{H}_2$. Its simple eigenvalues are given by Equation* (39) *and the eigenfunctions in* (40). *The relation* (39) *can be written in the form*

$$|\varrho|^2 r^{2(k+1)} = 1, \quad k = 0, 1, \ldots. \tag{41}$$

Proof. Find an operator A^* adjoint to the operator A by the definition of the scalar product (5)

$$(Af, g) = (f, A^* g). \tag{42}$$

It will be shown that $A = A^*$.

We have

$$(f, g) = \sum_{k=0}^{\infty} f_k \bar{g}_k r^{2k}. \tag{43}$$

Using (35), calculate

$$(Af, g) = \left(r^2 f\left(r^2 z\right), g(z)\right) = \sum_{k=0}^{\infty} f_k \bar{g}_k r^{2(k+1)} r^{2k}. \tag{44}$$

Equate the latter expression to $(f, A^* g)$

$$(A^* g)(z) = r^2 \sum_{k=0}^{\infty} g_k r^{2k} z^k = r^2 g\left(r^2 z\right). \tag{45}$$

Hence, $A = A^*$. Therefore, the operator A is self-adjoint.

It follows from the properties of the compact self-adjoint operators in the Hilbert space that the spectrum of the operator A consists of the eigenvalues in (39) satisfying the inequality

$$\lambda_0 > \lambda_1 > \lambda_2 > \dots \lambda_k > \dots. \tag{46}$$

One can see that $\lambda_k \to 0$, as $k \to \infty$, corresponds to the general theory. The eigenvalues given by (40) are simple since the monomials z^k are orthogonal with respect to the inner product (5), and form a basis for $\mathcal{H}^2(D_1)$.

The theorem is proved. □

4.2. Case of Different Contrast Parameters

In the present section, the parameter ϱ_1 is fixed and ϱ_2 is considered as a spectral parameter.

Introduce the operator

$$(B\psi_1)(z) = \overline{\varrho_1} r^2 \psi_1\left(r^2 z\right), \tag{47}$$

acting on $\psi_1(z)$ in the right part of (34).

Theorem 2. *The compact operator B defined by Equation* (47) *is self-adjoint in $\mathcal{H}_2$. Its eigenvalues are given by*

$$\lambda_k = \overline{\varrho_1} r^{2(k+1)} \quad (k = 0, 1, \dots), \tag{48}$$

the corresponding eigenfunctions z^k. The relation (48) *can be written in the form*

$$\varrho_2 \overline{\varrho_1} r^{2(k+1)} = 1 \quad (k = 0, 1, \dots). \tag{49}$$

Proof. Substitute the series (37) into Equation (34)

$$\lambda \sum_{k=0}^{\infty} \alpha_k z^k = \overline{\varrho_1} r^2 \sum_{k=0}^{\infty} \alpha_k r^{2k} z^k, \tag{50}$$

where $\lambda = \frac{1}{\varrho_2}$. In Equation (50), the coefficients with the same powers z are taken as

$$\lambda \alpha_k = \overline{\varrho_1} r^2 \alpha_k r^{2k}, \quad k = 0, 1, \dots \tag{51}$$

This yields Equation (48) and $\lambda_k \to 0$, as $k \to \infty$.

The operator $B = B^*$ is self-conjugate. It is proven in a similar way to Theorem 1 for the operator A.

The theorem is proved. □

Consider the particular case when $\varrho_1 > 0$. Then,

$$\lambda_0 > \lambda_1 > \lambda_2 > \ldots . \tag{52}$$

A numerical example of the vector field for $k = 1$ is presented in Figures 2 and 3. It is worth noting that the domains D_1 and D_2 are occupied by conducting and absorbing materials, respectively. This serves as the physical explanation of the non-vanishing field in the free-source medium. The vector field is expressed by the complex velocity. The gradient (22) can be calculated by the formula

$$\left(\frac{\partial u}{\partial x_1}(z), \frac{\partial u}{\partial x_2}(z)\right) = (\mathrm{Re}\ \psi(z), -\mathrm{Im}\ \psi(z)), \quad z \in D, \tag{53}$$

and

$$\left(\frac{\partial u_j}{\partial x_1}(z), \frac{\partial u_j}{\partial x_2}(z)\right) = \frac{2}{\sigma_j + 1}(\mathrm{Re}\ \psi_j(z), -\mathrm{Im}\ \psi(z)), \quad z \in D_j \quad (j = 1, 2). \tag{54}$$

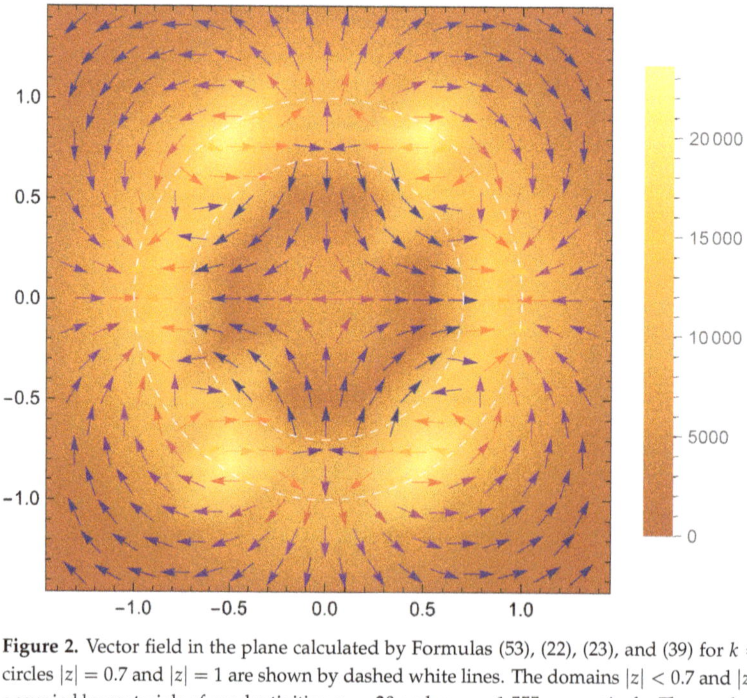

Figure 2. Vector field in the plane calculated by Formulas (53), (22), (23), and (39) for $k = 1$. The circles $|z| = 0.7$ and $|z| = 1$ are shown by dashed white lines. The domains $|z| < 0.7$ and $|z| > 1$ are occupied by materials of conductivities $\sigma_1 = 20$ and $\sigma_2 = -1.555$, respectively. The conductivity of the annulus $0.7 < |z| < 1$ is normalized to unity. The legend scale shows the intensity of the field, which is equal to $|\psi(z)|$ in D and $|\psi_j(z)|$ in D_j $(j = 1, 2)$.

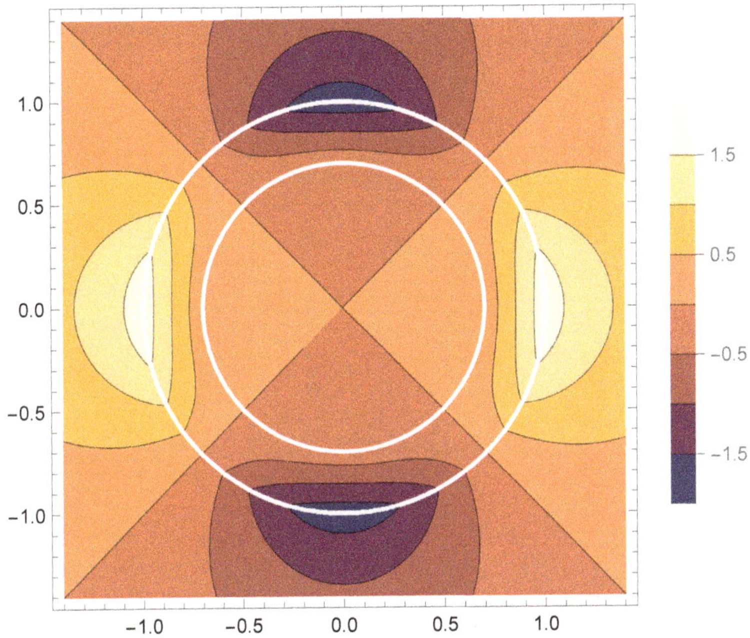

Figure 3. The contour plot illustrating the potentials $u(z) = \text{Re } \varphi(z)$ and $u_k(z) = \frac{2}{\sigma_k+1}\text{Re } \varphi_k(z)$ $(k = 1, 2)$ of the vector field from Figure 2.

4.3. Hashin–Shtrikman Assemblage

The Hashin–Shtrikman assemblage is not governed by the spectral problem considered above. The condition (29) means that the external flux at infinity is absent. Replace Equation (29) by the condition of the non-vanishing external flux parallel to the OX-axes:

$$\psi_2(\infty) = 1. \tag{55}$$

Then, we arrive at the Hashin–Shtrikman problem. Its solution is then written in terms of complex potentials in the corresponding domains:

$$\psi_1(z) = 1, \quad \psi_2(z) = 1, \quad \psi(z) = 1 + \frac{\varrho_2}{z^2}. \tag{56}$$

The analog of spectral relation (49) has the form

$$\varrho_2 = \varrho_1 r^2 \quad \Longleftrightarrow \quad \frac{\sigma_2 - 1}{\sigma_2 + 1} = \frac{\sigma_1 - 1}{\sigma_1 + 1} r^2. \tag{57}$$

Here, σ_1 and 1 are the conductivities of the core and the coating, respectively. In the framework of the Hashin–Shtrikman model, σ_2 yields the effective conductivity of neutral non-overlapping, coated inclusions distributed in the plane. It can be written in the form of the Hashin–Shtrikman bound and, equivalently, as the Clausius–Mossotti (Maxwell) approximation [2] (Chapter 7). It is worth noting that condition (57) can be fulfilled for positive conductivities σ_1 and σ_2.

5. Conclusions and Discussion

The coated inclusions in the theory of composites distinguish many exceptional properties. In the present paper, we consider a composite, displayed in Figure 1, with complex-valued conductivities under the absence of any sources and sinks in the whole plane, including infinity. It is established that a non-zero flux takes place under the conditions (49) or (41) on the contrast parameters ϱ_1 and ϱ_2 expressed through the conductivity of components by Equation (20). The relations (49) and (41) cannot be fulfilled for the traditional positive conductivities σ_1 and σ_2 and can hold only for negative and complex values. Such a case takes place for metamaterials [6,8].

A boundary value problem is stated with the spectral parameter on the boundary. Complex potentials are introduced in the considered domain through analytic functions. Next, the boundary value problem is reduced to a spectral problem for a functional equation that has been solved. Equal and different contrast parameters are considered. Using the classic theory of self-adjoint operators in the Hilbert space, the complete set of eigenvalues and eigenfunctions modeling special fields are written without external sources.

The classic spectral problem with the parameter λ in the partial differential equation, e.g., Laplace's equation, models wave processes for the special frequencies when an initial mode holds in a medium. In the considered model, Laplace's equation is not perturbed, which corresponds to a stationary process. However, the wave mode is hidden in the imaginary parts of the material constants σ_1 and σ_2 [9]. Hence, the main physical mechanism, wave mode, is the main reason for the occurrence of unusual phenomena in materials. It is interesting to study the general geometric configurations to check this hypothesis.

The Hashin–Shtrikman assemblage discussed in Section 4.3 is an example of neutral inclusions [2,3] with the classical case of positive material constants σ_1 and σ_2. It is worth noting that the complex velocity is bounded at infinity, e.g., Equation (55). At the same time, the complex potential $\varphi_2(z)$ defined by Equation (16) has a pole at infinity. This observation distinguishes the neutral inclusion from others.

The results obtained highlight the significance of the eigenvalue problem in metamaterial theory. This approach helps refine the range of specific metamaterial constants, contributing to a more precise characterization.

Author Contributions: Conceptualization and methodology, V.M.; software, validation, formal analysis, and investigation, V.M. and Z.Z.; writing—original draft preparation, and review and editing, V.M.; funding acquisition, Z.Z. All authors have read and agreed to the published version of the manuscript.

Funding: This work was supported by the Science Committee of the Ministry of Science and Higher Education of the Republic of Kazakhstan (grant no. AP23486576).

Institutional Review Board Statement: Not applicable.

Informed Consent Statement: Not applicable.

Data Availability Statement: Data are contained within the article.

Acknowledgments: We are grateful to the reviewers for the comments and suggestions that improve the presentation of the paper.

Conflicts of Interest: The authors declare no conflicts of interest.

References

1. Hashin, Z. The elastic moduli of heterogeneous materials. *J. Appl. Mech.* **1962**, *29*, 143–150. [CrossRef]
2. Milton, G.W. *The Theory of Composites*; Cambridge University Press: Cambridge, UK, 2002.
3. Jarczyk, P.; Mityushev, V. Neutral coated inclusions of finite conductivity. *Proc. R. Soc. Math. Phys. Eng. Sci.* **2012**, *468*, 954–970. [CrossRef]
4. Cherkaev, A. Optimal three-material wheel assemblage of conducting and elastic composites. *Int. J. Eng. Sci.* **2012**, *59*, 27–39. [CrossRef]
5. Cherkaev, A.; Pruss, A.D. Effective conductivity of spiral and other radial symmetric assemblages. *Mech. Mater.* **2013**, *65*, 103–109. [CrossRef]

6. McPhedran, R.C.; Stout, B. 'Killing Mie softly': Analytic integrals for complex resonant states. *Q. J. Mech. Appl. Math.* **2020**, *73*, 119–139. [CrossRef]
7. Capolino, F. *Theory and Phenomena of Metamaterials*; CRC Press: Boca Raton, FL, USA, 2017.
8. Veselago, V.G. The electrodynamics of substances with simultaneously negative values of permittivity and permeability. *Sov. Phys. Uspekhi* **1968**, *10*, 509–514. [CrossRef]
9. Mityushev, V.; Gric, T.; Kycia, R.; Rylko, N. *Anisotropy of Metamaterials: Beyond Conventional Paradigms*; CRC Press: Boca Raton, FL, USA, 2025.
10. Abatangelo, L.; Lena, C.; Musolino, P. Asymptotic behavior of generalized capacities with applications to eigenvalue perturbations: The higher dimensional case. *Nonlinear Anal.* **2024**, *238*, 113391. [CrossRef]
11. Demuth, M. Partial differential equations and spectral theory. In *Birkhäuser*; Springer: Berlin/Heidelberg, Germany, 2011.
12. Laugesen, R.S. Spectral theory of partial differential equations. *Contemp. Math* **2018**, **720**, 23–55.
13. Ruzhansky, M.; Sadybekov, M.; Suragan, D. *Spectral Geometry of Partial Differential Operators*; Taylor & Francis: Abingdon, UK, 2020; p. 378.
14. Ugurlu, E.; Bairamov, E. Spectral analysis of eigenparameter dependent boundary value transmission problems. *J. Math. Anal. Appl.* **2014**, *413*, 482–494. [CrossRef]
15. Chavel, I. *Eigenvalues in Riemannian Geometry*; Academic Press: New York, NY, USA, 1984.
16. Colbois, B.; Girouard, A.; Gordon, C.; Sher, D. Some recent developments on the Steklov eigenvalue problem. *Rev. MatemáTica Complut.* **2024**, *37*, 1–161. [CrossRef]
17. de Cristoforis, M.L.; Musolino, P.; Taskinen, J. A real analyticity result for symmetric functions of the eigenvalues of a quasiperiodic spectral problem for the Dirichlet Laplacian. *J. Oper. Theory* **2021**, *86*, 439–467. [CrossRef]
18. Ferrero, A.; Lamberti, P.D. Spectral stability for a class of fourth order Steklov problems under domain perturbations. *Calc. Var. Partial. Differ. Equ.* **2019**, *58*, 1–57. [CrossRef]
19. Ferrero, A.; Lamberti, P.D. Spectral stability of the Steklov problem. *Nonlinear Anal.* **2022**, *222*, 112989. [CrossRef]
20. Mityushev, V. Transport properties of doubly periodic arrays of circular cylinders and optimal design problems. *Appl. Math. Optim.* **2001**, **44**, 17–31. [CrossRef]
21. Gluzman, S.; Mityushev, V.; Nawalaniec, W. *Computational Analysis of Structured Media*; Academic Press: New York, NY, USA, 2017.
22. Andrianov, I.V.; Awrejcewicz, J.; Starushenko, G. *Approximate Models of Mechanics of Composites: An Asymptotic Approach*; CRC Press: Boca Raton, FL, USA, 2023.
23. Andrianov, I.V.; Awrejcewicz, J. *Asymptotic Methods for Engineers*; CRC Press: Boca Raton, FL, USA, 2024.
24. Schiffer, M. *Fredholm Eigen Values of Multiply-Connected Domains*; Applied Mathematics and Statistics Laboratory, Stanford University: Stanford, CA, USA, 1958.
25. Kuczma, M.; Choczewski, B.; Ger, R. *Iterative Functional Equations*; Cambridge University Press: Cambridge, UK, 1990.
26. Koosis, P. *Introduction to H_p Spaces*; Cambridge University Press: Cambridge, UK, 1998.
27. Chipot, M. *Elliptic Equations: An Introductory Course*; Birkhäuser: Basel, Switzerland, 2009.
28. Shapiro, J.H. *Composition Operators and Classical Function Theory*; Springer: Berlin/Heidelberg, Germany, 2011; 223p.

mathematics MDPI

Article

Line Defects in One-Dimensional Hexagonal Quasicrystals

Markus Lazar

Institute for Mechanics, Technical University of Darmstadt, D-64287 Darmstadt, Germany; markus.lazar1@tu-darmstadt.de

Abstract: Using the eight-dimensional framework of the integral formalism of one-dimensional quasicrystals, the analytical expressions for the displacement fields and stress functions of line defects, which are dislocations and line forces, in one-dimensional hexagonal quasicrystals of Laue class 10 are derived. The self-energy of a straight dislocation, the self-energy of a line force, the Peach–Koehler force between two straight dislocations, and the Cherepanov force between two straight line forces in one-dimensional hexagonal quasicrystals of Laue class 10 are calculated. In addition, the two-dimensional Green tensor of one-dimensional hexagonal quasicrystals of Laue class 10 is given within the framework of the integral formalism.

Keywords: line defects; dislocations; line forces; anisotropic elasticity; integral formalism; Stroh formalism; quasicrystals

MSC: 52C23; 74Bxx; 74E10; 35Qxx

1. Introduction

Quasicrystals represent an interesting class of novel materials discovered by Shechtman et al. [1]. Quasicrystals are aperiodic crystals and possess long-range orientational order but no translational symmetry in quasiperiodic directions. Quasicrystals have unusual physical properties such as high hardness, low conductivity, resistivity that decreases with temperature, very low conductivity, small specific heat, low friction coefficients, and wear and oxidation resistance [2–4]. The basis of the continuum theory of solid quasicrystals consists of two elementary excitations, namely the phonons and the phasons [5,6]. In the atomistic picture, phonons are related to the translation of atoms, and phasons lead to local rearrangements of atoms in a cell. The generalized elasticity theory of quasicrystals including phonon and phason fields was developed by Ding et al. [7].

Dislocations are important line defects in quasicrystals that cause plasticity and influence the physical properties of quasicrystals (see, e.g., [6,8–14]). Dislocations have been observed in quasicrystals in many experimental studies (see, e.g., [15–21]). To understand the effects of dislocations on the physical properties of quasicrystals and to simulate the electron micrographs of dislocations in quasicrystals, the elastic fields of dislocations in quasicrystals are necessary. Therefore, the determination of the elastic fields of dislocations in quasicrystals is an important task in theory and experiment. For quasicrystals, the basic key equations of dislocations were given by Ding et al. [22,23] and Lazar and Agiasofitou [24]. The general expressions for the displacement fields induced by straight dislocations in quasicrystals in terms of the elastic Green tensor were given by Ding et al. [22,23]. Ding et al. [25] extended the Stroh formalism [26,27] and the integral formalism [28,29] for the displacement field of a dislocation in quasicrystals (see also [14,30]). Ding et al. [25] derived the integral formalism from the Stroh formalism for quasicrystals using the eigenvectors and eigenvalues of the Stroh formalism. Using the extended Stroh formalism, Li and Liu [31] studied a dislocation in an icosahedral quasicrystal and Radi and Mariano [32] investigated the steady-state propagation of dislocations in quasicrystals. Wang and Schiavone [33] have given the detailed structure of the extended Stroh formalism for quasicrystals. In the Stroh formalism, the fields of line defects are given as complex form solutions of an eigenvalue problem and in the integral formalism, the fields of line defects are given as real-form solutions of a matrix partial differential equation of the first order (see [34]).

Lazar [35] recently derived the extended integral formalism of line defects (straight dislocations and line forces) for the displacement fields and stress functions for quasicrystals in a straightforward manner without an unnecessary detour via the Stroh formalism. Lazar [35] extended the six-dimensional framework of the integral formalism for line defects in anisotropic elasticity towards a $2n$-dimensional integral formalism for line defects in quasicrystals ($n = 4, 5, 6$ for one-, two-, three-dimensional quasicrystals). For generalized plane strain in quasicrystals, the appropriate two-dimensional matrix partial differential equation of the first order and its solution of the integral formalism have been given in [35]. For line defects in quasicrystals, the solution gives the $2n$-dimensional vector of the n-dimensional displacement vector $\boldsymbol{U}$ and the n-dimensional stress function vector $\boldsymbol{\Phi}$ of a straight dislocation with Burgers vector $\boldsymbol{b}$ and a line force with strength $\boldsymbol{F}$. The integral formalism provides suitable expressions for both analytical and numerical modelling of line defects in quasicrystals.

Explicit analytical solutions require analytical expressions for the integrals in the integral formalism. In the general case, only numerical solutions are possible using numerical integration. For quasicrystals with high symmetry and symmetrical orientation of the defect line, the complexity is reduced and analytical solutions become possible because the integrals in the integral formalism can be computed analytically. In the case of lower symmetry of quasicrystals and of the orientation of the line defect, numerical treatment is required in the integral formalism. The successful derivation of analytic solutions is of particular value because it confirms and manifests the theoretical prediction. The form of the analytic solutions shows their functional dependence on all relevant parameters. Analytic solutions are always useful as benchmark solutions with which to verify numerical results, and they are useful for degenerate cases (special orientations or ratios of elastic constants) where the numerics may diverge [36].

The most suitable type of quasicrystals for testing the integral formalism are one-dimensional hexagonal quasicrystals of Laue class 10. The Laue class 10 is the Laue class with the highest symmetry for one-dimensional quasicrystals. A one-dimensional quasicrystal is defined as a three-dimensional body which is periodic in the x_1x_2-plane and quasiperiodic in the x_3-direction. One-dimensional quasicrystals are quasicrystals with crystallographic symmetries, namely 31 point groups and 10 Laue classes [13,30,37].

In this work, analytical solutions of line defects in one-dimensional quasicrystals are derived for the first time using the integral formalism. In particular, the closed-form expressions for the displacement fields and the stress functions of line defects in one-dimensional hexagonal quasicrystals of Laue class 10, where the defect line is parallel to the aperiodic direction, are the aim of this work. Within the framework of the integral formalism, the self-energies of a straight dislocation and a line force, the Peach–Koehler force between two straight dislocations, and the Cherepanov force between two line forces in one-dimensional hexagonal quasicrystals of Laue class 10 are calculated. In addition, the two-dimensional Green tensor of one-dimensional hexagonal quasicrystals of Laue class 10 is derived within the framework of the integral formalism.

2. Generalized Elasticity of One-Dimensional Quasicrystals

We consider one-dimensional quasicrystals which are periodic in the x_1x_2-plane and quasiperiodic in the x_3-direction. A one-dimensional quasicrystal can be obtained by projecting a four-dimensional periodic structure onto the three-dimensional physical space. The four-dimensional hyperspace E^4 can be decomposed into the direct sum of two orthogonal subspaces:

$$E^4 = E^3_\parallel \oplus E^1_\perp\,, \tag{1}$$

where $E^3_\parallel$ is the three-dimensional physical or parallel space of the phonon fields and $E^1_\perp$ is the one-dimensional perpendicular space of the phason field with quasiperiodicity in the x_3-direction. Throughout the text, phonon fields will be denoted $(\cdot)^\parallel$ and phason fields $(\cdot)^\perp$. Note that all quantities (phonon and phason fields) depend on the so-called material space coordinates $\boldsymbol{x} \in \mathbb{R}^3$. Indices in the parallel space are denoted by small Latin letters i, j, k, l with $i, j, k, l = 1, 2, 3$.

In the theory of generalized elasticity of one-dimensional quasicrystals, the (elastic) *phonon and phason distortion tensors*, $\beta^\parallel_{kl}$ and $\beta^\perp_{3l}$, are defined as the gradients of the *phonon displacement vector* $u^\parallel_k \in E^3_\parallel$ and the *phason displacement field* $u^\perp_3 \in E^1_\perp$, respectively,

$$\beta^\parallel_{kl} = \partial_l u^\parallel_k\,, \tag{2}$$

$$\beta^\perp_{3l} = \partial_l u^\perp_3\,, \tag{3}$$

where $\partial_l = \partial/\partial x_l$ denotes the partial differentiation with respect to x_l. Therefore, the phason displacement vector has only one component $u^\perp_3$. The *constitutive relations* for a one-dimensional quasicrystal are given by

$$\sigma^\parallel_{ij} = C_{ijkl}\beta^\parallel_{kl} + D_{ij3l}\beta^\perp_{3l}\,, \tag{4}$$

$$\sigma^\perp_{3j} = D_{kl3j}\beta^\parallel_{kl} + E_{3j3l}\beta^\perp_{3l}\,, \tag{5}$$

where $\sigma^\parallel_{ij}$ and $\sigma^\perp_{3j}$ are the *phonon and phason stress tensors*, respectively. Note that the phonon stress tensor is symmetric, $\sigma^\parallel_{ij} = \sigma^\parallel_{ji}$, whereas $\sigma^\perp_{3j} \neq \sigma^\perp_{j3}$ (see [24,38]). C_{ijkl} is the *tensor of the elastic moduli of phonons*, E_{3j3l} is the *tensor of the elastic moduli of phasons*, and D_{ij3l} is the *tensor of the elastic moduli of the phonon–phason coupling*. The constitutive tensors possess the following symmetries [7,24]:

$$C_{ijkl} = C_{klij} = C_{ijlk} = C_{jikl}\,, \quad D_{ij3l} = D_{ji3l}\,, \quad E_{3j3l} = E_{3l3j}\,. \tag{6}$$

Using the hyperspace notation of quasicrystals given by Lazar and Agiasofitou [24], the phonon and phason fields can be unified in the corresponding extended fields in the hyperspace. The components of the extended fields will be denoted by capital letters, e.g., $I, K = 1, 2, 3, 4$. In the hyperspace notation, we have the *extended displacement vector* $U_K \in E^3_\parallel \oplus E^1_\perp$:

$$U_K = \begin{cases} u^\parallel_k\,, & K = 1,2,3\,, \\ u^\perp_3\,, & K = 4\,, \end{cases} \tag{7}$$

the *extended elastic distortion tensor* $B_{Kl} \in \left(E_{\parallel}^3 \oplus E_{\perp}^1\right) \otimes E_{\parallel}^3$:

$$B_{Kl} = \begin{cases} \beta_{kl}^{\parallel}, & K = 1,2,3, \\ \beta_{3l}^{\perp}, & K = 4, \end{cases} \tag{8}$$

the *extended stress tensor* $\Sigma_{Ij} \in \left(E_{\parallel}^3 \oplus E_{\perp}^1\right) \otimes E_{\parallel}^3$:

$$\Sigma_{Ij} = \begin{cases} \sigma_{ij}^{\parallel}, & I = 1,2,3, \\ \sigma_{3j}^{\perp}, & I = 4, \end{cases} \tag{9}$$

and the *tensor of the extended elastic moduli* $C_{IjKl} \in \left(E_{\parallel}^3 \oplus E_{\perp}^1\right) \otimes E_{\parallel}^3 \otimes \left(E_{\parallel}^3 \oplus E_{\perp}^1\right) \otimes E_{\parallel}^3$:

$$C_{IjKl} = \begin{cases} C_{ijkl}, & I = 1,2,3; \quad K = 1,2,3, \\ D_{ij3l}, & I = 1,2,3; \quad K = 4, \\ D_{kl3j}, & I = 4; \quad K = 1,2,3, \\ E_{3j3l}, & I = 4; \quad K = 4. \end{cases} \tag{10}$$

The tensor C_{IjKl} retains the *major symmetry*:

$$C_{IjKl} = C_{KlIj}. \tag{11}$$

In matrix form, Equation (10) can be written as

$$C_{IjKl} = \begin{pmatrix} C_{ijkl} & D_{ij3l} \\ D_{kl3j} & E_{3j3l} \end{pmatrix}. \tag{12}$$

Using the hyperspace notation, Equations (2) and (3) reduce to

$$B_{Kl} = \partial_l U_K \tag{13}$$

and the constitutive relations (4) and (5) become

$$\Sigma_{Ij} = C_{IjKl} B_{Kl} = C_{IjKl} \partial_l U_K, \tag{14}$$

which is the extended Hooke law for quasicrystals.

In the absence of external forces, the extended stress tensor (9) can be written in terms of an *extended stress function tensor* $\Phi_{Im} \in \left(E_{\parallel}^3 \oplus E_{\perp}^1\right) \otimes E_{\parallel}^3$:

$$\Sigma_{Ij} = \epsilon_{jkm} \partial_k \Phi_{Im} \tag{15}$$

with

$$\Phi_{Im} = \begin{cases} \Phi_{im}^{\parallel}, & I = 1,2,3, \\ \Phi_{3m}^{\perp}, & I = 4, \end{cases} \tag{16}$$

where $\Phi_{im}^{\parallel}$ and $\Phi_{3m}^{\perp}$ are the *phonon and phason stress function tensors*, respectively. Here, ϵ_{jkm} denotes the three-dimensional Levi–Civita tensor. Substituting Equation (15) into the left hand side of the extended Hooke law (14), we obtain

$$\epsilon_{jkm} \partial_k \Phi_{Im} = C_{IjKl} \partial_l U_K, \tag{17}$$

which is nothing but the extended Hooke law (14) written in terms of U_K and Φ_{Im}.

3. Integral Formalism for One-Dimensional Quasicrystals

The integral formalism for one-dimensional quasicrystals is based on the framework of generalized plane strain of one-dimensional quasicrystals. In generalized plane strain, all fields are independent of the variable x_3. The extended displacement fields depend only on x_1 and x_2, and $\boldsymbol{x} \in \mathbb{R}^2$, but with index in the hyperspace $U_I(x_1, x_2)$, $I = 1, 2, 3, 4$.

We consider a Cartesian coordinate system in the parallel space. Using a unit vector $\boldsymbol{m}$ in the x_1-direction and a unit vector $\boldsymbol{n}$ in the x_2-direction ($\boldsymbol{m} \equiv \boldsymbol{e}_1$, $\boldsymbol{n} \equiv \boldsymbol{e}_2$) and the notation

$$(a\, b)_{IK} = a_j C_{IjKl} b_l\,, \tag{18}$$

which is a 4×4 matrix in the hyperspace, it follows from the major symmetry (11) of the extended constitutive tensor C_{IjKl} that

$$(a\, b)_{IK} = (b\, a)_{KI}\,. \tag{19}$$

The extended constitutive relation (14) reads for generalized plane strain:

$$\Sigma_{I1} = (m\, m)_{IK} B_{K1} + (m\, n)_{IK} B_{K2} \tag{20}$$
$$\Sigma_{I2} = (n\, m)_{IK} B_{K1} + (n\, n)_{IK} B_{K2} \tag{21}$$

with $I, K = 1, 2, 3, 4$. For a one-dimensional quasicrystal, the characteristic 4×4 blocks of the constitutive tensor C_{IjKl} entering the extended constitutive relations (20) and (21) are given by

$$(m\, m)_{IK} = \begin{pmatrix} C_{1111} & C_{1121} & C_{1131} & D_{1131} \\ C_{2111} & C_{2121} & C_{2131} & D_{2131} \\ C_{3111} & C_{3121} & C_{3131} & D_{3131} \\ D_{1131} & D_{2131} & D_{3131} & E_{3131} \end{pmatrix} = C_{I1K1} \tag{22}$$

$$(m\, n)_{IK} = \begin{pmatrix} C_{1112} & C_{1122} & C_{1132} & D_{1132} \\ C_{2112} & C_{2122} & C_{2132} & D_{2132} \\ C_{3112} & C_{3122} & C_{3132} & D_{3132} \\ D_{1231} & D_{2231} & D_{3231} & E_{3132} \end{pmatrix} = C_{I1K2} \tag{23}$$

$$(n\, n)_{IK} = \begin{pmatrix} C_{1212} & C_{1222} & C_{1232} & D_{1232} \\ C_{2212} & C_{2222} & C_{2232} & D_{2232} \\ C_{3212} & C_{3222} & C_{3232} & D_{3232} \\ D_{1232} & D_{2232} & D_{3232} & E_{3232} \end{pmatrix} = C_{I2K2}\,. \tag{24}$$

Due to the major symmetry of C_{IjKl}, $(m\, m)_{IK}$ and $(n\, n)_{IK}$ are symmetric.

For generalized plane strain, the extended stress tensor can be written in terms of an extended stress function vector $\Phi_{I3} \equiv \Phi_I$, and Equation (15) reduces to

$$\Sigma_{I1} = \partial_2 \Phi_I \tag{25}$$
$$\Sigma_{I2} = -\partial_1 \Phi_I\,. \tag{26}$$

Substituting Equations (13), (25) and (26) into the extended Hooke law (20) and (21), we obtain

$$\partial_2\Phi_I = (m\,m)_{IK}\partial_1 U_K + (m\,n)_{IK}\partial_2 U_K \tag{27}$$

$$-\partial_1\Phi_I = (n\,m)_{IK}\partial_1 U_K + (n\,n)_{IK}\partial_2 U_K\,. \tag{28}$$

Equations (27) and (28), which have the meaning of the extended Hooke law (14) written in terms of U_K and Φ_I for generalized plane strain, are the two components of Equation (17) for generalized plane strain with $j = 1, 2$.

After rearrangement of Equations (27) and (28), we obtain the following matrix partial differential equation of the first order

$$[\boldsymbol{I}\,\partial_2 - \boldsymbol{N}\partial_1]\begin{pmatrix}\boldsymbol{U}(x_1,x_2)\\ \boldsymbol{\Phi}(x_1,x_2)\end{pmatrix} = \begin{pmatrix}0\\0\end{pmatrix}. \tag{29}$$

This is the *matrix differential equation* for generalized plane strain of one-dimensional quasicrystals. We have introduced the 8-vector of the extended displacement vector and the extended stress function vector for generalized plane strain:

$$\begin{pmatrix}\boldsymbol{U}\\ \boldsymbol{\Phi}\end{pmatrix} \equiv \begin{pmatrix}U_I\\ \Phi_I\end{pmatrix}, \qquad I = 1,2,3,4 \tag{30}$$

and the 8×8 real matrix is defined by its 4×4 blocks

$$\boldsymbol{N} = -\begin{pmatrix}(n\,n)^{-1}\,(n\,m) & (n\,n)^{-1}\\ (m\,n)\,(n\,n)^{-1}\,(n\,m) - (m\,m) & (m\,n)\,(n\,n)^{-1}\end{pmatrix}. \tag{31}$$

Here, $\boldsymbol{I}$ is the 8×8 identity matrix. The 8×8 matrix $\boldsymbol{N}$ is the *fundamental elasticity matrix for one-dimensional quasicrystals* depending on the elastic constants of quasicrystals.

Now, we choose two orthogonal unit vectors $\boldsymbol{m}$ and $\boldsymbol{n}$, which are orthogonal to $\boldsymbol{t}$ such that $(\boldsymbol{m},\boldsymbol{n},\boldsymbol{t})$ forms a right-handed Cartesian basis in $E^3_\parallel$. This basis is rotated around $\boldsymbol{t}$ by an angle ϕ against another fixed $(\boldsymbol{m}_0,\boldsymbol{n}_0,\boldsymbol{t})$ basis in $E^3_\parallel$, such that

$$\begin{pmatrix}\boldsymbol{m}(\phi)\\ \boldsymbol{n}(\phi)\\ \boldsymbol{t}\end{pmatrix} = \begin{pmatrix}\cos\phi & \sin\phi & 0\\ -\sin\phi & \cos\phi & 0\\ 0 & 0 & 1\end{pmatrix}\begin{pmatrix}\boldsymbol{m}_0\\ \boldsymbol{n}_0\\ \boldsymbol{t}\end{pmatrix} \tag{32}$$

as shown in Figure 1. It yields $\boldsymbol{m} \equiv \boldsymbol{m}(\phi)$ and $\boldsymbol{n} \equiv \boldsymbol{n}(\phi)$. Only the independent variable x_j is transformed, but not the dependent variables U_I and Φ_I. Using the rotation given in Equation (32) and polar coordinates (r,ϕ), the matrix partial differential Equation (29) becomes

$$\left[\boldsymbol{I}\,\frac{1}{r}\,\partial_\phi - \boldsymbol{N}(\phi)\,\partial_r\right]\begin{pmatrix}\boldsymbol{U}(r,\phi)\\ \boldsymbol{\Phi}(r,\phi)\end{pmatrix} = \begin{pmatrix}0\\0\end{pmatrix}. \tag{33}$$

Here, the 8×8 matrix $\boldsymbol{N}(\phi)$ is defined by contraction of the elastic constants with orthogonal unit vectors $\boldsymbol{m}$ and $\boldsymbol{n}$ according to Equation (18). As mentioned above, the vectors $\boldsymbol{m}(\phi)$ and $\boldsymbol{n}(\phi)$ are turned against the $\boldsymbol{m}_0$, $\boldsymbol{n}_0$ coordinate system by an angle ϕ (see Figure 1), so that $\boldsymbol{N}$ depends on the angle ϕ

$$\mathbf{N}(\phi) = -\begin{pmatrix} (n\,n)^{-1}\,(n\,m) & (n\,n)^{-1} \\ (m\,n)\,(n\,n)^{-1}\,(n\,m) - (m\,m) & (m\,n)\,(n\,n)^{-1} \end{pmatrix}. \tag{34}$$

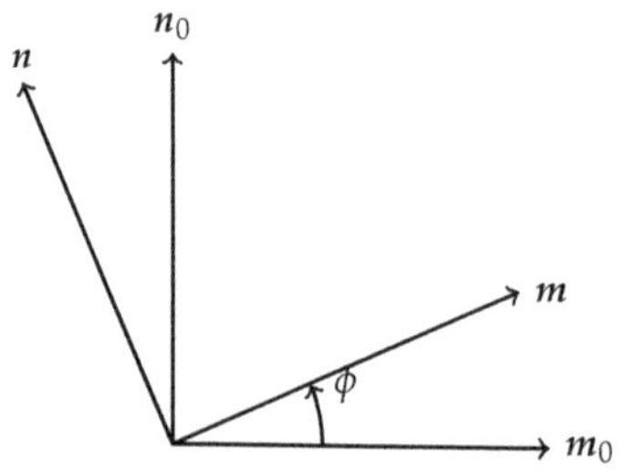

Figure 1. The unit vectors $\boldsymbol{m}$ and $\boldsymbol{n}$ are to be turned anticlockwise from $\boldsymbol{m}_0$ and $\boldsymbol{n}_0$ by an angle ϕ.

4. Line Defects in One-Dimensional Quasicrystals

Here, we consider a line defect in a one-dimensional quasicrystal, namely a straight dislocation with extended Burgers vector $\boldsymbol{b}$ and a line force with extended strength $\boldsymbol{F}$ in a one-dimensional quasicrystal located at the origin of the coordinate system. The defect line runs along the axis $\boldsymbol{t}$, which is parallel to the quasiperiodic direction. The fields of the straight dislocation and line force are the extended displacement vector $\boldsymbol{U}$ and the extended stress function vector $\boldsymbol{\Phi}$. In the hyperspace notation, the *extended Burgers vector* of a straight dislocation in a one-dimensional quasicrystal is given by

$$b_I = \begin{cases} b_i^{\|}\,, & I = 1,2,3\,, \\ b_3^{\perp}\,, & I = 4\,, \end{cases} \tag{35}$$

where $b_i^{\|}$ is the phonon component and $b_3^{\perp}$ is the phason component of the extended Burgers vector, $b_I \in E_{\|}^3 \oplus E_{\perp}^1$, and the *extended body force vector* of a line force in a one-dimensional quasicrystal reads

$$F_I = \begin{cases} F_i^{\|}\,, & I = 1,2,3\,, \\ F_3^{\perp}\,, & I = 4\,, \end{cases} \tag{36}$$

where $F_i^{\|}$ is the phonon component and $F_3^{\perp}$ is the phason component of the extended strength $F_I \in E_{\|}^3 \oplus E_{\perp}^1$.

For one-dimensional quasicrystals, the solution of Equation (33) for a straight dislocation with extended Burgers vector $\boldsymbol{b}$ and a line force with extended strength $\boldsymbol{F}$ reads (see [35])

$$\begin{pmatrix} \boldsymbol{U}(r,\phi) \\ \boldsymbol{\Phi}(r,\phi) \end{pmatrix} = -\frac{1}{(2\pi)^2}\left[\boldsymbol{I}\ln r + \int_0^{\phi} \mathbf{N}(\omega)\,\mathrm{d}\omega\right]\int_0^{2\pi} \mathbf{N}(\omega)\,\mathrm{d}\omega \begin{pmatrix} \boldsymbol{b} \\ -\boldsymbol{F} \end{pmatrix}. \tag{37}$$

Equation (37) is the 8-vector of the solution of the extended displacement vector $\boldsymbol{U}$ and the extended stress function vector $\boldsymbol{\Phi}$ for a straight dislocation with extended Burgers vector $\boldsymbol{b}$ and a line force with extended strength $\boldsymbol{F}$ in a one-dimensional quasicrystal. Using

$$\widetilde{\mathbf{N}}(\phi) = \frac{1}{2\pi}\int_0^{\phi} \mathbf{N}(\omega)\,\mathrm{d}\omega \tag{38}$$

and $\widetilde{N}(2\pi) = \widetilde{N}(\phi = 2\pi)$, the solution (37) can be written in compact form as (see also [35,39])

$$\begin{pmatrix} U(r,\phi) \\ \Phi(r,\phi) \end{pmatrix} = -\left[\frac{1}{2\pi} I \ln r + \widetilde{N}(\phi)\right] \widetilde{N}(2\pi) \begin{pmatrix} b \\ -F \end{pmatrix}. \tag{39}$$

The 4×4 matrices S_{IK}, Q_{IK} and B_{IK}, which are tensors of rank two in the hyperspace, are the 4×4 blocks of the 8×8 matrix $\widetilde{N}$, Equation (38), with the block structure (34)

$$\widetilde{N}(\phi) = \begin{pmatrix} S(\phi) & Q(\phi) \\ B(\phi) & S^{\mathrm{T}}(\phi) \end{pmatrix} \tag{40}$$

and

$$Q(\phi) = -\frac{1}{2\pi} \int_0^{\phi} (n\,n)^{-1}\,\mathrm{d}\omega\,, \tag{41}$$

$$S(\phi) = -\frac{1}{2\pi} \int_0^{\phi} (n\,n)^{-1}(n\,m)\,\mathrm{d}\omega\,, \tag{42}$$

$$S^{\mathrm{T}}(\phi) = -\frac{1}{2\pi} \int_0^{\phi} (m\,n)(n\,n)^{-1}\,\mathrm{d}\omega\,, \tag{43}$$

$$B(\phi) = -\frac{1}{2\pi} \int_0^{\phi} \Big[(m\,n)(n\,n)^{-1}(n\,m) - (m\,m)\Big]\,\mathrm{d}\omega\,, \tag{44}$$

where S, Q, B and S^{T} (the transpose of S) are 4×4 matrices resulting from the four blocks in N (see Equations (34) and (38)).

Equation (39) can be decomposed into its four pieces, which are four vectors in the four-dimensional hyperspace (see [35]):

- The extended displacement vector of a straight dislocation with extended Burgers vector b_I:

$$U_K(r,\phi) = -b_I\left[\frac{1}{2\pi} S_{KI}(2\pi)\ln r + Q_{KM}(\phi)B_{MI}(2\pi) + S_{KM}(\phi)S_{MI}(2\pi)\right]. \tag{45}$$

- The extended displacement vector of a line force with extended strength F_I:

$$U_K(r,\phi) = F_I\left[\frac{1}{2\pi} Q_{KI}(2\pi)\ln r + S_{KM}(\phi)Q_{MI}(2\pi) + Q_{KM}(\phi)S^{\mathrm{T}}_{MI}(2\pi)\right]. \tag{46}$$

- The extended stress function vector of a straight dislocation with extended Burgers vector b_K:

$$\Phi_I(r,\phi) = -b_K\left[\frac{1}{2\pi} B_{IK}(2\pi)\ln r + B_{IM}(\phi)S_{MK}(2\pi) + S^{\mathrm{T}}_{IM}(\phi)B_{MK}(2\pi)\right]. \tag{47}$$

- The extended stress function vector of a line force with extended strength F_K:

$$\Phi_I(r,\phi) = F_K\left[\frac{1}{2\pi} S^{\mathrm{T}}_{IK}(2\pi)\ln r + B_{IM}(\phi)Q_{MK}(2\pi) + S^{\mathrm{T}}_{IM}(\phi)S^{\mathrm{T}}_{MK}(2\pi)\right]. \tag{48}$$

Using the extended stress function vector of a straight dislocation (47) for $r = r_0/R$ and $\phi = 0$, the elastic self-energy of a straight dislocation per unit length reads [35]

$$\mathcal{E} = \frac{1}{2}\, b_I \Phi_I(r_0/R, 0)$$
$$= \frac{1}{4\pi}\, b_I B_{IK}(2\pi) b_K \ln(R/r_0)\,, \tag{49}$$

where r_0 and R are the inner and outer cutoff radii. Using the extended displacement vector of a line force (48) for $r = r_0/R$ and $\phi = 0$, the elastic self-energy of a line force per unit length reads [35]

$$\mathcal{E} = \frac{1}{2}\, F_K U_K(r_0/R, 0)$$
$$= -\frac{1}{4\pi}\, F_I Q_{IK}(2\pi) F_K \ln(R/r_0)\,. \tag{50}$$

The Peach–Koehler force per unit length between a straight dislocation with Burgers vector $b_I^{(A)}$ at position (r, ϕ) in the stress field $\Sigma_{Ij}^{(B)}$ produced by another dislocation with Burgers vector $b_K^{(B)}$ located at the position $(0, 0)$ is given by (see [35])

$$\mathcal{F}_l^{\text{PK}(AB)} = \epsilon_{lj3}\, \frac{b_I^{(A)} b_K^{(B)}}{2\pi r} \Big[n_j B_{IK}(2\pi) + m_j \big[(m\,n)(n\,n)^{-1}(n\,m) - (m\,m)\big]_{IM} S_{MK}(2\pi) + m_j \big[(m\,n)(n\,n)^{-1}\big]_{IM} B_{MK}(2\pi) \Big]\,. \tag{51}$$

In polar coordinates, the two non-vanishing components of the Peach–Koehler force (51) read

$$\mathcal{F}_r^{\text{PK}(AB)} = \frac{b_I^{(A)} b_K^{(B)}}{2\pi r}\, B_{IK}(2\pi)\,, \tag{52}$$
$$\mathcal{F}_\phi^{\text{PK}(AB)} = -\frac{b_I^{(A)} b_K^{(B)}}{2\pi r} \Big[\big[(m\,n)(n\,n)^{-1}(n\,m) - (m\,m)\big]_{IM} S_{MK}(2\pi) + \big[(m\,n)(n\,n)^{-1}\big]_{IM} B_{MK}(2\pi) \Big]\,. \tag{53}$$

The radial component (52) is connected with the matrix $\boldsymbol{B}$, and the tangential component (53) is connected with the matrices $\boldsymbol{S}$ and $\boldsymbol{B}$. Note that the ϕ-dependence in Equation (53) is due to the ϕ-dependence of $\boldsymbol{m}$ and $\boldsymbol{n}$.

The Cherepanov force per unit length between a line force with strength $F_K^{(A)}$ at position (r, ϕ) in the elastic distortion field $B_{KI}^{(B)}$ produced by another line force with strength $F_I^{(B)}$ located the the position $(0, 0)$ is given by (see [35])

$$\mathcal{F}_l^{\text{C}(AB)} = \frac{F_K^{(A)} F_I^{(B)}}{2\pi r} \Big[m_l Q_{KI}(2\pi) - n_l (n\,n)^{-1}_{KN} \big(S^{\text{T}}_{NI}(2\pi) + (n\,m)_{NM} Q_{MI}(2\pi)\big) \Big]\,. \tag{54}$$

In polar coordinates, the two non-vanishing components of the Cherepanov force (54) read

$$\mathcal{F}_r^{\text{C}(AB)} = \frac{F_K^{(A)} F_I^{(B)}}{2\pi r}\, Q_{KI}(2\pi)\,, \tag{55}$$
$$\mathcal{F}_\phi^{\text{C}(AB)} = -\frac{F_K^{(A)} F_I^{(B)}}{2\pi r}\, (n\,n)^{-1}_{KN} \Big[S^{\text{T}}_{NI}(2\pi) + (n\,m)_{NM} Q_{MI}(2\pi) \Big]\,. \tag{56}$$

The radial component (55) is connected with the matrix $\boldsymbol{Q}$, and the tangential component (56) is connected with the matrices $\boldsymbol{S}^{\text{T}}$ and $\boldsymbol{Q}$. The ϕ-dependence in Equation (56) is due to the ϕ-dependence of $\boldsymbol{m}$ and $\boldsymbol{n}$.

5. Line Defects in One-Dimensional Hexagonal Quasicrystals of Laue Class 10

We consider one-dimensional hexagonal quasicrystals of Laue class 10 characterized by the following elastic constants (see [37]), namely five elastic moduli of phonons:

$$\begin{aligned}
&C_{1111} = C_{2222} = C_{11}\,, \qquad C_{1122} = C_{2211} = C_{12}\,,\\
&C_{1133} = C_{2233} = C_{3311} = C_{3322} = C_{13}\,, \qquad C_{3333} = C_{33}\,,\\
&C_{2323} = C_{2332} = C_{3223} = C_{3232} = C_{1313} = C_{1331} = C_{3113} = C_{3131} = C_{44}\,,\\
&C_{1212} = C_{1221} = C_{2112} = C_{2121} = C_{66} = (C_{11} - C_{12})/2\,,
\end{aligned} \tag{57}$$

three elastic moduli of phonon–phason coupling:

$$D_{1133} = D_{2233} = R_1\,, \quad D_{3333} = R_2\,, \quad D_{3232} = D_{2332} = D_{3131} = D_{1331} = R_3 \tag{58}$$

and two elastic moduli of phasons:

$$E_{3333} = K_1\,, \quad E_{3131} = E_{3232} = K_2\,. \tag{59}$$

The conditions of positive definiteness for the elastic constants of one-dimensional hexagonal quasicrystals of Laue class 10 to ensure a positive elastic energy density read [40,41]

$$C_{11} > |C_{12}|\,, \qquad (C_{11}+C_{12})C_{33} - 2C_{13}^2 > 0\,, \qquad C_{44} > 0\,, \tag{60}$$

$$C_{44}K_2 - R_3^2 > 0\,, \tag{61}$$

$$\big[(C_{11}+C_{12})C_{33} - 2C_{13}^2\big]K_1 + 2(2C_{13}R_2 - C_{33}R_1)R_1 - (C_{11}+C_{12})R_2^2 > 0\,. \tag{62}$$

A one-dimensional hexagonal quasicrystal of Laue class 10 is a transversely isotropic medium, concerning generalized elasticity of quasicrystals, namely, a one-dimensional hexagonal quasicrystal of Laue class 10 is isotropic in the basal x_1x_2-plane (see [38]). Laue class 10 consists of the following point groups [13]: 62_h2_h, $6mm$, $\bar{6}m2_h$, $6/m_hmm$. Therefore, the Laue class 10 is the Laue class with the highest symmetry for one-dimensional quasicrystals. In order to keep equations as simple as possible, we may use $C_{66} = (C_{11} - C_{12})/2$ instead of C_{12} in the formulas. For one-dimensional hexagonal quasicrystals of Laue class 10, Equations (22)–(24) reduce to

$$(m\,m)_{IK} = \begin{pmatrix} C_{1111} & 0 & 0 & 0\\ 0 & C_{2121} & 0 & 0\\ 0 & 0 & C_{3131} & D_{3131}\\ 0 & 0 & D_{3131} & E_{3131} \end{pmatrix} = \begin{pmatrix} C_{11} & 0 & 0 & 0\\ 0 & C_{66} & 0 & 0\\ 0 & 0 & C_{44} & R_3\\ 0 & 0 & R_3 & K_2 \end{pmatrix} \tag{63}$$

$$(m\,n)_{IK} = \begin{pmatrix} 0 & C_{1122} & 0 & 0\\ C_{2112} & 0 & 0 & 0\\ 0 & 0 & 0 & 0\\ 0 & 0 & 0 & 0 \end{pmatrix} = \begin{pmatrix} 0 & C_{12} & 0 & 0\\ C_{66} & 0 & 0 & 0\\ 0 & 0 & 0 & 0\\ 0 & 0 & 0 & 0 \end{pmatrix} \tag{64}$$

$$(n\,n)_{IK} = \begin{pmatrix} C_{1212} & 0 & 0 & 0\\ 0 & C_{2222} & 0 & 0\\ 0 & 0 & C_{3232} & D_{3232}\\ 0 & 0 & D_{3232} & E_{3232} \end{pmatrix} = \begin{pmatrix} C_{66} & 0 & 0 & 0\\ 0 & C_{11} & 0 & 0\\ 0 & 0 & C_{44} & R_3\\ 0 & 0 & R_3 & K_2 \end{pmatrix}. \tag{65}$$

Therefore, for generalized plane strain, only three elastic moduli of phonons $C_{11} = C_{1111} = C_{2222}$, $C_{12} = C_{1122}$, $C_{44} = C_{3232} = C_{3131}$, $C_{66} = (C_{11} - C_{12})/2 = (C_{1111} - C_{1122})/2 = C_{2121} = C_{2112} = C_{1212}$; one elastic modulus of phonon–phason coupling $R_3 = D_{3232} = D_{3131}$; and one elastic modulus of phasons $K_2 = E_{3131} = E_{3232}$ are relevant.

The four 4×4 blocks of the 8×8 matrix $\boldsymbol{N}$ read

$$(n\,n)^{-1} = \begin{pmatrix} 1/C_{66} & 0 & 0 & 0 \\ 0 & 1/C_{11} & 0 & 0 \\ 0 & 0 & \dfrac{K_2}{C_{44}K_2 - R_3^2} & -\dfrac{R_3}{C_{44}K_2 - R_3^2} \\ 0 & 0 & -\dfrac{R_3}{C_{44}K_2 - R_3^2} & -\dfrac{C_{44}}{C_{44}K_2 - R_3^2} \end{pmatrix} \tag{66}$$

$$(n\,n)^{-1}(n\,m) = \begin{pmatrix} 0 & 1 & 0 & 0 \\ C_{12}/C_{11} & 0 & 0 & 0 \\ 0 & 0 & 0 & 0 \\ 0 & 0 & 0 & 0 \end{pmatrix} \tag{67}$$

$$(m\,n)(n\,n)^{-1} = \begin{pmatrix} 0 & C_{12}/C_{11} & 0 & 0 \\ 1 & 0 & 0 & 0 \\ 0 & 0 & 0 & 0 \\ 0 & 0 & 0 & 0 \end{pmatrix} \tag{68}$$

and

$$[(m\,n)(n\,n)^{-1}(n\,m) - (m\,m)] = -\begin{pmatrix} \alpha & 0 & 0 & 0 \\ 0 & 0 & 0 & 0 \\ 0 & 0 & C_{44} & R_3 \\ 0 & 0 & R_3 & K_2 \end{pmatrix}, \tag{69}$$

where

$$\alpha = \frac{C_{11}^2 - C_{12}^2}{C_{11}}. \tag{70}$$

The angular behaviour of the 4×4 matrices $\boldsymbol{Q}$, $\boldsymbol{S}$, and $\boldsymbol{B}$ is obtained by the transformations (see also [42,43])

$$\begin{aligned} \boldsymbol{Q}(\phi) &= -\frac{1}{2\pi}\int_0^{\phi} (n\,n)^{-1}(\omega)\,\mathrm{d}\omega \\ &= -\frac{1}{2\pi}\int_0^{\phi} \boldsymbol{R}^{\mathrm{T}}(\omega)\,(n\,n)^{-1}(0)\,\boldsymbol{R}(\omega)\,\mathrm{d}\omega\,, \end{aligned} \tag{71}$$

$$\begin{aligned} \boldsymbol{S}(\phi) &= -\frac{1}{2\pi}\int_0^{\phi} [(n\,n)^{-1}(n\,m)](\omega)\,\mathrm{d}\omega \\ &= -\frac{1}{2\pi}\int_0^{\phi} \boldsymbol{R}^{\mathrm{T}}(\omega)\,[(n\,n)^{-1}(n\,m)](0)\,\boldsymbol{R}(\omega)\,\mathrm{d}\omega\,, \end{aligned} \tag{72}$$

$$\begin{aligned} \boldsymbol{B}(\phi) &= -\frac{1}{2\pi}\int_0^{\phi} [(m\,n)(n\,n)^{-1}(n\,m) - (m\,m)](\omega)\,\mathrm{d}\omega \\ &= -\frac{1}{2\pi}\int_0^{\phi} \boldsymbol{R}^{\mathrm{T}}(\omega)\,[(m\,n)(n\,n)^{-1}(n\,m) - (m\,m)](0)\,\boldsymbol{R}(\omega)\,\mathrm{d}\omega\,, \end{aligned} \tag{73}$$

where

$$\boldsymbol{R} = \begin{pmatrix} \cos\omega & \sin\omega & 0 & 0 \\ -\sin\omega & \cos\omega & 0 & 0 \\ 0 & 0 & 1 & 0 \\ 0 & 0 & 0 & 1 \end{pmatrix} \tag{74}$$

is a 4×4 rotation matrix in the x_1x_2-plane.

Carrying out the integrations, we obtain

$$\boldsymbol{Q}(\phi) = -\frac{1}{4\pi} \begin{pmatrix} \beta\phi + \gamma\cos\phi\sin\phi & \gamma\sin^2\phi & 0 & 0 \\ \gamma\sin^2\phi & \beta\phi - \gamma\cos\phi\sin\phi & 0 & 0 \\ 0 & 0 & \dfrac{2K_2\phi}{C_{44}K_2 - R_3^2} & -\dfrac{2R_3\phi}{C_{44}K_2 - R_3^2} \\ 0 & 0 & -\dfrac{2R_3\phi}{C_{44}K_2 - R_3^2} & -\dfrac{2C_{44}\phi}{C_{44}K_2 - R_3^2} \end{pmatrix}, \tag{75}$$

where

$$\beta = \frac{1}{C_{66}} + \frac{1}{C_{11}} = \frac{C_{11} + C_{66}}{C_{11}C_{66}} \tag{76}$$

$$\gamma = \frac{1}{C_{66}} - \frac{1}{C_{11}} = \frac{C_{11} - C_{66}}{C_{11}C_{66}}, \tag{77}$$

$$\boldsymbol{B}(\phi) = \frac{1}{2\pi} \begin{pmatrix} \frac{\alpha}{2}(\phi + \cos\phi\sin\phi) & \frac{\alpha}{2}\sin^2\phi & 0 & 0 \\ \frac{\alpha}{2}\sin^2\phi & \frac{\alpha}{2}(\phi - \cos\phi\sin\phi) & 0 & 0 \\ 0 & 0 & C_{44}\phi & R_3\phi \\ 0 & 0 & R_3\phi & K_2\phi \end{pmatrix} \tag{78}$$

and

$$\boldsymbol{S}(\phi) = \frac{1}{2\pi} \begin{pmatrix} \dfrac{C_{11} - C_{66}}{C_{11}}\sin^2\phi & -\dfrac{C_{66}}{C_{11}}\phi - \dfrac{C_{11} - C_{66}}{C_{11}}\cos\phi\sin\phi & 0 & 0 \\ \dfrac{C_{66}}{C_{11}}\phi - \dfrac{C_{11} - C_{66}}{C_{11}}\cos\phi\sin\phi & -\dfrac{C_{11} - C_{66}}{C_{11}}\sin^2\phi & 0 & 0 \\ 0 & 0 & 0 & 0 \\ 0 & 0 & 0 & 0 \end{pmatrix}. \tag{79}$$

For $\phi = 2\pi$, the matrices (75), (78), and (79) reduce to

$$\boldsymbol{Q}(2\pi) = -\frac{1}{2} \begin{pmatrix} \beta & 0 & 0 & 0 \\ 0 & \beta & 0 & 0 \\ 0 & 0 & \dfrac{2K_2}{C_{44}K_2 - R_3^2} & -\dfrac{2R_3}{C_{44}K_2 - R_3^2} \\ 0 & 0 & -\dfrac{2R_3}{C_{44}K_2 - R_3^2} & \dfrac{2C_{44}}{C_{44}K_2 - R_3^2} \end{pmatrix}, \tag{80}$$

$$\boldsymbol{B}(2\pi) = \begin{pmatrix} \alpha/2 & 0 & 0 & 0 \\ 0 & \alpha/2 & 0 & 0 \\ 0 & 0 & C_{44} & R_3 \\ 0 & 0 & R_3 & K_2 \end{pmatrix} \tag{81}$$

and

$$S(2\pi) = \begin{pmatrix} 0 & -C_{66}/C_{11} & 0 & 0 \\ C_{66}/C_{11} & 0 & 0 & 0 \\ 0 & 0 & 0 & 0 \\ 0 & 0 & 0 & 0 \end{pmatrix}. \tag{82}$$

These three matrices (80)–(82) obey the relations

$$B(2\pi)\,S(2\pi) + S^{\mathrm{T}}(2\pi)\,B(2\pi) = \mathbf{0} \tag{83}$$

$$Q(2\pi)\,S^{\mathrm{T}}(2\pi) + S(2\pi)\,Q(2\pi) = \mathbf{0} \tag{84}$$

$$B(2\pi)\,Q(2\pi) + S^{\mathrm{T}}(2\pi)\,S^{\mathrm{T}}(2\pi) = -I \tag{85}$$

as in anisotropic elasticity (see [44,45]).

Now, substituting Equations (75), (79)–(82) into Equation (45), the extended displacement vector of a straight dislocation with extended Burgers vector b_I reads in matrix form:

$$2\pi U = b\phi + b\left(\frac{C_{11}-C_{66}}{C_{11}}\right)\begin{pmatrix} \sin\phi\cos\phi & \sin^2\phi & 0 & 0 \\ \sin^2\phi & -\sin\phi\cos\phi & 0 & 0 \\ 0 & 0 & 0 & 0 \\ 0 & 0 & 0 & 0 \end{pmatrix} - b\,\frac{C_{66}}{C_{11}}\ln r\begin{pmatrix} 0 & -1 & 0 & 0 \\ 1 & 0 & 0 & 0 \\ 0 & 0 & 0 & 0 \\ 0 & 0 & 0 & 0 \end{pmatrix} \tag{86}$$

and the four components of Equation (86) read explicitly:

$$u_1^{\parallel} = \frac{b_1^{\parallel}}{2\pi}\left(\arctan\frac{x_2}{x_1} + \frac{C_{11}-C_{66}}{C_{11}}\frac{x_1x_2}{r^2}\right) + \frac{b_2^{\parallel}}{2\pi}\left(\frac{C_{66}}{C_{11}}\ln r + \frac{C_{11}-C_{66}}{C_{11}}\frac{x_2^2}{r^2}\right), \tag{87}$$

$$u_2^{\parallel} = -\frac{b_1^{\parallel}}{2\pi}\left(\frac{C_{66}}{C_{11}}\ln r - \frac{C_{11}-C_{66}}{C_{11}}\frac{x_2^2}{r^2}\right) + \frac{b_2^{\parallel}}{2\pi}\left(\arctan\frac{x_2}{x_1} - \frac{C_{11}-C_{66}}{C_{11}}\frac{x_1x_2}{r^2}\right), \tag{88}$$

$$u_3^{\parallel} = \frac{b_3^{\parallel}}{2\pi}\arctan\frac{x_2}{x_1}, \tag{89}$$

$$u_3^{\perp} = \frac{b_3^{\perp}}{2\pi}\arctan\frac{x_2}{x_1}. \tag{90}$$

Equations (87)–(90) are the displacement components of a straight dislocation with extended Burgers vector $(b_1^{\parallel}, b_2^{\parallel}, b_3^{\parallel}, b_3^{\perp})$ in a one-dimensional hexagonal quasicrystal of Laue class 10. Equations (87) and (88) are the displacement components of an edge dislocation with Burgers vector $b_1^{\parallel}$ or $b_2^{\parallel}$. Equations (89) and (90) are the displacement components of a screw dislocation with Burgers vector components $b_3^{\parallel}$ and $b_3^{\perp}$. It can be seen that all four components of the Burgers vector are uncoupled. The displacement components of a straight dislocation in a one-dimensional hexagonal quasicrystal of Laue class 10 (87)–(90) obtained in the framework of the integral formalism are in agreement with the expressions given by Li and Fan [46], directly solving the field equations for the phonon and phason displacement fields (see also [47]). The only difference is an irrelevant constant displacement in Equation (88), leading to x_1^2/r^2 instead of $-x_2^2/r^2$ in [46]. The phonon displacement fields (87)–(89) take the same form as in hexagonal crystals.

Substituting Equations (75), (79), (80), and (82) into Equation (46), the extended displacement vector of a line force with extended strength F_I reads in matrix form:

$$4\pi \boldsymbol{U} = \boldsymbol{F}\gamma \begin{pmatrix} -\sin^2\phi & \sin\phi\cos\phi & 0 & 0 \\ \sin\phi\cos\phi & \sin^2\phi & 0 & 0 \\ 0 & 0 & 0 & 0 \\ 0 & 0 & 0 & 0 \end{pmatrix} - \boldsymbol{F}\ln r \begin{pmatrix} \beta & 0 & 0 & 0 \\ 0 & \beta & 0 & 0 \\ 0 & 0 & \dfrac{2K_2}{C_{44}K_2 - R_3^2} & -\dfrac{2R_3}{C_{44}K_2 - R_3^2} \\ 0 & 0 & -\dfrac{2R_3}{C_{44}K_2 - R_3^2} & \dfrac{2C_{44}}{C_{44}K_2 - R_3^2} \end{pmatrix} \tag{91}$$

and the four components of Equation (91) read explicitly:

$$u_1^{\parallel} = -\frac{F_1^{\parallel}}{4\pi}\left(\frac{C_{11}+C_{66}}{C_{11}C_{66}}\ln r + \frac{C_{11}-C_{66}}{C_{11}C_{66}}\frac{x_2^2}{r^2}\right) + \frac{F_2^{\parallel}}{4\pi}\frac{C_{11}-C_{66}}{C_{11}C_{66}}\frac{x_1x_2}{r^2}, \tag{92}$$

$$u_2^{\parallel} = \frac{F_1^{\parallel}}{4\pi}\frac{C_{11}-C_{66}}{C_{11}C_{66}}\frac{x_1x_2}{r^2} - \frac{F_2^{\parallel}}{4\pi}\left(\frac{C_{11}+C_{66}}{C_{11}C_{66}}\ln r - \frac{C_{11}-C_{66}}{C_{11}C_{66}}\frac{x_2^2}{r^2}\right), \tag{93}$$

$$u_3^{\parallel} = -\frac{F_3^{\parallel}K_2 - F_3^{\perp}R_3}{2\pi(C_{44}K_2 - R_3^2)}\ln r, \tag{94}$$

$$u_3^{\perp} = \frac{F_3^{\parallel}R_3 - F_3^{\perp}C_{44}}{2\pi(C_{44}K_2 - R_3^2)}\ln r. \tag{95}$$

Equations (92)–(95) are the displacement components of a line force with strength components $F_1^{\parallel}$, $F_2^{\parallel}$, $F_3^{\parallel}$ and $F_3^{\perp}$. In Equations (94) and (95), it can be seen that the phonon and phason components $F_3^{\parallel}$ and $F_3^{\perp}$ are coupled due to the elastic constant of the phonon–phason coupling R_3.

Substituting Equations (78), (79), (81), and (82) into Equation (47), the extended stress function vector of a straight dislocation with extended Burgers vector b_K reads in matrix form:

$$2\pi\boldsymbol{\Phi} = \boldsymbol{b}\,\frac{\alpha}{2}\begin{pmatrix} -\sin^2\phi & \sin\phi\cos\phi & 0 & 0 \\ \sin\phi\cos\phi & \sin^2\phi & 0 & 0 \\ 0 & 0 & 0 & 0 \\ 0 & 0 & 0 & 0 \end{pmatrix} - \boldsymbol{b}\ln r \begin{pmatrix} \alpha/2 & 0 & 0 & 0 \\ 0 & \alpha/2 & 0 & 0 \\ 0 & 0 & C_{44} & R_3 \\ 0 & 0 & R_3 & K_2 \end{pmatrix} \tag{96}$$

and the four components of Equation (96) read explicitly:

$$\Phi_1^{\parallel} = -\frac{C_{11}^2 - C_{12}^2}{2C_{11}}\left[\frac{b_1^{\parallel}}{2\pi}\left(\ln r + \frac{x_2^2}{r^2}\right) - \frac{b_2^{\parallel}}{2\pi}\frac{x_1x_2}{r^2}\right], \tag{97}$$

$$\Phi_2^{\parallel} = \frac{C_{11}^2 - C_{12}^2}{2C_{11}}\left[\frac{b_1^{\parallel}}{2\pi}\frac{x_1x_2}{r^2} - \frac{b_2^{\parallel}}{2\pi}\left(\ln r - \frac{x_2^2}{r^2}\right)\right], \tag{98}$$

$$\Phi_3^{\parallel} = -\frac{b_3^{\parallel}C_{44} + b_3^{\perp}R_3}{2\pi}\ln r, \tag{99}$$

$$\Phi_3^{\perp} = -\frac{b_3^{\parallel}R_3 + b_3^{\perp}K_2}{2\pi}\ln r. \tag{100}$$

Equations (97)–(100) are the stress function components of a straight dislocation with extended Burgers vector $(b_1^{\parallel}, b_2^{\parallel}, b_3^{\parallel}, b_3^{\perp})$ in a one-dimensional hexagonal quasicrystal of Laue class 10. Equations (97) and (98) are the stress function components of an edge dislocation with Burgers vector $b_1^{\parallel}$ or $b_2^{\parallel}$. Equations (99) and (100) are the stress function components of a screw dislocation with Burgers vector components $b_3^{\parallel}$ and $b_3^{\perp}$. In Equations (99) and (100), it can be seen that the phonon and phason components $b_3^{\parallel}$ and $b_3^{\perp}$ are coupled due to the elastic constant of the phonon–phason coupling R_3. It is interesting to note the similarity between the displacements of a line force (92)–(95) and the stress functions of a dislocation (97)–(100).

Substituting Equations (78)–(80) and (82) into Equation (48), the extended stress function vector of a line force with extended strength F_K reads in matrix form:

$$2\pi\boldsymbol{\Phi} = -\boldsymbol{F}\phi - \boldsymbol{F}\left(\frac{C_{11}-C_{66}}{C_{11}}\right)\begin{pmatrix} \sin\phi\cos\phi & \sin^2\phi & 0 & 0 \\ \sin^2\phi & -\sin\phi\cos\phi & 0 & 0 \\ 0 & 0 & 0 & 0 \\ 0 & 0 & 0 & 0 \end{pmatrix} + \boldsymbol{F}\frac{C_{66}}{C_{11}}\ln r\begin{pmatrix} 0 & 1 & 0 & 0 \\ -1 & 0 & 0 & 0 \\ 0 & 0 & 0 & 0 \\ 0 & 0 & 0 & 0 \end{pmatrix} \quad (101)$$

and the four components of Equation (101) read explicitly:

$$\Phi_1^{\parallel} = -\frac{F_1^{\parallel}}{2\pi}\left(\arctan\frac{x_2}{x_1} + \frac{C_{11}-C_{66}}{C_{11}}\frac{x_1x_2}{r^2}\right) + \frac{F_2^{\parallel}}{2\pi}\left(\frac{C_{66}}{C_{11}}\ln r - \frac{C_{11}-C_{66}}{C_{11}}\frac{x_2^2}{r^2}\right), \quad (102)$$

$$\Phi_2^{\parallel} = -\frac{F_1^{\parallel}}{2\pi}\left(\frac{C_{66}}{C_{11}}\ln r + \frac{C_{11}-C_{66}}{C_{11}}\frac{x_2^2}{r^2}\right) - \frac{F_2^{\parallel}}{2\pi}\left(\arctan\frac{x_2}{x_1} - \frac{C_{11}-C_{66}}{C_{11}}\frac{x_1x_2}{r^2}\right), \quad (103)$$

$$\Phi_3^{\parallel} = -\frac{F_3^{\parallel}}{2\pi}\arctan\frac{x_2}{x_1}, \quad (104)$$

$$\Phi_3^{\perp} = -\frac{F_3^{\perp}}{2\pi}\arctan\frac{x_2}{x_1}. \quad (105)$$

Equations (102)–(105) are the stress functions of a line force with strength components $F_1^{\parallel}$, $F_2^{\parallel}$, $F_3^{\parallel}$ and $F_3^{\perp}$. It can be seen that all four components of the strength components $F_1^{\parallel}$, $F_2^{\parallel}$, $F_3^{\parallel}$, and $F_3^{\perp}$ are uncoupled. The phonon stress functions (102)–(104) take the same form as in hexagonal crystals. It is interesting to note the similarity between the displacements of a dislocation (87)–(90) and the stress functions of a line force (102)–(105).

Substituting Equation (81) into Equation (49), the self-energy per unit length of a straight dislocation in a one-dimensional hexagonal quasicrystal of Laue class 10 is given by

$$\mathcal{E} = \frac{1}{4\pi}\left[\frac{\alpha}{2}\left(b_1^{\parallel}b_1^{\parallel} + b_2^{\parallel}b_2^{\parallel}\right) + C_{44}b_3^{\parallel}b_3^{\parallel} + 2R_3b_3^{\parallel}b_3^{\perp} + K_2b_3^{\perp}b_3^{\perp}\right]\ln(R/r_0). \quad (106)$$

Substituting Equation (80) into Equation (50), the self-energy per unit length of a line force in a one-dimensional hexagonal quasicrystal of Laue class 10 is given by

$$\mathcal{E} = \frac{1}{8\pi}\left[\beta\left(F_1^{\parallel}F_1^{\parallel} + F_2^{\parallel}F_2^{\parallel}\right) + \frac{2}{C_{44}K_2 - R_3^2}\left(K_2F_3^{\parallel}F_3^{\parallel} - 2R_3F_3^{\parallel}F_3^{\perp} + C_{44}F_3^{\perp}F_3^{\perp}\right)\right]\ln(R/r_0). \quad (107)$$

Using Equations (81) and (82), the non-vanishing components of the Peach–Koehler force per unit length between two straight dislocations, Equations (52) and (53), read for a one-dimensional hexagonal quasicrystal of Laue class 10:

$$\mathcal{F}_r^{\mathrm{PK}(AB)} = \frac{1}{2\pi r}\left[\frac{\alpha}{2}\left(b_1^{\parallel(A)}b_1^{\parallel(B)} + b_2^{\parallel(A)}b_2^{\parallel(B)}\right) + C_{44}b_3^{\parallel(A)}b_3^{\parallel(B)} + R_3b_3^{\parallel(A)}b_3^{\perp(B)} + R_3b_3^{\perp(A)}b_3^{\parallel(B)} + K_2b_3^{\perp(A)}b_3^{\perp(B)}\right], \quad (108)$$

$$\mathcal{F}_\phi^{\mathrm{PK}(AB)} = \frac{\alpha}{4\pi r}\left[\left(b_1^{\parallel(A)}b_1^{\parallel(B)} - b_2^{\parallel(A)}b_2^{\parallel(B)}\right)\sin 2\phi - \left(b_1^{\parallel(A)}b_2^{\parallel(B)} + b_2^{\parallel(A)}b_1^{\parallel(B)}\right)\cos 2\phi\right]. \quad (109)$$

Equation (108) contains phonon, phason and phonon–phason-coupling contributions, whereas Equation (109) contains only phonon contributions. If we consider only two parallel screw dislocations, Equations (108) and (109) reduce to the Peach–Kohler force given in [38,46]. In the isotropic limit (see below), the Peach–Koehler force, given in Equations (108) and (109), reduces to the Peach–Koehler force between straight dislocations in isotropic elasticity given in [48–51].

Using Equations (80) and (82), the non-vanishing components of the Cherepanov force per unit length of between two line forces, Equations (55) and (56), read for a one-dimensional hexagonal quasicrystal of Laue class 10

$$
\begin{aligned}
\mathcal{F}_r^{C(AB)} &= -\frac{1}{4\pi r}\Big[\beta\Big(F_1^{\|(A)}F_1^{\|(B)} + F_2^{\|(A)}F_2^{\|(B)}\Big) \\
&\qquad + \frac{2}{C_{44}K_2 - R_3^2}\Big(K_2F_3^{\|(A)}F_3^{\|(B)} - R_3F_3^{\|(A)}F_3^{\perp(B)} - R_3F_3^{\perp(A)}F_3^{\|(B)} + C_{44}F_3^{\perp(A)}F_3^{\perp(B)}\Big)\Big], \qquad (110) \\
\mathcal{F}_\phi^{C(AB)} &= -\frac{\beta}{4\pi r}\Big[\Big(F_1^{\|(A)}F_1^{\|(B)} - F_2^{\|(A)}F_2^{\|(B)}\Big)\sin 2\phi - \Big(F_1^{\|(A)}F_2^{\|(B)} + F_2^{\|(A)}F_1^{\|(B)}\Big)\cos 2\phi\Big]. \qquad (111)
\end{aligned}
$$

Equation (110) contains phonon, phason and phonon–phason-coupling contributions, whereas Equation (111) contains only phonon contributions. In the isotropic limit (see below), the Cherepanov force, given in Equations (110) and (111), reduces to the Cherepanov force between straight line forces in isotropic elasticity given in [52].

6. Green Tensor in the Integral Formalism

Within the framework of the integral formalism, the displacement field of a line force, Equations (92)–(95), gives the two-dimensional 4 × 4 Green tensor for quasicrystals in the hyperspace, namely

$$U_K = G_{KI}F_I \qquad (112)$$

with

$$G_{KI} = \begin{pmatrix} G_{11}^{\|\|} & G_{12}^{\|\|} & G_{13}^{\|\|} & G_{14}^{\|\perp} \\ G_{12}^{\|\|} & G_{22}^{\|\|} & G_{23}^{\|\|} & G_{24}^{\|\perp} \\ G_{13}^{\|\|} & G_{23}^{\|\|} & G_{33}^{\|\|} & G_{34}^{\|\perp} \\ G_{14}^{\|\perp} & G_{24}^{\|\perp} & G_{34}^{\|\perp} & G_{44}^{\perp\perp} \end{pmatrix}. \qquad (113)$$

Therefore, the non-vanishing components of the two-dimensional Green tensor for one-dimensional hexagonal quasicrystals of Laue class 10 read

$$
\begin{aligned}
G_{11}^{\|\|} &= -\frac{1}{4\pi}\left(\frac{C_{11}+C_{66}}{C_{11}C_{66}}\ln r + \frac{C_{11}-C_{66}}{C_{11}C_{66}}\frac{x_2^2}{r^2}\right), && (114) \\
G_{12}^{\|\|} &= \frac{1}{4\pi}\frac{C_{11}-C_{66}}{C_{11}C_{66}}\frac{x_1x_2}{r^2}, && (115) \\
G_{22}^{\|\|} &= -\frac{1}{4\pi}\left(\frac{C_{11}+C_{66}}{C_{11}C_{66}}\ln r - \frac{C_{11}-C_{66}}{C_{11}C_{66}}\frac{x_2^2}{r^2}\right), && (116) \\
G_{33}^{\|\|} &= -\frac{K_2}{2\pi(C_{44}K_2 - R_3^2)}\ln r, && (117) \\
G_{34}^{\|\perp} &= \frac{R_3}{2\pi(C_{44}K_2 - R_3^2)}\ln r, && (118) \\
G_{44}^{\perp\perp} &= -\frac{C_{44}}{2\pi(C_{44}K_2 - R_3^2)}\ln r. && (119)
\end{aligned}
$$

Note that the three-dimensional Green tensor of one-dimensional hexagonal quasicrystals of Laue class 10 has been recently given by Lazar et al. [40].

7. Isotropic Limit

Now, we obtain the isotropic limit of the phonon displacement fields and stress functions of a straight dislocation and a straight line force:

$u_i^{\|} \to u_i$, $\Phi_i^{\|} \to \Phi_i$, $b_i^{\|} \to b_i$ and $F_i^{\|} \to F_i$.

To verify the isotropic limit of Equations (87)–(89), we need $\frac{C_{11}-C_{66}}{C_{11}} \to \frac{1}{2(1-\nu)}$, $\frac{C_{66}}{C_{11}} \to \frac{1-2\nu}{2(1-\nu)}$, where ν denotes the Poisson ratio; we obtain for the three components of the displacement vector of a straight dislocation in an isotropic medium

$$u_1 = \frac{b_1}{2\pi}\left(\arctan\frac{x_2}{x_1} + \frac{1}{2(1-\nu)}\,\frac{x_1x_2}{r^2}\right) + \frac{b_2}{4\pi(1-\nu)}\left((1-2\nu)\,\ln r + \frac{x_2^2}{r^2}\right), \tag{120}$$

$$u_2 = -\frac{b_1}{4\pi(1-\nu)}\left((1-2\nu)\,\ln r - \frac{x_2^2}{r^2}\right) + \frac{b_2}{2\pi}\left(\arctan\frac{x_2}{x_1} - \frac{1}{2(1-\nu)}\,\frac{x_1x_2}{r^2}\right), \tag{121}$$

$$u_3 = \frac{b_3}{2\pi}\,\arctan\frac{x_2}{x_1}, \tag{122}$$

which are in agreement with deWit [53] and Mura [54].

To verify the isotropic limit of Equations (92)–(94), we need $R_3 \to 0$, $\frac{C_{11}+C_{66}}{C_{11}} \to \frac{3-4\nu}{2(1-\nu)}$, $\frac{1}{C_{66}} \to \frac{1}{\mu}$ and $\frac{1}{C_{44}} \to \frac{1}{\mu}$, where μ denotes the shear modulus, and we obtain for the three components of the displacement vector of a line force in an isotropic medium

$$u_1 = -\frac{F_1}{8\pi\mu(1-\mu)}\left((3-4\nu)\,\ln r + \frac{x_2^2}{r^2}\right) + \frac{F_2}{8\pi\mu(1-\mu)}\,\frac{x_1x_2}{r^2}, \tag{123}$$

$$u_2 = \frac{F_1}{8\pi\mu(1-\mu)}\,\frac{x_1x_2}{r^2} - \frac{F_2}{8\pi\mu(1-\mu)}\left((3-4\nu)\,\ln r - \frac{x_2^2}{r^2}\right), \tag{124}$$

$$u_3 = -\frac{F_3}{2\pi\mu}\,\ln r, \tag{125}$$

which are in agreement with Lazar and Agiasofitou [52].

To verify the isotropic limit of Equations (97)–(99), we need $R_3 \to 0$, $\frac{C_{11}^2-C_{12}^2}{2C_{11}} \to \frac{\mu}{1-\nu}$, $C_{44} \to \mu$, and we obtain for the three stress functions of a straight dislocation in an isotropic medium

$$\Phi_1 = -\frac{\mu}{1-\nu}\left[\frac{b_1}{2\pi}\left(\ln r + \frac{x_2^2}{r^2}\right) - \frac{b_2}{2\pi}\,\frac{x_1x_2}{r^2}\right], \tag{126}$$

$$\Phi_2 = \frac{\mu}{1-\nu}\left[\frac{b_1}{2\pi}\,\frac{x_1x_2}{r^2} - \frac{b_2}{2\pi}\left(\ln r - \frac{x_2^2}{r^2}\right)\right], \tag{127}$$

$$\Phi_3 = -\frac{\mu\, b_3}{2\pi}\,\ln r, \tag{128}$$

which are in agreement with Ni and Nemat-Nasser [55].

To verify the isotropic limit of Equations (102)–(104), we need $\frac{C_{11}-C_{66}}{C_{11}} \to \frac{1}{2(1-\nu)}$, $\frac{C_{66}}{C_{11}} \to \frac{1-2\nu}{2(1-\nu)}$, and we obtain for the three stress functions of a line force in an isotropic medium

$$\Phi_1 = -\frac{F_1}{2\pi}\left(\arctan\frac{x_2}{x_1} + \frac{1}{2(1-\nu)}\,\frac{x_1x_2}{r^2}\right) + \frac{F_2}{4\pi(1-\nu)}\left((1-2\nu)\,\ln r - \frac{x_2^2}{r^2}\right), \tag{129}$$

$$\Phi_2 = -\frac{F_1}{4\pi(1-\nu)}\left((1-2\nu)\,\ln r + \frac{x_2^2}{r^2}\right) - \frac{F_2}{2\pi}\left(\arctan\frac{x_2}{x_1} - \frac{1}{2(1-\nu)}\,\frac{x_1x_2}{r^2}\right), \tag{130}$$

$$\Phi_3 = -\frac{F_3}{2\pi}\arctan\frac{x_2}{x_1}. \tag{131}$$

Note that the components (129) and (130) are in agreement with Ni and Nemat-Nasser [55]. It is obvious that the component Φ_3 given in Ni and Nemat-Nasser [55] has a mistaken pre-factor.

8. Conclusions

Using the eight-dimensional integral formalism for one-dimensional quasicrystals, we have derived exact analytical solutions of the displacement fields and the stress functions of line defects, namely straight dislocations and line forces, parallel to the aperiodic direction in a one-dimensional hexagonal quasicrystal of Laue class 10. Moreover, we have calculated the self-energies of a straight dislocation and a line force, the Peach–Koehler force between two straight dislocations, and the Cherepanov force between two straight line forces. It has been shown that in the limit to isotropic elasticity, well-known results for the displacement fields and the stress functions of straight dislocations and line forces are recovered. The two-dimensional Green tensor of one-dimensional hexagonal quasicrystals of Laue class 10 has been derived within the framework of the integral formalism.

Funding: This research was funded by the Deutsche Forschungsgemeinschaft, grant number LA1974/5-1.

Data Availability Statement: The original contributions presented in this study are included in the article.

Acknowledgments: Markus Lazar gratefully acknowledges the grant from the Deutsche Forschungsgemeinschaft (Grant number LA1974/5-1). The author wishes to thank Eleni Agiasofitou for their many useful and stimulating discussions on the fascinating subject of quasicrystals.

Conflicts of Interest: The author declares no conflicts of interest.

References

1. Shechtman, D.; Blech, I.; Gratias, D.; Cahn, J.-W. Metallic phase with long-range orientational order and no translational symmetry. *Phys. Rev. Lett.* **1984**, *53*, 1951–1953. [CrossRef]
2. Maciá-Barber, E. *Quasicrystals: Fundamentals and Applications*; CRC Press: Boca Raton, FL, USA, 2021.
3. Maciá Barber, E. Chemical bonding and physical properties in quasicrystals and their related approximant phases: Known facts and current perspectives. *Appl. Sci.* **2019**, *9*, 2132. [CrossRef]
4. Maciá, E. The role of aperiodic order in science and technology. *Rep. Prog. Phys.* **2006**, *69*, 397–441. [CrossRef]
5. Bak, P. Symmetry, stability, and elastic properties of icosahedral incommensurate crystals. *Phys. Rev. B* **1985**, *32*, 5764–5772. [CrossRef]
6. Levine, D.; Lubensky, T.C.; Ostlund, S.; Ramaswamy, S.; Steinhardt, P.J.; Toner, J. Elasticity and dislocations in pentagonal and icosahedral quasicrystals. *Phys. Rev. Lett.* **1985**, *54*, 1520–1523. [CrossRef]
7. Ding, D.-H.; Wang, W.; Hu, C.; Yang, R. Generalized elasticity theory of quasicrystals. *Phys. Rev. B* **1993**, *48*, 7003–7010. [CrossRef]
8. Socolar, J.E.S.; Lubensky, T.C.; Steinhardt, P.J. Phonons, phasons, and dislocations in quasicrystals. *Phys. Rev. B* **1986**, *34*, 3345–3360. [CrossRef] [PubMed]
9. Lubensky, T.C.; Ramaswamy, S.; Toner, J. Dislocation motion in quasicrystals and implications for macroscopic properties. *Phys. Rev. B* **1986**, *33*, 7715–7719. [CrossRef]
10. Edagawa, K. Dislocations in quasicrystals. *Mater. Sci. Eng. A* **2001**, *309–310*, 528–538. [CrossRef]

11. Edagawa, K.; Takeuchi, S. Elasticity, dislocation and their motion in quasicrystals. In *Dislocations in Solids*; Nabarro, F.R.N., Hirth, J.P., Eds.; Elsevier B.V.: Amsterdam, The Netherlands, 2007; Volume 13, pp. 365–417.
12. Kleman, M. Phasons and the plastic deformation of quasicrystals. *Eur. Phys. J. B* **2003**, *31*, 315–325. [CrossRef]
13. Fan, T.Y. *Mathematical Theory of Elasticity of Quasicrystals and Its Applications*; Science Press and Springer Science+Business Media Singapore: Beijing, China, 2011.
14. Wang, R.; Hu, C. Dislocations in Quasicrystals. In *Intermetallic Compounds—Principles and Practice: Progress*; Westbrook, J.H., Fleischer, R.L., Eds.; John Wiley & Sons, Ltd.: Chichester, UK, 2002; Volume 3, pp. 379–402.
15. Zhang, Z.; Urban, K. Transmission electron microscope observations of dislocations and stacking faults in a decagonal Al-Cu-Co alloy. *Philos. Mag. Lett.* **1989**, *60*, 97–102. [CrossRef]
16. Wollgarten, M.; Zhang, Z.; Urban, K. Diffraction contrast of quasicrystalline dislocations in the transmission electron microscope. *Philos. Mag. Lett.* **1992**, *65*, 1–6. [CrossRef]
17. Zhang, Z.; Zhuang, Y. A transmission electron microscopy study of dislocations in $Al_{70}Ni_{10}Co_{20}$ and $Al_{75}Ni_{10}Fe_{15}$ decagonal quasicrystals. *Philos. Mag. Lett.* **1992**, *65*, 203–209. [CrossRef]
18. Wollgarten, M.; Beyss, M.; Urban, K.; Liebertz, H.; Köster, K. Direct evidence for plastic deformation of quasicrystals by means of a dislocation mechanism. *Phys. Rev. Lett.* **1993**, *71*, 549–552. [CrossRef]
19. Urban, K.; Wollgarten, M.; Gratias, D.M.; Zhang, Z. Electron microscopy of dislocations in quasicrystals. *Phase Transit.* **1993**, *44*, 187–194. [CrossRef]
20. Wollgarten, M.; Bartsch, M.; Messerschmidt, U.; Feuerbacher, M.; Rosenfeld, R.; Beyss, M.; Urban, K. In-situ observation of dislocation motion in icosahedral Al-Pd-Mn single quasicrystals. *Philos. Mag. Lett.* **1995**, *71*, 99–105. [CrossRef]
21. Feuerbacher, M. Dislocations in icosahedral quasicrystals. *Chem. Soc. Rev.* **2012**, *41*, 6745–6759. [CrossRef] [PubMed]
22. Ding, D.; Yang, W.; Hu, C.; Wang, R. Linear elasticity theory of quasicrystals and defects in quasicrystals. In *Materials Science Forum*; Trans Tech Publications, Ltd.: Baech, Switzerland, 1994; Volumes 150–151, pp. 345–354.
23. Ding, D.-H.; Wang, R.; Yang, W.; Hu, C. General expressions for the elastic displacement fields induced by dislocations in quasicrystals. *J. Phys. Condens. Matter* **1995**, *7*, 5423–5436. [CrossRef]
24. Lazar, M.; Agiasofitou, E. Fundamentals in generalized elasticity and dislocation theory of quasicrystals: Green tensor, dislocation key-formulas and dislocation loops. *Philos. Mag.* **2014**, *94*, 4080–4101. [CrossRef]
25. Ding, D.-H.; Wang, R.; Yang, W.; Hu, C.; Qin, Y. Elasticity theory of straight dislocations in quasicrystals. *Philos. Mag. Lett.* **1995**, *72*, 353–359. [CrossRef]
26. Stroh, A.N. Dislocations and cracks in anisotropic elasticity. *Philos. Mag.* **1958**, *3*, 625–646. [CrossRef]
27. Stroh, A.N. Steady state problems in anisotropic elasticity. *J. Math. Phys.* **1962**, *41*, 77–103. [CrossRef]
28. Barnett, D.M.; Lothe, J. Synthesis of the sextic and the integral formalism for dislocations, Green's function and structure waves in anisotropic elastic solids. *Phys. Norv.* **1973**, *7*, 13–19.
29. Asaro, R.J.; Hirth, J.P.; Barnett, D.M.; Lothe, J. A further synthesis of sextic and integral theories for dislocations and line forces in anisotropic media. *Phys. Stat. Sol. (b)* **1973**, *60*, 261–271. [CrossRef]
30. Hu, C.; Wang, R.; Ding, D.-H. Symmetry groups, physical property tensors, elasticity and dislocations in quasicrystals. *Rep. Prog. Phys.* **2000**, *63*, 1–39. [CrossRef]
31. Li, L.H.; Liu, G.T. Stroh formalism for icosahedral quasicrystal and its application. *Phys. Lett. A* **2012**, *376*, 987–990. [CrossRef]
32. Radi, E.; Mariano, P.M. Steady-state propagation of dislocations in quasi-crystals. *Proc. R. Soc. A* **2011**, *467*, 3490–3508. [CrossRef]
33. Wang, X.; Schiavone, P. Dislocations, imperfect interfaces and interface cracks in anisotropic elasticity for quasicrystals. *Math. Mech. Complex Syst.* **2013**, *1*, 1–17. [CrossRef]
34. Ting, T.C.T. *Anisotropic Elasticity*; Oxford Science Publishers: Oxford, UK, 1996.
35. Lazar, M. Line defects in quasicrystals. *Crystals* **2025**, *15*, 275. [CrossRef]
36. Wu, M.S.; Kirchner, H.O.K. Line defects in the (110)-plane of a cubic crystal—An outstanding problem solved by the integral formalism. *Proc. R. Soc. Lond. A* **2003**, *459*, 2033–2047. [CrossRef]
37. Wang, R.; Yang, W.; Hu, C.; Ding, D.-H. Point and space groups and elastic behaviours of one-dimensional quasicrystals. *J. Phys. Condens. Matter* **1997**, *9*, 2411–2422. [CrossRef]
38. Lazar, M.; Agiasofitou, E. Eshelbian mechanics of novel materials: Quasicrystals. *J. Micromech. Mol. Phys.* **2016**, *1*, 1640008. [CrossRef]
39. Lazar, M.; Kirchner, H.O.K. Generalized plane strain embedded in three-dimensional anisotropic elasticity. *Philos. Mag.* **2021**, *101*, 2584–2598. [CrossRef]
40. Lazar, M.; Michelitsch, T.; Agiasofitou, E. Three-dimensional Green tensor of one-dimensional hexagonal quasicrystals. *Crystals* **2024**, *14*, 1034. [CrossRef]
41. Agiasofitou, E.; Lazar, M. On elastic stability and positive definiteness for one-dimensional quasicrystals. *Crystals* **2025**, *15*, 219. [CrossRef]

42. Ting, T.C.T. Some identities and the structure of N_i in the Stroh formalism of anisotropic elasticity. *Q. J. Mech. Appl. Math.* **1988**, *46*, 109–120.
43. Kirchner, H.O.K.; Bluemel, K.H. Elastic quasi-isotropy normal to the basal plane in the hexagonal system. *Phys. Stat. Sol. (b)* **1976**, *75*, 527–532. [CrossRef]
44. Bacon, D.J.; Barnett, D.M.; Scattergood, R.O. Anisotropic continuum theory of defects. *Prog. Mater. Sci.* **1980**, *23*, 51–262. [CrossRef]
45. Hirth, J.P.; Lothe, J. *Theory of Dislocations*, 2nd ed.; John Wiley: New York, NY, USA, 1982.
46. Li, X.F.; Fan, T.Y. A straight dislocation in one-dimensional hexagonal quasicrystals. *Phys. Stat. Sol. (b)* **1999**, *212*, 19–26. [CrossRef]
47. Fan, T.Y.; Li, X.F.; Sun, Y.F. A moving screw dislocation in a one-dimensional hexagonal quasicrystal. *Acta Phys. Sin.* **1999**, *8*, 288–295.
48. Lardner, R.W. *Mathematical Theory of Dislocations and Fracture*; University of Toronto Press: Toronto, ON, Canada, 1974.
49. Weertman, J.; Weertman, J.R. *Elementary Dislocation Theory*; Oxford University Press: Oxford, UK; New York, NY, USA, 1992.
50. Landau, L.D.; Lifshitz, E.M. *Theory of Elasticity*; Pergamon Press Ltd.: New York, NY, USA, 1970.
51. Agiasofitou, E.; Lazar, M. Micromechanics of dislocations in solids: ***J***-, *M*-, and ***L***-integrals and their fundamental relations. *Int. J. Eng. Sci.* **2017**, *114*, 16–40. [CrossRef]
52. Lazar, M.; Agiasofitou, E. ***J***-, *M*- and ***L***-integrals of line charges and line forces. *Acta Mech.* **2023**, *234*, 3131–3159. [CrossRef]
53. DeWit, R. Theory of disclinations IV. *J. Res. Natl. Bur. Stand.* **1973**, *77A*, 607–658. [CrossRef]
54. Mura, T. *Micromechanics of Defects in Solids*, 2nd ed.; Martinus Nijhoff: Dordrecht, The Netherlands, 1987.
55. Ni, L.; Nemat-Nasser, S. A general duality in elasticity. *Mech. Mater.* **1996**, *24*, 87–123. [CrossRef]

mathematics

Article

In-Plane Vibrations of Elastic Lattice Plates and Their Continuous Approximations

Noël Challamel [1,*], Huu Phu Nguyen [2], Chien Ming Wang [2] and Giuseppe Ruta [3]

1 Centre de Recherche, Université Bretagne Sud, IRDL (CNRS UMR 6027), Rue de Saint Maudé, BP92116, 56321 Lorient Cedex, France
2 School of Civil Engineering, University of Queensland, St Lucia, QLD 4072, Australia; huuphu.nguyen@uq.edu.au (H.P.N.); cm.wang@uq.edu.au (C.M.W.)
3 Dipartimento D'Ingegneria Strutturale e Geotecnica, Università "La Sapienza", 00185 Roma, Italy; giuseppe.ruta@uniroma1.it
* Correspondence: noel.challamel@univ-ubs.fr

Abstract: This paper presents an analytical study on the in-plane vibrations of a rectangular elastic lattice plate. The plane lattice is modelled considering central and angular interactions. The lattice difference equations are shown to coincide with a spatial finite difference scheme of the corresponding continuous plate. The considered lattice converges to a 2D linear isotropic elastic continuum at the asymptotic limit for a sufficiently small lattice spacing. This continuum has a free Poisson's ratio, which must be lower than that foreseen by the rare-constant theory, to preserve the definite positiveness of the associated discrete energy. Exact solutions for the in-plane eigenfrequencies and modes are analytically derived for the discrete plate. The stiffness characterising the lattice interactions at the boundary is corrected to preserve the symmetry properties of the discrete displacement field. Two classes of constraints are considered, i.e., sliding supports at the nodes, one normal and the other parallel to the boundary. For both boundary conditions, a single equation for the eigenfrequency spectrum is derived, with two families of eigenmodes. Such behaviour of the lattice plate is like that of the continuous plate, the eigenfrequency spectrum of which has been given by Rayleigh. The convergence of the spectrum of the lattice plate towards the spectrum of the continuous plate from below is confirmed. Two continuous size-dependent plate models, considering the strain gradient elasticity and non-local elasticity, respectively, are built from the lattice difference equations and are shown to approximate the plane lattice accurately.

Keywords: lattice/discrete elasticity; continuum elasticity; elastodynamics; difference equations; partial differential equations; in-plane vibrations; eigenvalue problem; discrete plate

MSC: 34A33; 34K31; 37K60; 39A36; 52C05; 74B05; 74H45; 70J10; 34C46

1. Introduction

This paper presents new solutions for the in-plane vibration problems of two-dimensional rectangular linear elastic lattices. The exact solution of this discrete problem is connected to that of a continuous plate, obtained at the asymptotic limit for an infinite number of particles in each direction. Bridging discrete and continuous elasticity is an old problem of mechanics that emerged during the 19th century with the so-called molecular foundation of elasticity (see the historical analysis of Capecchi et al., 2010, [1] Capecchi et al., 2011 [2] or Challamel et al., 2023 [3]). This problem of bridging discrete to continuous elasticity is still active nowadays, and the problem also includes the development of generalised continuum elasticity (Andrianov et al., 2021 [4]; dell'Isola and Steigmann, 2020 [5]; Wang et al., 2020 [6]).

Although many solutions are available for 1D lattices since the solution obtained by Lagrange (1759) [7,8] for the eigenfrequencies of a fixed–fixed lattice string (mathematically equivalent to a lattice rod), the exact solutions for the free vibration of 2D or 3D finite elastic lattices are not so well documented in the literature. There are reference solutions for the out-of-plane vibration of lattice membranes and plates. Rosenau (1987) [9], Andrianov and Awrejcewicz (2008) [10], Lombardo and Askes (2010) [11] and Hérisson et al. (2018) [12] obtained the exact wave dispersive equation in the infinite lattice membrane. Rosenau (1987) [9] also derived a lattice-based continuous nonlocal membrane model, by using the Padé approximant of the associated pseudodifferential operator of the discrete lattice. The eigenfrequencies of a fixed–fixed rectangular lattice membrane (finite lattice membrane) have been analytically found by Hérisson et al. (2018) [12] from the resolution of a linear difference eigenvalue problem. This approach is mathematically equivalent to the finite difference (FD) formulation of the continuous membrane problem, regarding which the exact discrete resolution was already presented by Chen (1971) [13] and Tong et al. (1971) [14]. Hérisson et al. (2018) [12] also investigated the efficiency of the nonlocal continuous membrane model pioneered by Rosenau (1987) [9] for approximating the eigenfrequency of the fixed–fixed lattice membrane.

Zhang et al. (2015) [15] studied analytically the free vibration of simply supported lattice rectangular plates (discrete Kirchhoff–Love plate model) and derived natural frequencies as exact solutions of the associated linear difference eigenvalue problem. As the mathematical problem is equivalent to the finite difference (FD) formulation of a continuous Kirchhoff–Love plate, the eigenfrequencies of the simply supported lattice plate coincide with those derived exactly from the FD scheme by Chen (1971) [13]. Wang et al. (2020) [6] gave exact solutions for lattice plates with various constraints (including Levy-type boundary conditions), thus generalising the solution of Chen (1971) [13] and Zhang et al. (2015) [15], that has been restricted to Navier-type boundary conditions. Challamel et al. (2016) [16] and Hache et al. (2017) [17] (see also Challamel et al., 2021 [18]) derived a gradient and a nonlocal continuum Kirchhoff–Love plate model from the lattice plate model and compared the higher-order solutions to the lattice one for simply supported boundary conditions. These two structural cases of out-of-plane vibration of a discrete membrane and plate are 2D one-field discrete problems, i.e., with only one kinematic field for the unknown displacement.

To the authors' knowledge, there is no analogous result for in-plane 2D/3D finite lattices described by a two-field or a three-field unknown displacement. Most results available in the literature for in-plane 2D or 3D lattices hold for infinite media, for which the dispersion response has been analytically characterised (Born and von Kármán, 1912 [19]; Gazis et al., 1960 [2]; Challamel et al., 2023 [3]). Born and von Kármán (1912) [19] first introduced the mixed differential-difference equations of a cubic lattice with central and non-central interactions. They also derived the wave dispersive property in an infinite Lagrange lattice, and in infinite 3D lattices [19]. The non-central interactions considered by Born and von Kármán [19] are of the shear type, which have been shown later to violate the rotation invariance principle (the conditions for rotational invariance of lattices were extensively studied during the 1960s by Lax, 1965 [20], Gazis and Wallis, 1966 [21], Keating, 1966 [22] and Keating, 1966 [23]). Gazis et al. (1960) [24] obtained the mixed differential-difference equations of a cubic lattice with central and non-central (angular-type) interactions, consistent with the rotation invariance principle. The mixed differential-difference equations derived by Gazis et al. [20] differ from the ones of Born and von Kármán [19], as shown by Challamel et al. [3] for 2D or 3D lattices. Gazis et al. [20] also analytically characterised the wave dispersive behaviour of such infinite lattice. To the authors' knowledge, the exact solutions for the vibrations of finite cubic lattices with central and angular interactions have not been derived in the literature.

The paper is focused on the analytical determination of the in-plane eigenfrequencies of a rectangular lattice (2D lattice). The lattice considered herein is the cubic lattice introduced by Gazis et al. (1960) [24]. This lattice exhibits central and non-central interactions that make it possible to calibrate the macroscopic elastic parameters for a wide range (see the discussion in Challamel et al., 2023 [3] and Challamel et al., 2024 [25]). Gazis et al. (1960) [24] already obtained the exact dispersive curves of their lattice in the case of infinite media. Mindlin (1968) [26] expanded the difference equations of Gazis et al.'s [23] lattice and obtained a lattice-based gradient elasticity model, if second-order terms are included. Mindlin (1970) [27] obtained exact solutions for Gazis' type of finite lattices for four problems: thickness-shear vibration of a plane lattice; face-shear and thickness-twist waves in a plate; axial shear vibration and, in the end, the torsion of a rectangular bar. The first three are reduced to a one-dimensional eigenfrequency problem, whereas the last one is a static problem. The present paper is restricted to a linear vibration analysis, due to the linearity behaviour in each periodic lattice cell. We do not explore the vibration behaviour of nonlinear elastic lattices, as initially studied by Fermi et al. (1955) [28] for a one-dimensional periodic lattice with a nonlinear restoring force. The nonlinear lattice of Fermi et al. (1955) [28] initiated some new research areas in the field of nonlinear wave propagation and soliton theory (Weissert, 1997 [29]; Dauxois et al., 2005 [30]; Dauxois and Peyrard, 2006 [31] or more recently Vainchtein, 2022 [32]). The papers of Potapov et al. (2002) [33] or Friesecke and Matthies (2003) [34] on solitary waves in 2D lattices may be mentioned for 2D nonlinear lattices, where the interactions have been restricted to the central type.

In this paper, the in-plane vibration of a rectangular linear elastic lattice according to the model by Gazis et al. [24], in which non-central interactions are of the angular type, is investigated to obtain the exact solutions for the eigen-frequencies and modes analytically, for simply supported boundary conditions (in-plane vibration problem with sliding supports). This lattice problem (discrete problem) has not been solved analytically in the literature. The free vibration characteristics of the rectangular lattice are obtained by posing and solving a linear two-field difference eigenvalue problem. The stiffness of the lattice interactions in the vicinity of the boundary is corrected to preserve the symmetry properties of the discrete displacement field along the boundaries. Two classes of constraints are considered, i.e., sliding supports at the nodes, once normal and then parallel to the boundary. For both boundary conditions, a single equation for the eigenfrequency spectrum is derived, with two families of eigenmodes. The lattice solution is compared to the available continuous one of the in-plane eigenfrequency for a plate, analytically derived by Rayleigh (1894) [35] (see also Rayleigh, 1889 [36]), and numerically obtained by Bardell et al. (1996) [37] and Gorman (2006) [38]. We also derived higher-order continuum plate models, under plane stress assumptions, which are asymptotically formulated from the lattice plate difference equations. It is possible to obtain the exact solutions of the eigenfrequencies for both higher-order plate models, a strain gradient elasticity plate model and a nonlocal plate model. The strain gradient plate model is similar to the one derived by Mindlin (1968) [26], here restricted to plane stress assumptions. Both higher-order continua are shown to accurately fit the behaviour of the lattice, including its scale-dependence behaviour, i.e., the dependence of the normalised eigenfrequency with respect to the number of cells in each direction. Both models converge towards the local plate solution derived by Rayleigh (1894) [35] when the number of cells in each direction is sufficiently large.

2. In-Plane Vibrations of a Discrete Rectangular Plate

We investigated the free vibration of a 2D linear elastic lattice exhibiting central and non-central angular interactions (see Figure 1). This lattice is the one pioneered by Gazis et al. (1960) [24] for cubic lattices with angular interactions, which has also been studied by Mindlin (1968) [26] and Mindlin (1970) [27]. This lattice, restricted to a 2D analysis in this study, is monatomic and square, composed of particles with an equal mass M and lattice spacing a. The linear elastic interactions among them consist of (a) nearest and next-to-nearest central forces, represented by axial springs of stiffness α and β between the nearest and next-to-nearest particles, respectively; (b) non-central interactions given by force couples, represented by rotational springs of stiffness $C = \gamma a^2$. α and β are the two parameters of the central interaction, whereas γ is the parameter of the non-central (or angular) interaction. The three stiffness parameters α, β and γ have the same dimensions. The governing equations of the lattice may be derived from the elastic potential energy.

$$
\begin{aligned}
U = \sum_i \sum_j \frac{\alpha}{4} & \left[\left(u_{i+1,j} - u_{i,j}\right)^2 + \left(u_{i+1,j+1} - u_{i,j+1}\right)^2 + \left(v_{i,j+1} - v_{i,j}\right)^2 + \left(v_{i+1,j+1} - v_{i+1,j}\right)^2 \right] \\
+ \frac{\beta}{4} & \left[\left(u_{i+1,j+1} - u_{i,j} + v_{i+1,j+1} - v_{i,j}\right)^2 + \left(u_{i+1,j} - u_{i,j+1} - v_{i+1,j} + v_{i,j+1}\right)^2 \right] \\
+ \frac{1}{2}\gamma & \left[\begin{array}{l} \left(u_{i,j+1} - u_{i,j} + v_{i+1,j} - v_{i,j}\right)^2 + \left(u_{i+1,j+1} - u_{i+1,j} + v_{i+1,j} - v_{i,j}\right)^2 \\ + \left(u_{i,j+1} - u_{i,j} + v_{i+1,j+1} - v_{i,j+1}\right)^2 + \left(u_{i+1,j+1} - u_{i+1,j} + v_{i+1,j+1} - v_{i,j+1}\right)^2 \end{array} \right]
\end{aligned}
\tag{1}
$$

where the stiffnesses α, β, γ are positive to ensure the positive definiteness of the energy; pure central interactions are attained if $\gamma = 0$ (Navier's, Poisson's or Cauchy's molecular assumption).

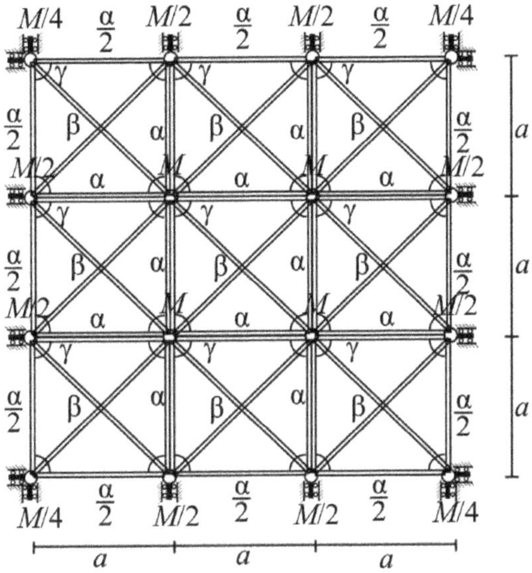

Figure 1. Lattice plate with central and angular interactions on sliding support (case A); $n = m = 3$; corrections of the elastic interactions at the border.

Since the lattice has concentrated masses only, its kinetic energy has the simple form:

$$
T = \sum_i \sum_j \frac{1}{2} M\, \dot{u}_{i,j}^2 + \frac{1}{2} M\, \dot{v}_{i,j}^2 \tag{2}
$$

where the superdot denotes the time derivative. Hamilton's principle applied to the Lagrangian $L = T - U$ yields the mixed differential-difference equations of the 2D Gazis et al.'s lattice (Challamel et al., 2022 [39]):

$$
\left\{
\begin{array}{l}
\alpha\left(u_{i+1,j} - 2u_{i,j} + u_{i-1,j}\right) + \frac{\beta}{2}\left(u_{i+1,j+1} + u_{i-1,j-1} + u_{i-1,j+1} + u_{i+1,j-1} - 4u_{i,j}\right) + \\
\left(\frac{\beta}{2} + \gamma\right)\left(v_{i+1,j+1} - v_{i-1,j+1} - v_{i+1,j-1} + v_{i-1,j-1}\right) + 4\gamma\left(u_{i,j+1} - 2u_{i,j} + u_{i,j-1}\right) = M\,\ddot{u}_{i,j} \\
\alpha\left(v_{i,j+1} - 2v_{i,j} + v_{i,j-1}\right) + \frac{\beta}{2}\left(v_{i+1,j+1} + v_{i-1,j-1} + v_{i-1,j+1} + v_{i+1,j-1} - 4v_{i,j}\right) + \\
\left(\frac{\beta}{2} + \gamma\right)\left(u_{i+1,j+1} - u_{i-1,j+1} - u_{i+1,j-1} + u_{i-1,j-1}\right) + 4\gamma\left(v_{i+1,j} - 2v_{i,j} + v_{i-1,j}\right) = M\,\ddot{v}_{i,j}
\end{array}
\right. \tag{3}
$$

If n, m are integers, we calculate the in-plane eigenfrequencies of rectangular lattices of length $L_1 = na$ along the 'horizontal' and width $L_2 = ma$ along the 'vertical' with sliding supports at their boundaries. In Figure 1, for instance, the lattice plate on sliding supports is represented with $n = m = 3$. Then, if ω is the circular frequency of the in-plane natural vibration, the mixed differential-difference equation, Equation (3), is reduced to a system of spatial difference equations. We must solve the linear eigenvalue problem resulting from the coupled system of difference equations

$$\begin{cases} \alpha\left(u_{i+1,j}-2u_{i,j}+u_{i-1,j}\right)+\frac{\beta}{2}\left(u_{i+1,j+1}+u_{i-1,j-1}+u_{i-1,j+1}+u_{i+1,j-1}-4u_{i,j}\right)+ \\ \left(\frac{\beta}{2}+\gamma\right)\left(v_{i+1,j+1}-v_{i-1,j+1}-v_{i+1,j-1}+v_{i-1,j-1}\right)+4\gamma\left(u_{i,j+1}-2u_{i,j}+u_{i,j-1}\right)+M\omega^2u_{i,j}=0 \\ \alpha\left(v_{i,j+1}-2v_{i,j}+v_{i,j-1}\right)+\frac{\beta}{2}\left(v_{i+1,j+1}+v_{i-1,j-1}+v_{i-1,j+1}+v_{i+1,j-1}-4v_{i,j}\right)+ \\ \left(\frac{\beta}{2}+\gamma\right)\left(u_{i+1,j+1}-u_{i-1,j+1}-u_{i+1,j-1}+u_{i-1,j-1}\right)+4\gamma\left(v_{i+1,j}-2v_{i,j}+v_{i-1,j}\right)+M\omega^2v_{i,j}=0 \end{cases} \quad (4)$$

We consider two cases of constraints (which can be referred to as simply supported boundary conditions) provided by sliding supports at the boundary particles, as discussed, for instance, by Gorman (2006) [38] for a continuum in the shape of a plate:

- In case A (Figure 1), the tangential displacement and the normal load at the lattice boundary vanish (sliding supports normal to the boundary). The displacement-based boundary conditions are as follows:

$$v_{0,j}=0 \text{ for } j\in\{0,1,\ldots,m\};\ v_{n,j}=0 \text{ for } j\in\{0,1,\ldots,m\};$$

$$u_{i,0}=0 \text{ for } i\in\{0,1,\ldots,n\};\ u_{i,m}=0 \text{ for } i\in\{0,1,\ldots,n\}; \quad (5)$$

The discrete Dirichlet boundary conditions in Equation (5) for the simply supported edges (case A) are completed by the discrete Neumann boundary conditions of vanishing normal force along the four edges. The same boundary conditions of case A can be considered for the continuous plate, as shown in Figure 2.

- In case B (Figure 3), the normal displacement and the tangential load at the lattice boundary vanish (sliding supports parallel to the boundary).

$$u_{0,j}=0 \text{ for } j\in\{0,1,\ldots,m\};\ u_{n,j}=0 \text{ for } j\in\{0,1,\ldots,m\};$$

$$v_{i,0}=0 \text{ for } i\in\{0,1,\ldots,n\};\ v_{i,m}=0 \text{ for } i\in\{0,1,\ldots,n\}; \quad (6)$$

Figure 2. Continuous plate on sliding support (for the in-plane continuous plate problem)—case A.

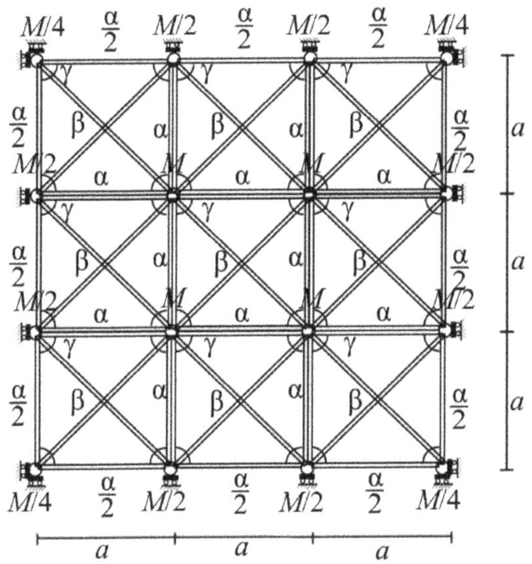

Figure 3. Lattice with central and angular interactions on sliding support (case B); $n = m = 3$; corrections of the elastic interactions at the border.

The discrete Dirichlet boundary conditions in Equation (6) of simply supported edges (case B) are completed by the discrete Neumann boundary conditions of vanishing of the tangential force along the four edges. The same boundary conditions of case B can be considered for the continuous plate, as shown in Figure 4.

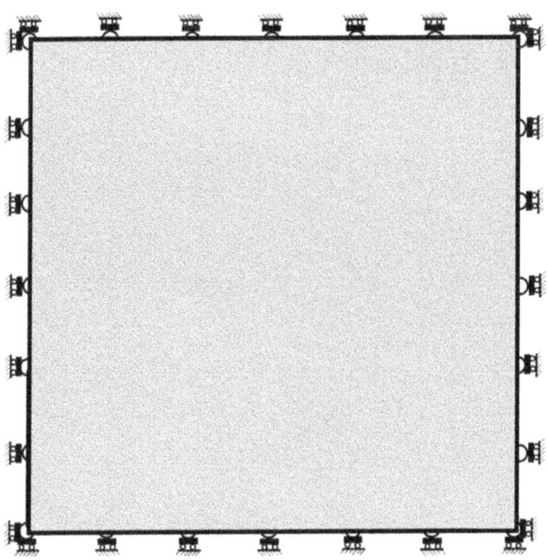

Figure 4. Continuous plate on sliding support (for the in-plane continuous plate problem)—case B.

As shown in Figures 1 and 3 for a square lattice with $n = m = 3$, the stiffness of border springs is half of that for the inner springs, as detailed by McHenry (1943) [40] and Hrennikoff (1941) [41] for 2D lattices with pure central interactions; we used the same approach in Challamel et al. (2023) [3] and Challamel et al. (2024) [25] to obtain the static responses of Gazis et al.'s lattices with central and non-central interactions. We follow here the same methodology for the free in-plane vibration of such lattices: the eigenfrequencies and eigenmodes derive from the linear algebraic problem associated with the discrete displacement (standard matrix eigenvalue problem), and the stiffness and the mass matrices are assembled from the difference equation, Equation (4).

Additionally, some exact solutions can be obtained for the eigenfrequencies and eigenmodes of the $(n \times m)$ rectangular lattice supported as in cases A and B. Using a methodology described by Mindlin (1970) [27] for the free shear vibration of Gazis et al.'s lattice, and posing the (k_1, k_2) mode numbers, the solution for case A simply supported edges (with sliding normal to the boundary) is thought to be in the form:

$$u_{i,j} = U\cos\left(\frac{k_1\pi a i}{L_1}\right)\sin\left(\frac{k_2\pi a j}{L_2}\right) = U\cos\left(\frac{k_1\pi i}{n}\right)\sin\left(\frac{k_2\pi j}{m}\right)$$

$$\text{for } i \in \{0,1,\ldots,n\} \text{ and } j \in \{0,1,\ldots,m\}$$

$$\text{and } v_{i,j} = V\sin\left(\frac{k_1\pi a i}{L_1}\right)\cos\left(\frac{k_2\pi a j}{L_2}\right) = V\sin\left(\frac{k_1\pi i}{n}\right)\cos\left(\frac{k_2\pi j}{m}\right)$$

$$\text{for } i \in \{0,1,\ldots,n\} \text{ and } j \in \{0,1,\ldots,m\} \tag{7}$$

Similarly, the solution for case B simply supported edges (with sliding support along the boundary) is thought to be in the form:

$$u_{i,j} = U\sin\left(\frac{k_1\pi a i}{L_1}\right)\cos\left(\frac{k_2\pi a j}{L_2}\right) = U\sin\left(\frac{k_1\pi i}{n}\right)\cos\left(\frac{k_2\pi j}{m}\right)$$

$$\text{for } i \in \{0,1,\ldots,n\} \text{ and } j \in \{0,1,\ldots,m\}$$

$$\text{and } v_{i,j} = V\cos\left(\frac{k_1\pi a i}{L_1}\right)\sin\left(\frac{k_2\pi a j}{L_2}\right) = V\cos\left(\frac{k_1\pi i}{n}\right)\sin\left(\frac{k_2\pi j}{m}\right)$$

$$\text{for } i \in \{0,1,\ldots,n\} \text{ and } j \in \{0,1,\ldots,m\} \tag{8}$$

For both simply supported lattice plate problems with case A and case B boundary conditions, the substitution of Equation (7) into Equation (4) or the substitution of Equation (8) into Equation (4) furnishes:

$$\left\{\begin{array}{l} \left[4\alpha\sin^2\left(\frac{k_1\pi}{2n}\right) + 2\beta\left[1-\cos\left(\frac{k_1\pi}{n}\right)\cos\left(\frac{k_2\pi}{m}\right)\right] + 16\gamma\sin^2\left(\frac{k_2\pi}{2m}\right) - M\omega^2\right]U \\ +\left[(2\beta+4\gamma)\sin\left(\frac{k_1\pi}{n}\right)\sin\left(\frac{k_2\pi}{m}\right)\right]V = 0 \\ \left[(2\beta+4\gamma)\sin\left(\frac{k_1\pi}{n}\right)\sin\left(\frac{k_2\pi}{m}\right)\right]U \\ +\left[4\alpha\sin^2\left(\frac{k_2\pi}{2m}\right) + 2\beta\left[1-\cos\left(\frac{k_1\pi}{n}\right)\cos\left(\frac{k_2\pi}{m}\right)\right] + 16\gamma\sin^2\left(\frac{k_1\pi}{2n}\right) - M\omega^2\right]V = 0 \end{array}\right. \tag{9}$$

which can be rewritten as:

$$\left\{\begin{array}{l} A_{11}(\omega^2)\,U + A_{12}V = 0 \\ A_{21}U + A_{22}(\omega^2)\,V = 0 \end{array}\right. \tag{10}$$

with:

$$\left\{\begin{array}{l} A_{11}(\omega^2) = 4\alpha\sin^2\left(\frac{k_1\pi}{2n}\right) + 2\beta\left[1-\cos\left(\frac{k_1\pi}{n}\right)\cos\left(\frac{k_2\pi}{m}\right)\right] + 16\gamma\sin^2\left(\frac{k_2\pi}{2m}\right) - M\omega^2 \\ A_{21} = A_{12} = (2\beta+4\gamma)\sin\left(\frac{k_1\pi}{n}\right)\sin\left(\frac{k_2\pi}{m}\right) \\ A_{22}(\omega^2) = 4\alpha\sin^2\left(\frac{k_2\pi}{2m}\right) + 2\beta\left[1-\cos\left(\frac{k_1\pi}{n}\right)\cos\left(\frac{k_2\pi}{m}\right)\right] + 16\gamma\sin^2\left(\frac{k_1\pi}{2n}\right) - M\omega^2 \end{array}\right. \tag{11}$$

The vanishing of the determinant of the system in Equation (9) or Equation (10), with the positions in Equation (11), gives a quartic characteristic equation for the in-plane eigenfrequencies of the rectangular lattice constrained according to cases A or B, which is then:

$$A_{11}\left(\omega^2\right)A_{22}\left(\omega^2\right) - A_{12}A_{21} = 0 \tag{12}$$

This quartic equation can be explicitly written as:

$$\begin{aligned}
&M^2\omega^4 \\
&-M\omega^2\left[(4\alpha+16\gamma)\left[\sin^2\left(\tfrac{k_1\pi}{2n}\right)+\sin^2\left(\tfrac{k_2\pi}{2m}\right)\right]+4\beta\left[1-\cos\left(\tfrac{k_1\pi}{n}\right)\cos\left(\tfrac{k_2\pi}{m}\right)\right]\right] \\
&+\left[4\alpha\sin^2\left(\tfrac{k_1\pi}{2n}\right)+2\beta\left[1-\cos\left(\tfrac{k_1\pi}{n}\right)\cos\left(\tfrac{k_2\pi}{m}\right)\right]+16\gamma\sin^2\left(\tfrac{k_2\pi}{2m}\right)\right] \\
&\left[4\alpha\sin^2\left(\tfrac{k_2\pi}{2m}\right)+2\beta\left[1-\cos\left(\tfrac{k_1\pi}{n}\right)\cos\left(\tfrac{k_2\pi}{m}\right)\right]+16\gamma\sin^2\left(\tfrac{k_1\pi}{2n}\right)\right] \\
&-(2\beta+4\gamma)^2\sin^2\left(\tfrac{k_1\pi}{n}\right)\sin^2\left(\tfrac{k_2\pi}{m}\right)=0
\end{aligned} \tag{13}$$

and admits two families of solutions: shearing modes (which we will see correspond to smaller roots) and the longitudinal modes (larger roots). For each boundary condition (simply supported edges with case A or B), the modes associated with the eigenfrequencies are selected to preserve the kinematic compatibility, as detailed below for the continuous plate problem.

3. Continuous Calibration and Numerical Results

The difference equations Equation (4) of Gazis et al.'s lattice asymptotically converge to Navier's partial differential equations for a sufficiently small spacing between pairs of particles. Between the discrete and the equivalent continuum connected at the nodal points $u_{i,j} = u(x = ai, y = aj)$ and $v_{i,j} = v(x = ai, y = aj)$, and for sufficiently smooth functions, it is:

$$u_{i+1,j+1} = u(x+a, y+a) = e^{a(\partial_x+\partial_y)}u(x,y) \text{ and } v_{i+1,j+1} = v(x+a, y+a) = e^{a(\partial_x+\partial_y)}v(x,y) \tag{14}$$

We shall make use of the technique developed by Cauchy (1828) [42], which allows the expansion of some difference operators in the Taylor series of differential operators by assuming a sufficiently smooth evolution function. The expansion of the difference operators in the governing equation, Equation (4), up to the first terms yields Navier's type of partial differential equations:

$$\begin{cases}
(\alpha+\beta)\frac{\partial^2 u}{\partial x^2}+(2\beta+4\gamma)\frac{\partial^2 v}{\partial x\partial y}+(4\gamma+\beta)\frac{\partial^2 u}{\partial y^2}+\omega^2\frac{M}{a^2}u=0 \\
(\alpha+\beta)\frac{\partial^2 v}{\partial y^2}+(2\beta+4\gamma)\frac{\partial^2 u}{\partial x\partial y}+(4\gamma+\beta)\frac{\partial^2 v}{\partial x^2}+\omega^2\frac{M}{a^2}v=0
\end{cases} \tag{15}$$

Navier's asymptotic continuous linear isotropic equations for plane stress are:

$$\begin{cases}
\frac{E}{1-v^2}\frac{\partial^2 u}{\partial x^2}+\frac{E}{2(1-v)}\frac{\partial^2 v}{\partial x\partial y}+\frac{E}{2(1+v)}\frac{\partial^2 u}{\partial y^2}+\omega^2\rho u=0 \\
\frac{E}{1-v^2}\frac{\partial^2 v}{\partial y^2}+\frac{E}{2(1-v)}\frac{\partial^2 u}{\partial x\partial y}+\frac{E}{2(1+v)}\frac{\partial^2 v}{\partial x^2}+\omega^2\rho v=0
\end{cases} \tag{16}$$

Comparing Equations (15) and (16) makes it possible to identify the elastic constants of the lattice, yielding the following micro/macro scaling relations (see also Challamel et al., 2023 [3]):

$$M=\rho a^2 h \text{ and } \begin{cases}
\alpha+\beta=\frac{Eh}{1-v^2} \\
2\beta+4\gamma=\frac{Eh}{2(1-v)} \\
4\gamma+\beta=\frac{Eh}{2(1+v)}
\end{cases} \Rightarrow \begin{cases}
\alpha=\frac{Eh}{1+v} \\
\beta=\frac{Ehv}{1-v^2} \\
\gamma=\frac{Eh(1-3v)}{8(1-v^2)}
\end{cases} \tag{17}$$

where h is the depth of the plane element, E the Young's modulus, v is the Poisson's ratio and ρ is the mass density of the material per unit volume.

The definite positivity of the potential energy of Gazis et al.'s lattice in plane stress implies the positivity of each central and non-central stiffness parameters, i.e.,

$$\alpha \geq 0,\ \beta \geq 0 \text{ and } \gamma \geq 0 \Rightarrow E \geq 0 \text{ and } v \in \left[0; \frac{1}{3}\right] \tag{18}$$

It is possible to include other non-central interactions, based on the surface potentials, to calibrate the lattice so as to have larger equivalent Poisson's ratios υ, as shown by Challamel et al. (2024) [25].

With these scaling coefficients, the difference equations to be solved for the free vibration of Gazis et al.'s (1960) lattice could be equivalently rewritten as (as obtained by Challamel et al., 2023 [3] or Challamel et al., 2024 [25]):

$$\begin{cases} \frac{E}{1+\upsilon}\left(u_{i+1,j}-2u_{i,j}+u_{i-1,j}\right)+\frac{E\upsilon}{2(1-\upsilon^2)}\left(u_{i+1,j+1}+u_{i-1,j-1}+u_{i-1,j+1}+u_{i+1,j-1}-4u_{i,j}\right)+ \\ \frac{E}{8(1-\upsilon)}\left(v_{i+1,j+1}-v_{i-1,j+1}-v_{i+1,j-1}+v_{i-1,j-1}\right)+\frac{E(1-3\upsilon)}{2(1-\upsilon^2)}\left(u_{i,j+1}-2u_{i,j}+u_{i,j-1}\right)+\omega^2\rho a^2 u_{i,j}=0 \\ \frac{E}{1+\upsilon}\left(v_{i,j+1}-2v_{i,j}+v_{i,j-1}\right)+\frac{E\upsilon}{2(1-\upsilon^2)}\left(v_{i+1,j+1}+v_{i-1,j-1}+v_{i-1,j+1}+v_{i+1,j-1}-4v_{i,j}\right)+ \\ \frac{E}{8(1-\upsilon)}\left(u_{i+1,j+1}-u_{i-1,j+1}-u_{i+1,j-1}+u_{i-1,j-1}\right)+\frac{E(1-3\upsilon)}{2(1-\upsilon^2)}\left(v_{i+1,j}-2v_{i,j}+v_{i-1,j}\right)+\omega^2\rho a^2 v_{i,j}=0 \end{cases} \tag{19}$$

In this way, the elastic lattice turns to a finite difference (FD) formulation of the continuous elastic plate (see Challamel et al. [3]). It is also possible to provide a non-dimensional formulation of the problem, starting from the introduction of the dimensionless frequency Ω:

$$\Omega=\frac{\omega L_1}{c_L}\text{ with }c_L=\sqrt{\frac{E}{\rho(1-\upsilon^2)}} \tag{20}$$

where c_L is the longitudinal wave speed of the asymptotic elastic linear isotropic continuum. The dimensionless frequency can be also defined with respect to the lattice parameters:

$$\Omega=\frac{\omega L_1}{c_L}\text{ with }c_L=\sqrt{\frac{\alpha+\beta}{\rho h}}=a\sqrt{\frac{\alpha+\beta}{M}} \tag{21}$$

By Equations (20) and (21), the difference equations, Equation (19), can be equivalently formulated as:

$$\begin{cases} (1-\upsilon)\left(u_{i+1,j}-2u_{i,j}+u_{i-1,j}\right)+\frac{\upsilon}{2}\left(u_{i+1,j+1}+u_{i-1,j-1}+u_{i-1,j+1}+u_{i+1,j-1}-4u_{i,j}\right)+ \\ \frac{1+\upsilon}{8}\left(v_{i+1,j+1}-v_{i-1,j+1}-v_{i+1,j-1}+v_{i-1,j-1}\right)+\frac{1-3\upsilon}{2}\left(u_{i,j+1}-2u_{i,j}+u_{i,j-1}\right)+\frac{\Omega^2}{n^2}u_{i,j}=0 \\ (1-\upsilon)\left(v_{i,j+1}-2v_{i,j}+v_{i,j-1}\right)+\frac{\upsilon}{2}\left(v_{i+1,j+1}+v_{i-1,j-1}+v_{i-1,j+1}+v_{i+1,j-1}-4v_{i,j}\right)+ \\ \frac{1+\upsilon}{8}\left(u_{i+1,j+1}-u_{i-1,j+1}-u_{i+1,j-1}+u_{i-1,j-1}\right)+\frac{1-3\upsilon}{2}\left(v_{i+1,j}-2v_{i,j}+v_{i-1,j}\right)+\frac{\Omega^2}{n^2}v_{i,j}=0 \end{cases} \tag{22}$$

The discrete eigenfrequency problem only depends on three dimensionless parameters, Poisson's ratio υ and the number of particles in each direction n and m:

$$\Omega=\Omega(\upsilon,n,m) \tag{23}$$

For a square plate, the eigenfrequency problem only depends on the Poisson's ratio and the number $n = m$ of particles, $\Omega=\Omega(\upsilon,n)$. At the asymptotic limit $n\to\infty$, i.e., for the asymptotic continuum, the eigenfrequency determination of the square continuous plate only depends on the Poisson's ratio υ.

The numerical results for the in-plane eigenfrequencies of square lattices composed of 10×10, 20×20 and 40×40 Gazis et al.'s cells with the Poisson's ratio $\upsilon = 0.3$ are presented in Tables 1 and 2 (case A), and in Tables 3 and 4 (case B). Tables 1 and 3 show a perfect agreement between the analytical solution obtained from the quartic equation and the direct numerical method based on matrix eigenvalues. The eigenfrequencies increase with respect to the number n of cells, and asymptotically converge to those of the continuous plate. It is also confirmed that the two cases, A and B, controlled by the same eigenfrequency equation, do not have the same spectra, due to the incompatibility of some modes with the given constraints. Consequently, the two boundary conditions yield distinct spectra for both the lattice and the asymptotic continuum. The eigenmode numbers associated with each mode are indicated in Tables 2 and 4 for cases A and B, respectively. The eigenmodes for both boundary conditions are shown in Figures 5 and 6: note that the mode shapes of both the lattice and continuum do not depend on the Poisson's ratio, as highlighted by the eigenmode equations, Equation (7) (case A) and Equation (8) (case B). The 10 first modes are shown for cases A and B for the 20 × 20 cells lattice: this confirms the physical difference between the two lattices with distinct constraints. The lattice appears well suited for a better understanding of the discrete modes according to the lattice cell deformations.

Table 1. Comparison of eigenfrequencies Ω between the lattice solution (analytical) and the numerical lattice solution from the algebraic eigenvalue computation—υ = 0.3—case A.

Mode	Numerical Solutions for Lattice Model, Ω			Analytical Solutions for Lattice Model, Ω			Analytical Solutions for Continuous Plate, Ω
	10 × 10 Cell Lattice	20 × 20 Cell Lattice	40 × 40 Cell Lattice	10 × 10 Cell Lattice	20 × 20 Cell Lattice	40 × 40 Cell Lattice	
1	1.851	1.857	1.858	1.851	1.857	1.858	1.859
2	1.851	1.857	1.858	1.851	1.857	1.858	1.859
3	2.620	2.626	2.628	2.620	2.626	2.628	2.628
4	3.656	3.702	3.713	3.656	3.702	3.713	3.717
5	3.656	3.702	3.713	3.656	3.702	3.713	3.717
6	4.104	4.143	4.153	4.104	4.143	4.153	4.156
7	4.104	4.143	4.153	4.104	4.143	4.153	4.156
8	4.391	4.430	4.440	4.391	4.430	4.440	4.443
9	5.188	5.240	5.253	5.188	5.240	5.253	5.257
10	5.372	5.524	5.563	5.372	5.524	5.563	5.576

Table 2. Eigenfrequency Ω and eigenmode lattice solution (numerical and analytical)—υ = 0.3—case A.

Mode	Eigenmode Numbers and Mode Type			Analytical and Numerical Solutions for Lattice Model, Ω			Analytical Solutions for Continuous Plate, Ω
	k_1	k_2	Mode	10 × 10 Cell Lattice	20 × 20 Cell Lattice	40 × 40 Cell Lattice	
1	0	1	shearing	1.851	1.857	1.858	1.859
2	1	0	shearing	1.851	1.857	1.858	1.859
3	1	1	shearing	2.620	2.626	2.628	2.628
4	0	2	shearing	3.656	3.702	3.713	3.717
5	2	0	shearing	3.656	3.702	3.713	3.717
6	1	2	shearing	4.104	4.143	4.153	4.156
7	2	1	shearing	4.104	4.143	4.153	4.156
8	1	1	longitudinal	4.391	4.430	4.440	4.443
9	2	2	shearing	5.188	5.240	5.253	5.257
10	0	3	shearing	5.372	5.524	5.563	5.576

Table 3. Comparison of eigenfrequencies Ω between the lattice solution (analytical) and the numerical lattice solution from the algebraic eigenvalue computation—υ = 0.3—case B.

Mode	Numerical Solutions for Lattice Model, Ω			Analytical Solutions for Lattice Model, Ω			Analytical Solutions for Continuous Plate, Ω
	10 × 10 Cell Lattice	20 × 20 Cell Lattice	40 × 40 Cell Lattice	10 × 10 Cell Lattice	20 × 20 Cell Lattice	40 × 40 Cell Lattice	
1	2.620	2.626	2.628	2.620	2.626	2.628	2.628
2	3.129	3.138	3.141	3.129	3.138	3.141	3.142
3	3.129	3.138	3.141	3.129	3.138	3.141	3.142
4	4.104	4.143	4.153	4.104	4.143	4.153	4.156
5	4.104	4.143	4.153	4.104	4.143	4.153	4.156
6	4.391	4.430	4.440	4.391	4.430	4.440	4.443
7	5.188	5.240	5.253	5.188	5.240	5.253	5.257
8	5.691	5.830	5.866	5.691	5.830	5.866	5.877
9	5.691	5.830	5.866	5.691	5.830	5.866	5.877
10	6.180	6.257	6.277	6.180	6.257	6.277	6.283

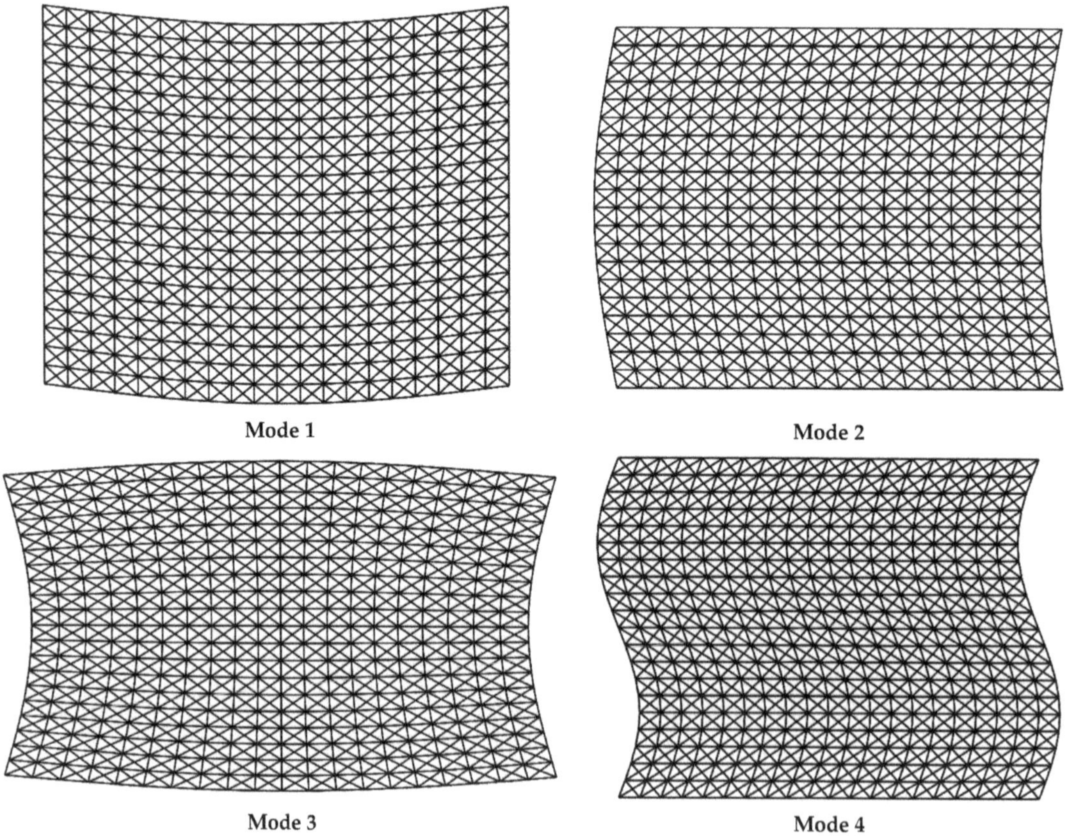

Figure 5. *Cont.*

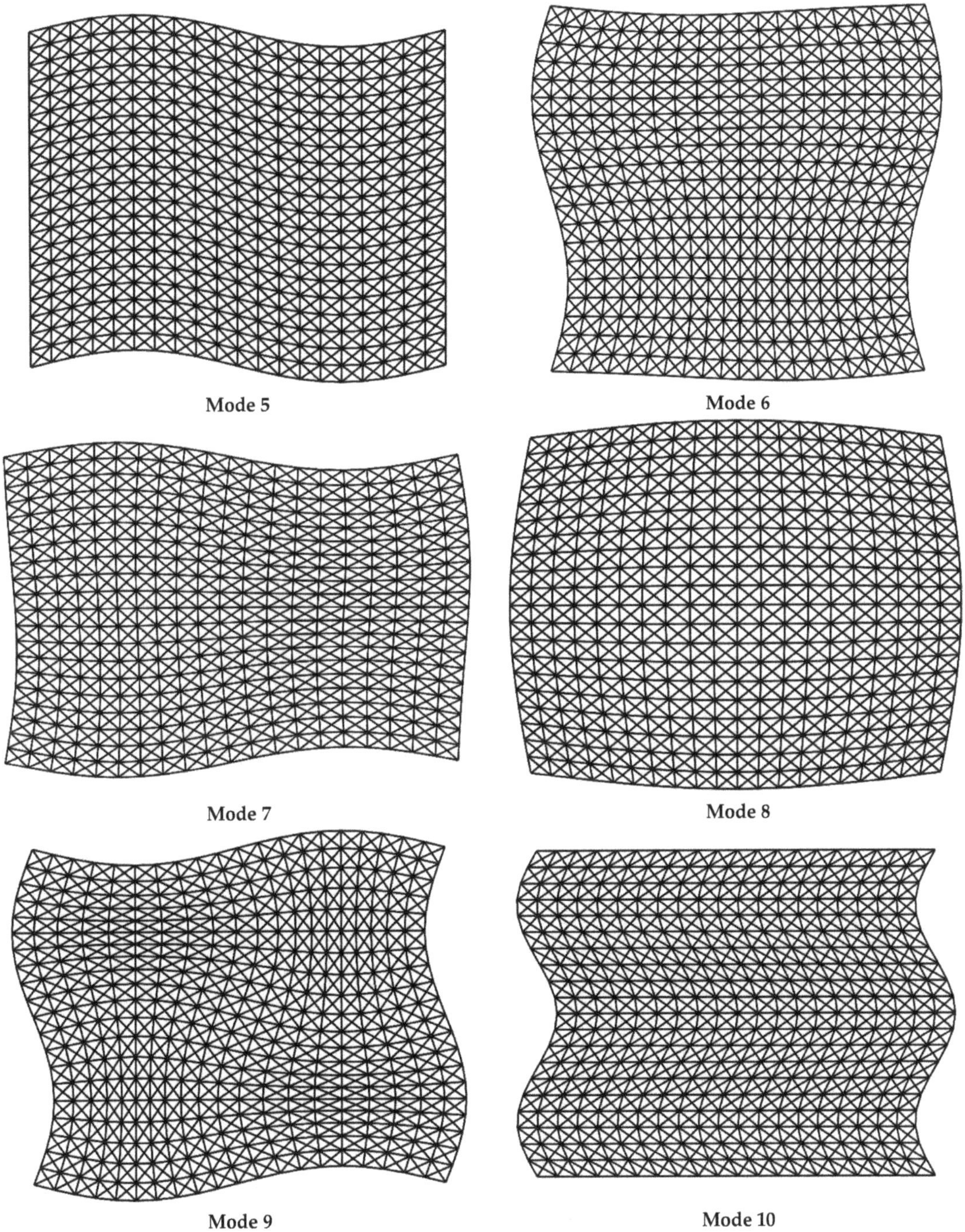

Figure 5. Eigenmodes of the lattice plate on sliding support (for the in-plane lattice plate problem)—20 × 20 cell square lattice (case A).

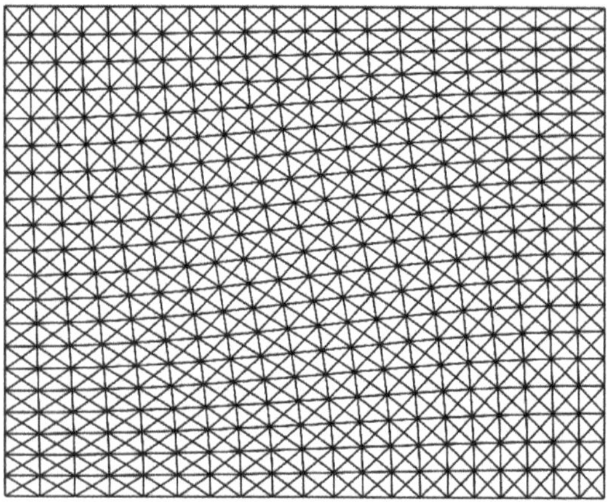

Mode 1

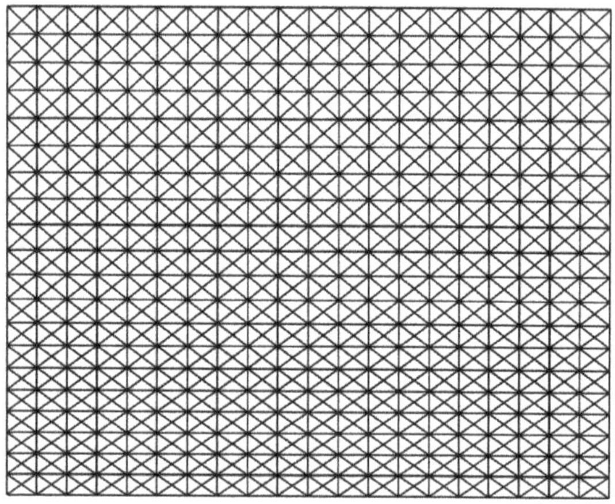

Mode 2

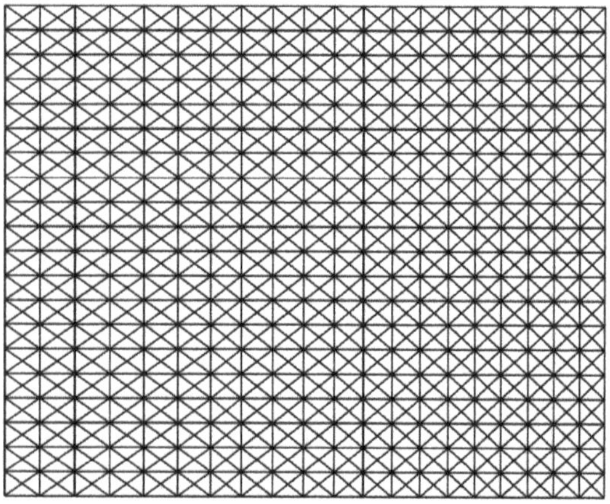

Mode 3

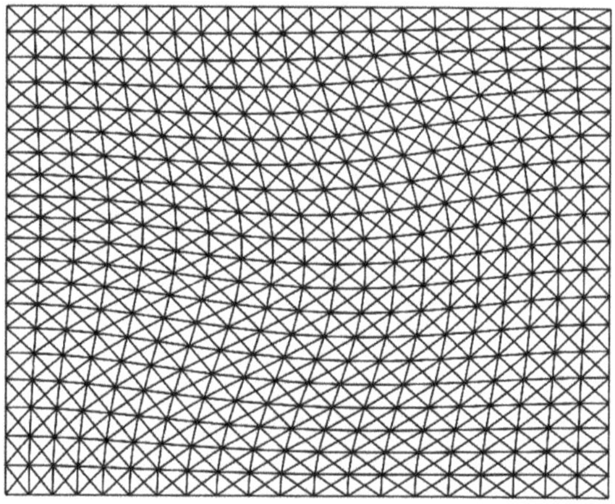

Mode 4

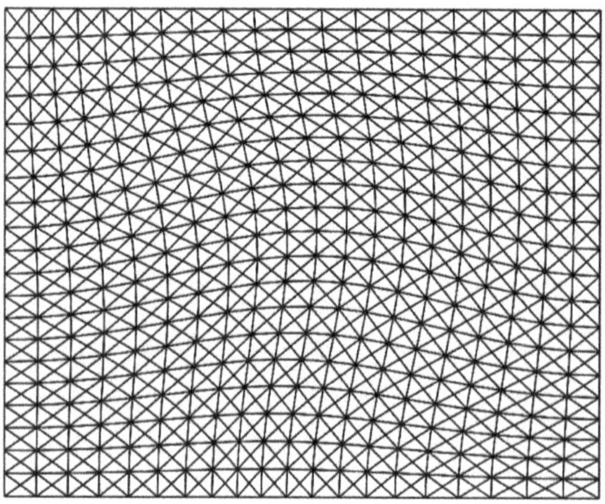

Mode 5

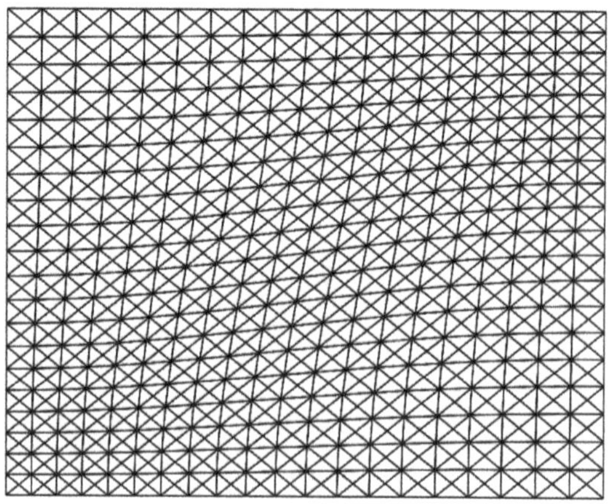

Mode 6

Figure 6. *Cont.*

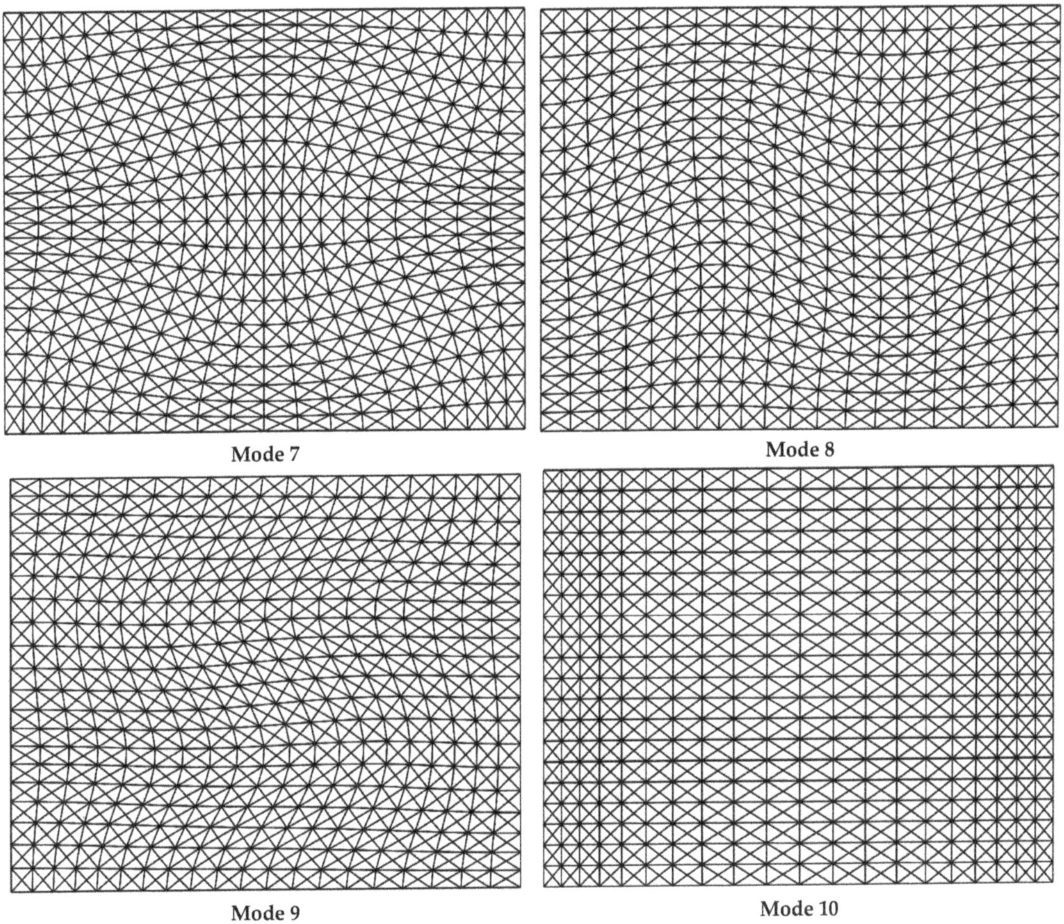

Figure 6. Eigenmodes of the lattice plate on sliding support (for the in-plane lattice plate problem)—20 × 20 cell square lattice (case B).

Table 4. Eigenfrequency Ω and eigenmode lattice solution (numerical and analytical)—υ = 0.3—case B.

Mode	Eigenmode Numbers and Mode Type			Analytical Solutions for Lattice Model, Ω			Analytical Solutions for Continuous Plate, Ω
	k_1	k_2	Mode	10 × 10 Cell Lattice	20 × 20 Cell Lattice	40 × 40 Cell Lattice	
1	1	1	shearing	2.620	2.626	2.628	2.628
2	0	1	longitudinal	3.129	3.138	3.141	3.142
3	1	0	longitudinal	3.129	3.138	3.141	3.142
4	1	2	shearing	4.104	4.143	4.153	4.156
5	2	1	shearing	4.104	4.143	4.153	4.156

Table 4. *Cont.*

Mode	Eigenmode Numbers and Mode Type			Analytical Solutions for Lattice Model, Ω			Analytical Solutions for Continuous Plate, Ω
	k_1	k_2	Mode	10 × 10 Cell Lattice	20 × 20 Cell Lattice	40 × 40 Cell Lattice	
6	1	1	longitudinal	4.391	4.430	4.440	4.443
7	2	2	shearing	5.188	5.240	5.253	5.257
8	1	3	shearing	5.691	5.830	5.866	5.877
9	3	1	shearing	5.691	5.830	5.866	5.877
10	0	2	longitudinal	6.180	6.257	6.277	6.283

4. In-Plane Vibrations of Continuous Plates—Rayleigh Solution

We now detail and comment on the exact solution of continuous plates compared to that of discrete lattices. Navier's partial differential equations for linear isotropic elastic plates, Equation (16), hold for plane stress and describe the in-plane vibration. As for the lattice constrained by sliding supports, we consider two boundary conditions (see also Gorman, 2006 [38]):

- In case A (Figure 2), the continuous tangential displacement and normal stress vanish (sliding supports normal to the boundary). The solution can be thought of in the same form considered for the lattice under the constraints of case A:

$$u(x,y) = U\cos\left(\frac{k_1\pi x}{L_1}\right)\sin\left(\frac{k_2\pi y}{L_2}\right) \text{ and } v(x,y) = V\sin\left(\frac{k_1\pi x}{L_1}\right)\cos\left(\frac{k_2\pi y}{L_2}\right) \quad (24)$$

- In case B (Figure 4), the continuous normal displacement and shear stress vanish (sliding supports along the boundary). The solution can be thought of in the same form considered for the lattice under the constraints of case B:

$$u(x,y) = U\sin\left(\frac{k_1\pi x}{L_1}\right)\cos\left(\frac{k_2\pi y}{L_2}\right) \text{ and } v(x,y) = V\cos\left(\frac{k_1\pi x}{L_1}\right)\sin\left(\frac{k_2\pi y}{L_2}\right) \quad (25)$$

The substitution of Equation (24) into Equation (16), or Equation (25) into Equation (16) gives the same eigenfrequency solution (for both simply supported boundary conditions) for the linear elastic 2D continuous medium in plane stress, which coincides with the expressions in Rayleigh (1894) [35]:

$$\begin{cases} \left[\frac{E}{1-v^2}\left(\frac{k_1\pi}{L_1}\right)^2 + \frac{E}{2(1+v)}\left(\frac{k_2\pi}{L_2}\right)^2 - \rho\omega^2\right]U + \frac{E}{2(1-v)}\left(\frac{k_1\pi}{L_1}\right)\left(\frac{k_2\pi}{L_2}\right)V = 0 \\ \frac{E}{2(1-v)}\left(\frac{k_1\pi}{L_1}\right)\left(\frac{k_2\pi}{L_2}\right)U + \left[\frac{E}{1-v^2}\left(\frac{k_2\pi}{L_2}\right)^2 + \frac{E}{2(1+v)}\left(\frac{k_1\pi}{L_1}\right)^2 - \rho\omega^2\right]V = 0 \end{cases} \quad (26)$$

From Equation (26), the eigenfrequencies are solutions of a quartic equation:

$$\begin{array}{l} \left[\frac{E}{1-v^2}\left(\frac{k_1\pi}{L_1}\right)^2 + \frac{E}{2(1+v)}\left(\frac{k_2\pi}{L_2}\right)^2 - \rho\omega^2\right]\left[\frac{E}{1-v^2}\left(\frac{k_2\pi}{L_2}\right)^2 + \frac{E}{2(1+v)}\left(\frac{k_1\pi}{L_1}\right)^2 - \rho\omega^2\right] \\ -\frac{E^2}{4(1-v)^2}\left(\frac{k_1\pi}{L_1}\right)^2\left(\frac{k_2\pi}{L_2}\right)^2 = 0 \end{array} \quad (27)$$

which may be presented in the equivalent form:

$$\left(\rho\omega^2\right)^2 - \frac{E(3-v)}{2(1-v^2)}\left[\left(\frac{k_1\pi}{L_1}\right)^2 + \left(\frac{k_2\pi}{L_2}\right)^2\right]\rho\omega^2 + \frac{E^2}{2(1+v)(1-v^2)}\left[\left(\frac{k_1\pi}{L_1}\right)^2 + \left(\frac{k_2\pi}{L_2}\right)^2\right]^2 = 0 \quad (28)$$

Equation (28) gives the two families of frequencies for the in-plane vibration in plane stress:

$$\rho\omega^2 = \frac{1}{4}\frac{E}{1-v^2}[3-v\pm(1+v)]\left[\left(\frac{k_1\pi}{L_1}\right)^2+\left(\frac{k_2\pi}{L_2}\right)^2\right] \tag{29}$$

The position in Equation (20) holds here also; thus, we obtain the non-dimensional frequency of the continuous plate for both boundary conditions:

$$\Omega^2 = \frac{1}{4}[3-v\pm(1+v)]\pi^2\left[k_1^2+k_2^2\left(\frac{L_1}{L_2}\right)^2\right] \tag{30}$$

which can be also written as:

$$\Omega^2 = \frac{1-v}{2}\pi^2\left[k_1^2+k_2^2\left(\frac{L_1}{L_2}\right)^2\right] \text{ for the shearing modes}$$

$$\text{or } \Omega^2 = \pi^2\left[k_1^2+k_2^2\left(\frac{L_1}{L_2}\right)^2\right] \text{ for the longitudinal modes} \tag{31}$$

Equation (31) was obtained by Rayleigh (page 405 of Rayleigh, 1894 [35]) and numerically confirmed by Bardell et al. (1996) [37] using a Rayleigh–Ritz approach. Rayleigh (1894) [35] implicitly gave the solution for a cylinder with an infinite radius that converges to a plate, with cyclic boundary conditions equivalent to simple supports.

For a square plate, $L_1 = L_2$, and we have for both boundary conditions:

$$\Omega = \sqrt{\frac{1-v}{2}}\pi\sqrt{k_1^2+k_2^2} \text{ for the shearing modes}$$

$$\text{or } \Omega = \pi\sqrt{k_1^2+k_2^2} \text{ for the longitudinal modes} \tag{32}$$

It is worth studying the situations when k_1 or k_2 vanishes; for instance, let $k_2 = 0$ in Equation (26):

$$\begin{cases} \left[\frac{E}{1-v^2}\left(\frac{k_1\pi}{L_1}\right)^2-\rho\omega^2\right]U=0 \\ \left[\frac{E}{2(1+v)}\left(\frac{k_1\pi}{L_1}\right)^2-\rho\omega^2\right]V=0 \end{cases} \tag{33}$$

For the boundary condition of free normal stress (case A), $k_2 = 0$ is associated with $U = 0$; then, only the shearing modes must be considered:

$$k_2 = 0 \Rightarrow \Omega = k_1\pi\sqrt{\frac{1-v}{2}} \tag{34}$$

An analogous result is obtained considering $k_1 = 0$; thus, for $k_1 = 0$ or $k_2 = 0$ and the boundary conditions of case A, only the shearing modes must be selected:

$$k_1 = 0 \Rightarrow \Omega = k_2\pi\sqrt{\frac{1-v}{2}} \text{ or } k_2 = 0 \Rightarrow \Omega = k_1\pi\sqrt{\frac{1-v}{2}} \tag{35}$$

For the boundary condition of no normal displacement (case B), $k_2 = 0$ is associated with $V = 0$; then, only the longitudinal modes must be considered:

$$k_2 = 0 \Rightarrow \Omega = k_1\pi \tag{36}$$

An analogous result is obtained considering $k_1 = 0$; thus, for $k_1 = 0$ or $k_2 = 0$ and the boundary conditions of case B, only the longitudinal modes must be selected:

$$k_1 = 0 \Rightarrow \Omega = k_2\pi \text{ or } k_2 = 0 \Rightarrow \Omega = k_1\pi \tag{37}$$

To summarise, the two boundary conditions of cases A and B are associated with the same spectrum (with different mode shapes), except for the limit situation in which $k_1k_2 = 0$. In such situation, we must consider only pure shear modes for case A and only pure longitudinal modes for case B. The numerical results are presented in Table 5 for a continuous square plate constrained as in cases A and B.

Table 5. Exact eigenfrequencies Ω of a simply supported square local continuous plate—in-plane vibrations; case A and case B boundary conditions; υ = 0.3.

(k_1, k_2)	Case A (as Analysed by Bardell et al., 1996 [37])	Case B
(0, 1)	π[(1 − υ)/2]^0.5 ≈ 1.859	
(1, 0)	π[(1 − υ)/2]^0.5 ≈ 1.859	
(1, 1)	π[1 − υ]^0.5 ≈ 2.628	π[1 − υ]^0.5 ≈ 2.628
(0, 1) †		π ≈ 3.142
(1, 0) †		π ≈ 3.142
(0, 2)	π[2*(1 − υ)]^0.5 ≈ 3.717	
(2, 0)	π[2*(1 − υ)]^0.5 ≈ 3.717	
(1, 2)	π[5*(1 − υ)/2]^0.5 ≈ 4.156	π[5*(1 − υ)/2]^0.5 ≈ 4.156
(2, 1)	π[5*(1 − υ)/2]^0.5 ≈ 4.156	π[5*(1 − υ)/2]^0.5 ≈ 4.156
(1, 1) †	π[2]^0.5 ≈ 4.443	π[2]^0.5 ≈ 4.443
(2, 2)	2π[1 − υ]^0.5 ≈ 5.257	2π[1 − υ]^0.5 ≈ 5.257

† These are longitudinal modes (all the others are shearing modes).

In conclusion, the eigenfrequencies for the square plate constrained as in case A are:

$$\Omega = \sqrt{\frac{1-\upsilon}{2}}\pi\sqrt{k_1^2 + k_2^2} \text{ for the shearing modes including } k_1k_2 = 0$$

$$\text{or } \Omega = \pi\sqrt{k_1^2 + k_2^2} \text{ for the longitudinal modes with } k_1k_2 \neq 0 \tag{38}$$

The eigenfrequencies for the square plate constrained as in case B are:

$$\Omega = \sqrt{\frac{1-\upsilon}{2}}\pi\sqrt{k_1^2 + k_2^2} \text{ for the shearing modes with } k_1k_2 \neq 0$$

$$\text{or } \Omega = \pi\sqrt{k_1^2 + k_2^2} \text{ for the longitudinal modes including } k_1k_2 = 0 \tag{39}$$

The same conclusions hold for the plane lattice for both boundary conditions of cases A and B. Note that the modes of the lattice and of the continuous plate coincide, while the eigenfrequencies differ between the discrete and the continuous problem. As highlighted by Tables 1 and 3, the eigenfrequencies of the plane lattice converge to those of the continuous plate from below, as classically obtained from Finite Difference Methods (FDMs) applied to homogeneous structural dynamic problems (see, for instance, Collatz, 1960 [43]; Challamel et al., 2015 [44]; Wang et al., 2020 [6]). Challamel et al. (2015) [44] commented on the lower status of FDM for one-dimensional continuous eigenvalue problems, such as the free vibration of finite rods or strings. The FDM has also a lower bound status for the eigenfrequency calculation of continuous membranes with the so-called lumped mass method (see also Polya, 1952 [45] or Tong et al., 1971 [14] for FD linear 2D homogeneous eigenvalue problems such as that of the membrane). It is worth mentioning that the lower bound or upper bound status of the discrete approach is strongly sensitive to the mass repartition in the lattice, as analysed by Polya (1952) [45] for the discrete membrane problem or Challamel et al. (2016) [46] for the discrete rod or string problem (lumped mass method as opposed to the consistent distributed mass method).

5. Higher-Order Elasticity and Nonlocal Elasticity

By using the continualisation procedure presented in Equation (14), the difference eigenvalue problem in Equation (22) can be approximated via a higher-order elasticity continuous eigenvalue problem, by considering second-order terms in the asymptotic expansion of the difference operators:

$$\left\{\begin{array}{l} \frac{\partial^2 u}{\partial x^2}+\left(\frac{1+v}{2}\right)\frac{\partial^2 v}{\partial x\partial y}+\left(\frac{1-v}{2}\right)\frac{\partial^2 u}{\partial y^2} \\ +\frac{a^2}{12}\left[\frac{\partial^4 u}{\partial x^4}+\left(\frac{1-v}{2}\right)\frac{\partial^4 u}{\partial y^4}+6v\frac{\partial^4 u}{\partial x^2\partial y^2}+(1+v)\left(\frac{\partial^4 v}{\partial x^3\partial y}+\frac{\partial^4 v}{\partial x\partial y^3}\right)\right]+\frac{\Omega^2}{L_1^2}u=0 \\ \frac{\partial^2 v}{\partial y^2}+\left(\frac{1+v}{2}\right)\frac{\partial^2 u}{\partial x\partial y}+\left(\frac{1-v}{2}\right)\frac{\partial^2 v}{\partial x^2} \\ +\frac{a^2}{12}\left[\frac{\partial^4 v}{\partial y^4}+\left(\frac{1-v}{2}\right)\frac{\partial^4 v}{\partial x^4}+6v\frac{\partial^4 v}{\partial x^2\partial y^2}+(1+v)\left(\frac{\partial^4 u}{\partial x^3\partial y}+\frac{\partial^4 u}{\partial x\partial y^3}\right)\right]+\frac{\Omega^2}{L_1^2}v=0 \end{array}\right. \tag{40}$$

Mindlin (1968) [26] adopted the same reasoning to derive cubic strain gradient elasticity equations from the 3D lattice model of Gazis et al. [20]. Equation (40) corresponds to the partial differential equations of a strain gradient elastic continuum under plane stress assumptions, with additional terms proportional to $a^2/12$ that account for the lattice's small-scale effects. Equation (40) was also derived by Challamel et al. (2023) [3] and Challamel et al. (2024) [25].

The substitution of Equation (24) into Equation (40) or Equation (25) into Equation (40) gives the same linear system (for both simply supported boundary conditions) for the linear strain gradient elastic 2D medium:

$$\left\{\begin{array}{l} B_{11}(\Omega^2)\,U+B_{12}V=0 \\ B_{21}U+B_{22}(\Omega^2)\,V=0 \end{array}\right. \tag{41}$$

with:

$$\left\{\begin{array}{l} B_{11}(\Omega^2)=\left(\frac{k_1\pi}{L_1}\right)^2\left[1-\frac{a^2}{12}\left(\frac{k_1\pi}{L_1}\right)^2\right] \\ +\left(\frac{1-v}{2}\right)\left(\frac{k_2\pi}{L_2}\right)^2\left[1-\frac{a^2}{12}\left(\frac{k_2\pi}{L_2}\right)^2\right]-\frac{a^2}{2}v\left(\frac{k_1\pi}{L_1}\right)^2\left(\frac{k_2\pi}{L_2}\right)^2-\frac{\Omega^2}{L_1^2} \\ B_{21}=B_{12}=\left(\frac{1+v}{2}\right)\left(\frac{k_1\pi}{L_1}\right)\left(\frac{k_2\pi}{L_2}\right)-\frac{a^2}{12}(1+v)\left(\frac{k_1\pi}{L_1}\right)\left(\frac{k_2\pi}{L_2}\right)\left[\left(\frac{k_1\pi}{L_1}\right)^2+\left(\frac{k_2\pi}{L_2}\right)^2\right] \\ B_{22}(\Omega^2)=\left(\frac{k_2\pi}{L_2}\right)^2\left[1-\frac{a^2}{12}\left(\frac{k_2\pi}{L_2}\right)^2\right] \\ +\left(\frac{1-v}{2}\right)\left(\frac{k_1\pi}{L_1}\right)^2\left[1-\frac{a^2}{12}\left(\frac{k_1\pi}{L_1}\right)^2\right]-\frac{a^2}{2}v\left(\frac{k_1\pi}{L_1}\right)^2\left(\frac{k_2\pi}{L_2}\right)^2-\frac{\Omega^2}{L_1^2} \end{array}\right. \tag{42}$$

The quartic equation for the in-plane eigenfrequencies of the lattice-based rectangular strain gradient elastic plate (with case A or case B boundary conditions) is then obtained from:

$$B_{11}\left(\Omega^2\right)B_{22}\left(\Omega^2\right)-B_{12}B_{21}=0 \tag{43}$$

A nonlocal plate model is obtained by multiplying the partial differential equations, Equation (40), by the differential operator $(1-a^2\Delta/12)$, where Δ is the 2D Laplacian operator, and limiting to the terms in a^2 to avoid higher-order differential operators (see also Challamel et al., 2022 [39] who applied this methodology for deriving 2D lattice-based nonlocal elasticity equations in plane strain, or Zhang et al., 2021 [47] for alternative 2D lattices in plane stress). The nonlocal elasticity equations in plane stress are:

$$\left\{\begin{array}{l}\frac{\partial^2 u}{\partial x^2}+\left(\frac{1+\upsilon}{2}\right)\frac{\partial^2 v}{\partial x \partial y}+\left(\frac{1-\upsilon}{2}\right)\frac{\partial^2 u}{\partial y^2}\\+\frac{a^2}{24}\left[(-3+13\upsilon)\frac{\partial^4 u}{\partial x^2 \partial y^2}+(1+\upsilon)\left(\frac{\partial^4 v}{\partial x^3 \partial y}+\frac{\partial^4 v}{\partial x \partial y^3}\right)\right]+\frac{\Omega^2}{L_1^2}\left(1-\frac{a^2}{12}\Delta\right)u=0\\\frac{\partial^2 v}{\partial y^2}+\left(\frac{1+\upsilon}{2}\right)\frac{\partial^2 u}{\partial x \partial y}+\left(\frac{1-\upsilon}{2}\right)\frac{\partial^2 v}{\partial x^2}\\+\frac{a^2}{24}\left[(-3+13\upsilon)\frac{\partial^4 v}{\partial x^2 \partial y^2}+(1+\upsilon)\left(\frac{\partial^4 u}{\partial x^3 \partial y}+\frac{\partial^4 u}{\partial x \partial y^3}\right)\right]+\frac{\Omega^2}{L_1^2}\left(1-\frac{a^2}{12}\Delta\right)v=0\end{array}\right. \tag{44}$$

This lattice-based nonlocal plate model has the advantage of avoiding higher-order differential operators and the discussion of higher-order variationally based boundary conditions. As already commented by Zhang et al. (2021) [47] or Challamel et al. (2022) [39], these lattice-based partial differential equations differ from the in-plane Eringen's nonlocal plate model (based on Eringen's differential law—Eringen, 1983 [48] applied to plate mechanics), due to the additional spatial coupling terms in Equation (44) (this phenomenon was already commented on for the out-of-plane responses of lattice plates in Challamel et al., 2021 [18]). The substitution of Equation (24) into Equation (44), or Equation (25) into Equation (44) gives the same linear system (for both simply supported boundary conditions) for the linear lattice-based nonlocal elastic 2D medium:

$$\left\{\begin{array}{l}C_{11}(\Omega^2)\,U+C_{12}V=0\\C_{21}U+C_{22}(\Omega^2)\,V=0\end{array}\right. \tag{45}$$

with:

$$\left\{\begin{array}{l}C_{11}(\Omega^2)=\left(\frac{k_1\pi}{L_1}\right)^2+\left(\frac{1-\upsilon}{2}\right)\left(\frac{k_2\pi}{L_2}\right)^2-\frac{a^2}{24}(-3+13\upsilon)\left(\frac{k_1\pi}{L_1}\right)^2\left(\frac{k_2\pi}{L_2}\right)^2\\-\frac{\Omega^2}{L_1^2}\left[1+\frac{a^2}{12}\left[\left(\frac{k_1\pi}{L_1}\right)^2+\left(\frac{k_2\pi}{L_2}\right)^2\right]\right]\\C_{21}=C_{12}=\left(\frac{1+\upsilon}{2}\right)\left(\frac{k_1\pi}{L_1}\right)\left(\frac{k_2\pi}{L_2}\right)\left[1-\frac{a^2}{12}\left[\left(\frac{k_1\pi}{L_1}\right)^2+\left(\frac{k_2\pi}{L_2}\right)^2\right]\right]\\C_{22}(\Omega^2)=\left(\frac{k_2\pi}{L_2}\right)^2+\left(\frac{1-\upsilon}{2}\right)\left(\frac{k_1\pi}{L_1}\right)^2-\frac{a^2}{24}(-3+13\upsilon)\left(\frac{k_1\pi}{L_1}\right)^2\left(\frac{k_2\pi}{L_2}\right)^2\\-\frac{\Omega^2}{L_1^2}\left[1+\frac{a^2}{12}\left[\left(\frac{k_1\pi}{L_1}\right)^2+\left(\frac{k_2\pi}{L_2}\right)^2\right]\right]\end{array}\right. \tag{46}$$

The quartic equation for the in-plane eigenfrequencies of the simply supported lattice-based rectangular nonlocal elastic plate (with case A or case B boundary conditions), is then obtained from:

$$C_{11}\left(\Omega^2\right)C_{22}\left(\Omega^2\right)-C_{12}C_{21}=0 \tag{47}$$

The lattice-based gradient elasticity and the nonlocal elasticity solutions for the plate are then compared to the exact lattice plate solution in Tables 6 and 7 for a 10 × 10 cell square lattice (with an asymptotic Poisson's ratio υ equal to 0.3), for both boundary conditions. As expected, an excellent agreement can be noticed between the second-order (gradient or nonlocal) elastic plate models and the lattice one, especially for the lower modes. For both models (gradient and nonlocal plate models), and for both simply supported boundary conditions, the fundamental eigenfrequencies of the enriched continuous models coincide with the lattice fundamental eigenfrequencies up to three digits. The difference tends to increase with the mode numbers. This is also confirmed by the curves in Figures 7–9 also valid for the n × n cell square lattice with the asymptotic Poisson's ratio υ equal to 0.3. The dependence of the dimensionless fundamental frequency $\Omega_{1,0}=\Omega_{0,1}$ with respect to the number n of cells in each direction (with case A simply supported boundary conditions) is compared in Figure 7 for the lattice, the strain gradient and the nonlocal elastic plate models. For the three models, the convergence towards the local solution (Rayleigh solution) is from below. Additionally, it is shown that both the gradient elasticity and the nonlocal elasticity solutions give accurate results for small values of n such as $n = 4$. As expected, the higher-order solutions which are obtained from asymptotic analysis for sufficiently large values of n become more precise with an increasing value of n. The same tendencies are observed in Figure 8 for the dimensionless fundamental frequency $\Omega_{1,1}$ with respect to the number n of cells in each direction (with case B simply supported boundary conditions). For both fundamental eigenfrequencies, the gradient elasticity model furnishes slightly better results than the nonlocal elastic plate model. Finally, the dimensionless eigenfrequency $\Omega_{2,2}$ for both sliding supports—case A (which corresponds to mode 9) or Case B (which corresponds to mode 7) (for the in-plane lattice plate problem)—is plotted versus the number n of cells. The higher-order solutions are less accurate for this higher frequency regime, but they still constitute a relevant approximation of the lattice solution, especially for the lattice-based strain gradient plate solution.

Table 6. Eigenfrequencies Ω of the gradient elasticity and nonlocal plate models compared to the lattice ones—$\upsilon = 0.3$—case A.

Mode	Eigenmode Numbers and Mode Type			Analytical and Numerical Solutions for Lattice Model, Ω			Local Continuous Plate, Ω
	k_1	k_2	Mode	10 × 10 Cell Lattice	Gradient Elasticity	Nonlocal Elasticity	
1	0	1	shearing	1.851	1.851	1.851	1.859
2	1	0	shearing	1.851	1.851	1.851	1.859
3	1	1	shearing	2.620	2.620	2.620	2.628
4	0	2	shearing	3.656	3.656	3.658	3.717
5	2	0	shearing	3.656	3.656	3.658	3.717
6	1	2	shearing	4.104	4.103	4.105	4.156
7	2	1	shearing	4.104	4.103	4.105	4.156
8	1	1	longitudinal	4.391	4.390	4.391	4.443
9	2	2	shearing	5.188	5.189	5.193	5.257
10	0	3	shearing	5.372	5.366	5.380	5.576

Table 7. Eigenfrequencies Ω of the gradient elasticity and nonlocal plate models compared to the lattice ones—$\upsilon = 0.3$—case B.

Mode	Eigenmode Numbers and Mode Type			Analytical and Numerical Solutions for Lattice Model, Ω			Local Continuous Plate, Ω
	k_1	k_2	Mode	10 × 10 Cell Lattice	Gradient Elasticity	Nonlocal Elasticity	
1	1	1	shearing	2.620	2.620	2.620	2.628
2	0	1	longitudinal	3.129	3.129	3.129	3.142
3	1	0	longitudinal	3.129	3.129	3.129	3.142
4	1	2	shearing	4.104	4.103	4.105	4.156
5	2	1	shearing	4.104	4.103	4.105	4.156
6	1	1	longitudinal	4.391	4.390	4.391	4.443
7	2	2	shearing	5.188	5.189	5.193	5.257
8	1	3	shearing	5.691	5.683	5.698	5.877
9	3	1	shearing	5.691	5.683	5.698	5.877
10	0	2	longitudinal	6.180	6.179	6.182	6.283

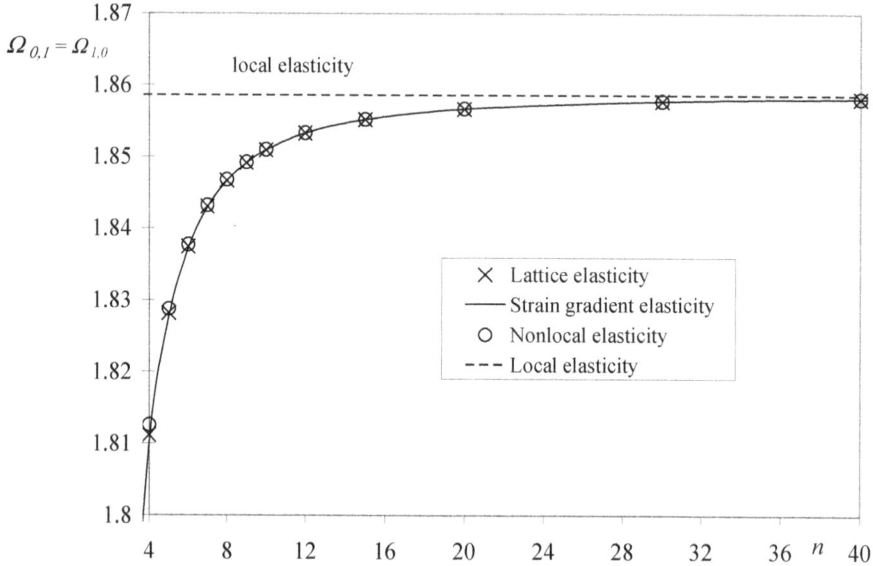

Figure 7. Evolution of the fundamental eigenfrequency $\Omega_{1,0} = \Omega_{0,1}$ of the $n \times n$ cell lattice plate on sliding support—case A (for the in-plane lattice plate problem)—comparison between the exact lattice plate, the strain gradient elastic plate, the nonlocal elastic plate and the continuous local plate models; $\upsilon = 0.3$.

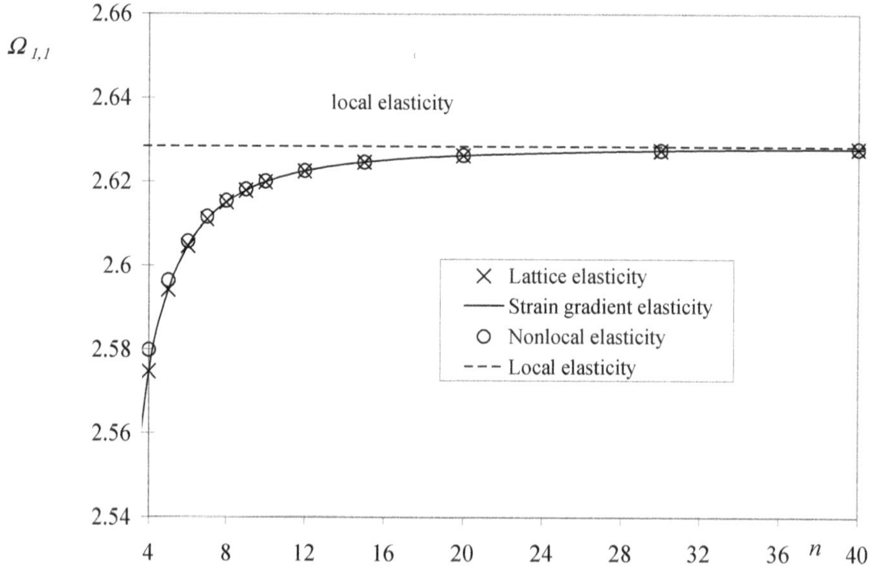

Figure 8. Evolution of the fundamental eigenfrequency $\Omega_{1,1}$ of the $n \times n$ cell lattice plate on sliding support—case B (for the in-plane lattice plate problem)—comparison between the exact lattice plate, the strain gradient elastic plate, the nonlocal elastic plate and the continuous local plate models; $\upsilon = 0.3$.

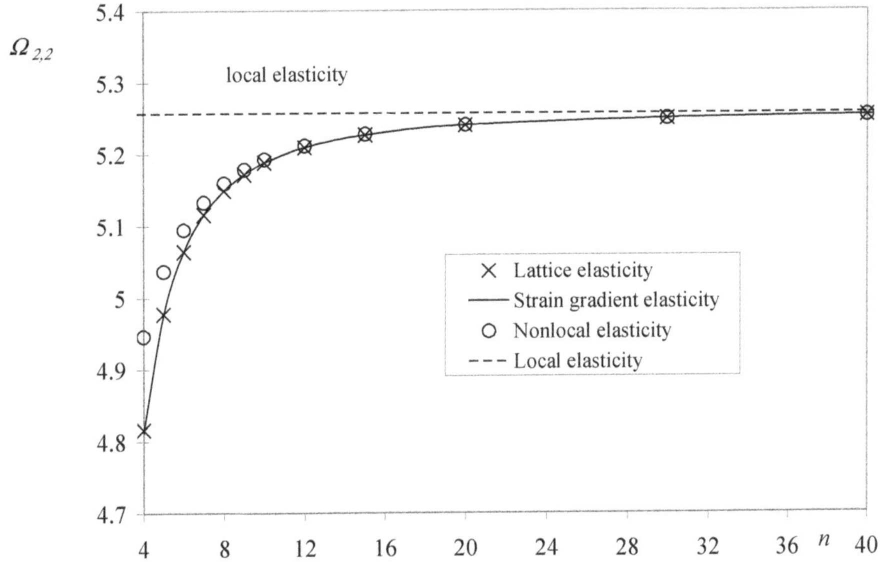

Figure 9. Evolution of the eigenfrequency $\Omega_{2,2}$ of the $n \times n$ cell lattice plate on sliding support—case A (mode 9) or Case B (mode 7) (for the in-plane lattice plate problem)—comparison between the exact lattice plate, the strain gradient elastic plate, the nonlocal elastic plate and the continuous local plate models; $\upsilon = 0.3$.

6. Remarks—Plane Strain

The 2D problem investigated in this paper can be alternatively analysed in plane strain, by comparing the lattice difference equations with those valid for linear elastic isotropy in plane strain:

$$\begin{cases} (\lambda+2\mu)\frac{\partial^2 u}{\partial x^2}+(\lambda+\mu)\frac{\partial^2 v}{\partial x\partial y}+\mu\frac{\partial^2 u}{\partial y^2}+\rho\omega^2 u=0 \\ (\lambda+2\mu)\frac{\partial^2 v}{\partial y^2}+(\lambda+\mu)\frac{\partial^2 u}{\partial x\partial y}+\mu\frac{\partial^2 v}{\partial x^2}+\rho\omega^2 v=0 \end{cases} \tag{48}$$

We identify the micro–macro relations for the plane strain analysis:

$$\begin{cases} \alpha = 2\mu h \\ \beta = \lambda h \\ \gamma = \frac{\mu-\lambda}{4}h \end{cases} \tag{49}$$

The eigenfrequencies of the rectangular lattice in plane strain constrained as in cases A and B can be calculated using the same methodology applied in a plane stress framework. For the continuum problem based on Equation (48), the displacement fields in Equation (24) and Equation (25) are assumed for both boundary conditions of cases A and B. Inserting Equation (24) or Equation (25) into Equation (48) gives the eigenfrequency problem for the linear elastic 2D medium in plane strain:

$$\begin{cases} \left[(\lambda+2\mu)\left(\frac{k_1\pi}{L_1}\right)^2+\mu\left(\frac{k_2\pi}{L_2}\right)^2-\rho\omega^2\right]U+(\lambda+\mu)\left(\frac{k_1\pi}{L_1}\right)\left(\frac{k_2\pi}{L_2}\right)V=0 \\ (\lambda+\mu)\left(\frac{k_1\pi}{L_1}\right)\left(\frac{k_2\pi}{L_2}\right)U+\left[(\lambda+2\mu)\left(\frac{k_2\pi}{L_2}\right)^2+\mu\left(\frac{k_1\pi}{L_1}\right)^2-\rho\omega^2\right]V=0 \end{cases} \tag{50}$$

From Equation (50), the eigenfrequencies are solutions of a quartic equation:

$$\left[(\lambda+2\mu)\left(\frac{k_1\pi}{L_1}\right)^2+\mu\left(\frac{k_2\pi}{L_2}\right)^2-\rho\omega^2\right]\left[(\lambda+2\mu)\left(\frac{k_2\pi}{L_2}\right)^2+\mu\left(\frac{k_1\pi}{L_1}\right)^2-\rho\omega^2\right] \\ -(\lambda+\mu)^2\left(\frac{k_1\pi}{L_1}\right)^2\left(\frac{k_2\pi}{L_2}\right)^2=0 \quad (51)$$

which may be presented in the equivalent form:

$$\left(\rho\omega^2\right)^2-(\lambda+3\mu)\left[\left(\frac{k_1\pi}{L_1}\right)^2+\left(\frac{k_2\pi}{L_2}\right)^2\right]\rho\omega^2+\mu(\lambda+2\mu)\left[\left(\frac{k_1\pi}{L_1}\right)^2+\left(\frac{k_2\pi}{L_2}\right)^2\right]^2=0 \quad (52)$$

Equation (52) gives the two families of frequencies in plane strain:

$$\rho\omega^2=\frac{1}{2}[\lambda+3\mu\pm(\lambda+\mu)]\left[\left(\frac{k_1\pi}{L_1}\right)^2+\left(\frac{k_2\pi}{L_2}\right)^2\right] \quad (53)$$

as obtained by Mindlin (2006) [49]. Equation (53) can be also rewritten as:

$$\frac{\omega^2}{c_{shear}^2}=\left[\left(\frac{k_1\pi}{L_1}\right)^2+\left(\frac{k_2\pi}{L_2}\right)^2\right] \text{ or } \frac{\omega^2}{c_{long}^2}=\left[\left(\frac{k_1\pi}{L_1}\right)^2+\left(\frac{k_2\pi}{L_2}\right)^2\right] \quad (54)$$

where we have used the following wave speed valid for the plane strain analysis:

$$c_{shear}=\sqrt{\frac{\mu}{\rho}};\ c_{long}=\sqrt{\frac{\lambda+2\mu}{\rho}} \quad (55)$$

7. Long-Range Interactions

The present analysis, valid for the in-plane vibration of a plane lattice composed of Gazis et al.'s [20] cells, exhibiting central and non-central interactions, can be extended to the vibration of lattices composed of cells exhibiting both short- and long-range interactions, as already investigated by Challamel et al. (2024) [50] for the wave propagation problem in infinite media:

$$\alpha_1\left(u_{i+1,j}-2u_{i,j}+u_{i-1,j}\right)+\alpha_2\left(u_{i+2,j}-2u_{i,j}+u_{i-2,j}\right)+ \\ \frac{\beta_1}{2}\left(u_{i+1,j+1}+u_{i-1,j-1}+u_{i-1,j+1}+u_{i+1,j-1}-4u_{i,j}\right)+\frac{\beta_2}{2}\left(u_{i+2,j+2}+u_{i-2,j-2}+u_{i-2,j+2}+u_{i+2,j-2}-4u_{i,j}\right)+ \\ \left(\frac{\beta_1}{2}+\gamma_1\right)\left(v_{i+1,j+1}-v_{i-1,j+1}-v_{i+1,j-1}+v_{i-1,j-1}\right)+\left(\frac{\beta_2}{2}+\gamma_2\right)\left(v_{i+2,j+2}-v_{i-2,j+2}-v_{i+2,j-2}+v_{i-2,j-2}\right)+ \\ 4\gamma_1\left(u_{i,j+1}-2u_{i,j}+u_{i,j-1}\right)+4\gamma_2\left(u_{i,j+2}-2u_{i,j}+u_{i,j-2}\right)+M\omega^2u_{i,j}=0$$

and

$$\alpha_1\left(v_{i,j+1}-2v_{i,j}+v_{i,j-1}\right)+\alpha_2\left(v_{i,j+2}-2v_{i,j}+v_{i,j-2}\right)+ \\ \frac{\beta_1}{2}\left(v_{i+1,j+1}+v_{i-1,j-1}+v_{i-1,j+1}+v_{i+1,j-1}-4v_{i,j}\right)+\frac{\beta_2}{2}\left(v_{i+2,j+2}+v_{i-2,j-2}+v_{i-2,j+2}+v_{i+2,j-2}-4v_{i,j}\right)+ \\ \left(\frac{\beta_1}{2}+\gamma_1\right)\left(u_{i+1,j+1}-u_{i-1,j+1}-u_{i+1,j-1}+u_{i-1,j-1}\right)+\left(\frac{\beta_2}{2}+\gamma_2\right)\left(u_{i+2,j+2}-u_{i-2,j+2}-u_{i+2,j-2}+u_{i-2,j-2}\right)+ \\ 4\gamma_1\left(v_{i+1,j}-2v_{i,j}+v_{i-1,j}\right)+4\gamma_2\left(v_{i+2,j}-2v_{i,j}+v_{i-2,j}\right)+M\omega^2v_{i,j}=0 \quad (56)$$

where the short-range stiffnesses $\alpha_1, \beta_1, \gamma_1$ are positive to ensure the positive definiteness of the associated short-range energy, and the long-range stiffnesses $\alpha_2, \beta_2, \gamma_2$ are also positive to ensure the positive definiteness of the associated long-range energy. This model can also be understood as a two-cell 2D interaction lattice.

For this generalised lattice problem with two generic interactions in plane stress, we have the possible material calibration:

$$\begin{cases} \alpha_1 = \frac{Eh}{1+\upsilon}\xi_1 \\ \beta_1 = \frac{Eh\upsilon}{1-\upsilon^2}\xi_1 \\ \gamma_1 = \frac{Eh(1-3\upsilon)}{8(1-\upsilon^2)}\xi_1 \end{cases} \text{and} \begin{cases} \alpha_2 = \frac{Eh}{1+\upsilon}\frac{\xi_2}{4} \\ \beta_2 = \frac{Eh\upsilon}{1-\upsilon^2}\frac{\xi_2}{4} \\ \gamma_2 = \frac{Eh(1-3\upsilon)}{8(1-\upsilon^2)}\frac{\xi_2}{4} \end{cases} \text{and } \xi_1 + \xi_2 = 1;\ \xi_1 \geq 0;\ \xi_2 \geq 0 \quad (57)$$

where (ξ_1, ξ_2) are weighting coefficients which control the proportion of short-range to long-range interactions. (ξ_1, ξ_2) = (1, 0) corresponds to a pure short-range interaction problem, as previously studied, whereas (ξ_1, ξ_2) = (0, 1) corresponds to a pure long-range interaction problem.

To investigate its vibration, it is possible for both boundary conditions of cases A and B to assume the modes in Equations (7) and (8), thus leading to the coupled system of linear equations:

$$\begin{cases} \begin{bmatrix} 4\alpha_1\sin^2\left(\frac{k_1\pi}{2n}\right) + 4\alpha_2\sin^2\left(\frac{k_1\pi}{n}\right) + 2\beta_1\left[1-\cos\left(\frac{k_1\pi}{n}\right)\cos\left(\frac{k_2\pi}{m}\right)\right] \\ +2\beta_2\left[1-\cos\left(\frac{2k_1\pi}{n}\right)\cos\left(\frac{2k_2\pi}{m}\right)\right] + 16\gamma_1\sin^2\left(\frac{k_2\pi}{2m}\right) + 16\gamma_2\sin^2\left(\frac{k_2\pi}{m}\right) - M\omega^2 \end{bmatrix} U \\ +\left[(2\beta_1+4\gamma_1)\sin\left(\frac{k_1\pi}{n}\right)\sin\left(\frac{k_2\pi}{m}\right) + (2\beta_2+4\gamma_2)\sin\left(\frac{2k_1\pi}{n}\right)\sin\left(\frac{2k_2\pi}{m}\right)\right] V = 0 \\ \left[(2\beta_1+4\gamma_1)\sin\left(\frac{k_1\pi}{n}\right)\sin\left(\frac{k_2\pi}{m}\right) + (2\beta_2+4\gamma_2)\sin\left(\frac{2k_1\pi}{n}\right)\sin\left(\frac{2k_2\pi}{m}\right)\right] U \\ + \begin{bmatrix} 4\alpha_1\sin^2\left(\frac{k_2\pi}{2m}\right) + 4\alpha_2\sin^2\left(\frac{k_2\pi}{m}\right) + 2\beta_1\left[1-\cos\left(\frac{k_1\pi}{n}\right)\cos\left(\frac{k_2\pi}{m}\right)\right] \\ 2\beta_2\left[1-\cos\left(\frac{2k_1\pi}{n}\right)\cos\left(\frac{2k_2\pi}{m}\right)\right] + 16\gamma_1\sin^2\left(\frac{k_1\pi}{2n}\right) + 16\gamma_2\sin^2\left(\frac{k_1\pi}{n}\right) - M\omega^2 \end{bmatrix} V = 0 \end{cases} \quad (58)$$

The vanishing of the determinant of the system Equation (58) gives the quartic equation for the in-plane eigenfrequencies of rectangular lattices composed of cells with long-range interactions:

$$D_{11}\left(\omega^2\right)D_{22}\left(\omega^2\right) - D_{12}D_{21} = 0 \quad (59)$$

with

$$\begin{cases} D_{11}(\omega^2) = 4\alpha_1\sin^2\left(\frac{k_1\pi}{2n}\right) + 4\alpha_2\sin^2\left(\frac{k_1\pi}{n}\right) + 2\beta_1\left[1-\cos\left(\frac{k_1\pi}{n}\right)\cos\left(\frac{k_2\pi}{m}\right)\right] \\ +2\beta_2\left[1-\cos\left(\frac{2k_1\pi}{n}\right)\cos\left(\frac{2k_2\pi}{m}\right)\right] + 16\gamma_1\sin^2\left(\frac{k_2\pi}{2m}\right) + 16\gamma_2\sin^2\left(\frac{k_2\pi}{m}\right) - M\omega^2 \\ D_{12} = (2\beta_1+4\gamma_1)\sin\left(\frac{k_1\pi}{n}\right)\sin\left(\frac{k_2\pi}{m}\right) + (2\beta_2+4\gamma_2)\sin\left(\frac{2k_1\pi}{n}\right)\sin\left(\frac{2k_2\pi}{m}\right) \\ D_{21} = D_{12} = (2\beta_1+4\gamma_1)\sin\left(\frac{k_1\pi}{n}\right)\sin\left(\frac{k_2\pi}{m}\right) + (2\beta_2+4\gamma_2)\sin\left(\frac{2k_1\pi}{n}\right)\sin\left(\frac{2k_2\pi}{m}\right) \\ D_{22}(\omega^2) = 4\alpha_1\sin^2\left(\frac{k_2\pi}{2m}\right) + 4\alpha_2\sin^2\left(\frac{k_2\pi}{m}\right) + 2\beta_1\left[1-\cos\left(\frac{k_1\pi}{n}\right)\cos\left(\frac{k_2\pi}{m}\right)\right] \\ 2\beta_2\left[1-\cos\left(\frac{2k_1\pi}{n}\right)\cos\left(\frac{2k_2\pi}{m}\right)\right] + 16\gamma_1\sin^2\left(\frac{k_1\pi}{2n}\right) + 16\gamma_2\sin^2\left(\frac{k_1\pi}{n}\right) - M\omega^2 \end{cases} \quad (60)$$

Equation (60) is a quartic equation for long-range, also dubbed as multi-cell, interactions, which generalises the one valid for the in-plane vibration of the rectangular lattices composed of Gazis et al.'s cells [20] exhibiting only short-range, also dubbed as cell-to-cell, interactions.

8. Conclusions

Studied herein is the in-plane vibration problem of rectangular elastic lattice plates. The lattice considered here is a 2D model pioneered by Gazis et al. (1960) [24] and accounts for central and non-central interactions, the latter of the angular type. For sufficiently small spacing, the lattice converges to a linear isotropic elastic continuum, at the asymptotic limit. The rectangular lattice plate is assumed to be constrained by sliding supports along its boundary. The elastic interaction in the vicinity of the boundaries is corrected, as described by Hrennikoff (1941) [41] and McHenry (1943) [40] for central lattices. The exact solutions in terms of eigenfrequencies and eigenmodes are analytically derived for this in-plane discrete plate problem from the resolution of a linear difference eigenvalue problem. We show how the eigenfrequency spectrum of the rectangular lattice converges to the continuous one (provided by Rayleigh [35]) from below, when the lattice spacing decreases, as usually observed for such spatial finite difference eigenvalue problems. We also developed two lattice-based higher-order continuous plate models to account for the specific scale effects associated with the lattice response. The gradient elasticity model derived herein is built by asymptotic expansion of the lattice difference equations, including central and non-central interactions. The gradient elasticity equations coincide with the ones of Mindlin (1968) [26], even if Mindlin [26] developed the 3D gradient elasticity model from the 3D Gazis et al. lattice model [20], whereas in the present paper we apply the same methodology for the plane stress problem. An exact solution for the eigenfrequencies of the simply supported gradient elasticity plate is also presented, and is shown to accurately fit the lattice response. We also present the exact solutions for a lattice-based nonlocal plate model which avoids the used of higher-order differential operators. The exact eigenfrequency solution for this alternative nonlocal plate model is also derived and compared to the lattice model. Both size-dependent lattice-based plate models accurately capture the scale effects of the in-plane lattice. A possible extension of this work would be to consider more general long-range elastic interactions for the lattice. It would be also interesting to include some additional inner masses (2D or 3D mass-in-mass lattices), as recently studied by Karampour et al. (2024) [51] for simply supported mass-in-mass rectangular lattices.

Author Contributions: Methodology, N.C.; Validation, H.P.N.; Writing—review & editing, C.M.W. and G.R. All authors have read and agreed to the published version of the manuscript.

Funding: This research received no external funding.

Data Availability Statement: Data is contained within the article.

Acknowledgments: Noël Challamel would like to thank Francesco Dell'Isola for stimulating discussions on discrete to continuous modelling in structural mechanics. Giuseppe Ruta acknowledges the financial support of the national grant PRIN PNRR 2022 P2022ATTAR.

Conflicts of Interest: The authors declare no conflict of interest.

References

1. Capecchi, D.; Ruta, G.; Trovalusci, P. From classical to Voigt's molecular models in elasticity. *Arch. Hist. Exact Sci.* **2010**, *64*, 525–559. [CrossRef]
2. Capecchi, D.; Ruta, G.; Trovalusci, P. Voigt and Poincaré's mechanistic-energetic approaches to linear elasticity and suggestions for multiscale modelling. *Arch. Appl. Mech.* **2011**, *81*, 1573–1584. [CrossRef]
3. Challamel, N.; Zhang, Y.P.; Wang, C.M.; Ruta, G.; Dell'Isola, F. Discrete and continuous models of linear elasticity: History and connections. *Contin. Mech. Thermodyn.* **2023**, *35*, 347–391. [CrossRef]
4. Andrianov, I.V.; Awrejcewicz, J.; Danishevskyy, V.V. *Linear and Nonlinear Waves in Microstructured Solids—Homogenization and Asymptotic Approaches*; CRC Press: Boca Raton, FL, USA, 2021.
5. Dell'Isola, F.; Steigmann, D.J. (Eds.) *Discrete and Continuum Models for Complex Metamaterials*; Cambridge University Press: Cambridge, UK, 2020.
6. Wang, C.M.; Zhang, H.; Challamel, N.; Pan, W. *Hencky-Bar-Chain/Net for Structural Analysis*; World Scientific: Singapore, 2020.
7. Lagrange, J.L. Recherches sur la nature et la propagation du son. In *Miscellanea Philosophico-Mathematica Societatis Privatae Taurinensis I*; 2rd Pagination, i-112, 1759 (see also Œuvres, Tome 1, 39-148); Ex Typographia Regia: Turin, Italy, 1759.
8. Lagrange, J.L. *Mécanique Analytique, Paris, 1788*, 3rd ed.; Mallet-Bachelier, Ed.; Imprimeur-Libraire du Bureau des Longitudes, de L'école Polytechnique, de L'école Centrale des Arts et Manufactures: France, Paris, 1853.
9. Rosenau, P. Dynamics of dense lattices. *Phys. Rev. B* **1987**, *36*, 5868–5876. [CrossRef] [PubMed]
10. Andrianov, I.; Awrejcewicz, J. Continuous models for 2D discrete media valid for higher-frequency domain. *Comput. Struct.* **2008**, *86*, 140–144. [CrossRef]
11. Lombardo, M.; Askes, H. Elastic wave dispersion in microstructured membrane. *Proc. R. Soc. Lond. A: Math. Phys. Eng. Sci.* **2010**, *466*, 1789–1807. [CrossRef]
12. Hérisson, B.; Challamel, N.; Picandet, V.; Perrot, A.; Wang, C.M. Static and dynamic behaviors of microstructured membranes within nonlocal mechanics. *J. Eng. Mech.* **2018**, *144*, 04017155. [CrossRef]
13. Chen, F.Y. On modeling and direct solution of certain free vibration systems. *J. Sound Vib.* **1971**, *14*, 57–79. [CrossRef]

14. Tong, P.; Pian, T.H.H.; Bucciarblli, L.L. Mode shapes and frequencies by finite element method using consistent and lumped masses. *Comput. Struct.* **1971**, *1*, 623–638. [CrossRef]
15. Zhang, Z.; Wang, C.M.; Challamel, N. Eringen's length scale coefficient for vibration and buckling of nonlocal rectangular plates with simply supported edges. *J. Eng. Mech.* **2015**, *141*, 04014117. [CrossRef]
16. Challamel, N.; Hache, F.; Elishakoff, I.; Wang, C.M. Buckling and vibration of microstructured rectangular plates considering phenomenological and lattice-based nonlocal continuum models. *Compos. Struct.* **2016**, *149*, 145–156. [CrossRef]
17. Hache, F.; Challamel, N.; Elishakoff, I.; Wang, C.M. Comparison of nonlocal continualization schemes for lattice beams and plates. *Arch. Appl. Mech.* **2017**, *87*, 1105–1138. [CrossRef]
18. Challamel, N.; Wang, C.M.; Zhang, H.; Elishakoff, I. Lattice-based nonlocal elastic structural models. In *Size-Dependent Continuum Mechanics Approaches: Theory & Applications*; Ghavanloo, E., Fazelzadeh, S.A., Marotti de Sciarra, F., Eds.; Springer: Berlin/Heidelberg, Germany, 2021; pp. 1–50.
19. Born, M.; von Kármán, T. On fluctuations in spatial grids. *Phys. Z.* **1912**, *13*, 297–309.
20. Lax, M. E1. The relation between microscopic and macroscopic theories of elasticity. In *Lattice Dynamics*; Wallis, R.F., Ed.; Pergamon Press: New York, NY, USA, 1965; pp. 583–596.
21. Gazis, D.C.; Wallis, R.F. Conditions for rotational invariance of a harmonic lattice. *Phys. Rev.* **1966**, *151*, 578–580, Erratum in *Phys. Rev.* **1967**, *156*, 1038. [CrossRef]
22. Keating, P.N. Effect of invariance requirements on the elastic strain energy of crystals with application to the diamond structure. *Phys. Rev.* **1966**, *145*, 637–645. [CrossRef]
23. Keating, P.N. Relationship between themacroscopic and microscopic theory of crystal elasticity. I. Primitive crystals. *Phys. Rev.* **1966**, *152*, 774–779. [CrossRef]
24. Gazis, D.C.; Herman, R.; Wallis, R.F. Surface elastic waves in cubic crystals. *Phys. Rev.* **1960**, *119*, 533–544. [CrossRef]
25. Challamel, N.; Ruta, G.; Nguyen, H.P.; Wang, C.M.; Zhang, Y.P. On angular and surface interactions in two-dimensional elastic lattices. *Eur. J. Mech. A/Solids* **2024**, *104*, 105176. [CrossRef]
26. Mindlin, R.D. Theories of elastic continua and crystal lattice theories. In *Mechanics of Generalized Continua: Proceedings of the IUTAM-Symposium on the Generalized Cosserat Continuum and the Continuum Theory of dislocations with applications, Freudenstadt and Stuttgart, Germany, 1967*; Kröner, E., Ed.; Springer: Berlin/Heidelberg, Germany, 1968; pp. 312–320.
27. Mindlin, R.D. Lattice theory of shear modes of vibration and torsional equilibrium of simple-cubic crystal plates and bars. *Int. J. Solids Struct.* **1970**, *6*, 725–738. [CrossRef]
28. Fermi, E.; Pasta, P.; Ulam, S.; Tsingou, M. *Studies of the Nonlinear Problems, Technical Report*; Los Alamos National Lab. (LANL): Los Alamos, NM, USA, 1955.
29. Weissert, T.P. *The Genesis of Simulation in Dynamics: Pursuing the Fermi-Pasta-Ulam Problem*; Springer Science & Business Media: Berlin/Heidelberg, Germany, 1997.
30. Dauxois, T.; Peyrard, M.; Ruffo, S. The Fermi–Pasta–Ulam 'numerical experiment': History and pedagogical perspectives. *Eur. J. Phys.* **2005**, *26*, S3. [CrossRef]
31. Dauxois, T.; Peyrard, M. *Physics of Solitons*; Cambridge University Press: Cambridge, UK, 2006.
32. Vainchtein, A. Solitary waves in FPU-type lattices. *Phys. D* **2022**, *434*, 133252. [CrossRef]
33. Potapov, A.I.; Pavlov, I.S.; Gorshkov, K.A.; Maugin, G.A. Nonlinear interaction of solitary waves in a 2D lattice. *Wave Motion* **2002**, *34*, 83–95. [CrossRef]
34. Friesecke, G.; Matthies, K. Geometric solitary waves in a 2D mass-spring lattice. *Discret. Contin. Dyn. Syst.* **2003**, *1*, 105–114. [CrossRef]
35. Strutt, J.W. *The Theory of Sound*; Dover: New York, NY, USA, 1894; Volume I, p. 405.
36. Strutt, J.W. On the free vibration of an infinite plate of homogeneous isotropic elastic matter. *Proc. Lond. Math. Soc.* **1889**, *20*, 225–234.
37. Bardell, N.S.; Langley, R.S.; Dunsdon, J.M. On the free in-plane vibration of isotropic rectangular plates. *J. Sound Vib.* **1996**, *191*, 459–467. [CrossRef]
38. Gorman, D.J. Exact solutions for the free in-plane vibration of rectangular plates with two opposite edges simply supported. *J. Sound Vib.* **2006**, *294*, 131–161. [CrossRef]
39. Challamel, N.; Atanackovic, T.; Zhang, Y.P.; Wang, C.M. A fractional nonlocal elastic model for lattice wave analysis. *Mech. Res. Comm.* **2022**, *126*, 103999. [CrossRef]
40. McHenry, D. A lattice analogy for the solution of stress problems. *J. Inst. Civ. Eng.* **1943**, *2*, 59–82. [CrossRef]
41. Hrennikoff, A. Solutions of problems of elasticity by the framework method. *J. Appl. Mech.* **1941**, *8*, A169–A175. [CrossRef]
42. Cauchy, A. Sur les différences finies et les intégrales aux différences des fonctions entières d'une on de plusieurs variables. *Exerc. De Mathématiques* **1828**, 155–159.
43. Collatz, L. *The Numerical Treatment of Differential Equations*; Springer: Berlin/Heidelberg, Germany, 1960.
44. Challamel, N.; Picandet, V.; Collet, B.; Michelitsch, T.; Elishakoff, I.; Wang, C.M. Revisiting finite difference and finite element methods applied to structural mechanics within enriched continua. *Eur. J. Mech. A/Solids* **2015**, *53*, 107–120. [CrossRef]
45. Polya, G. Sur une interprétation de la méthode des différences finies qui peut fournir des bornes supérieures ou inférieures. *C.R. Acad. Sci.* **1952**, *235*, 995–997.

46. Challamel, N.; Wang, C.M.; Elishakoff, I. Nonlocal or gradient elasticity macroscopic models: A question of concentrated or distributed microstructure. *Mech. Res. Commun.* **2016**, *71*, 25–31. [CrossRef]
47. Zhang, Y.P.; Challamel, N.; Wang, C.M. Modelling nano-plane structures with body force using Hencky bar-grid model, continualised nonlocal model and Eringen nonlocal model. *Contin. Mech. Thermodyn.* **2021**, *33*, 2453–2480. [CrossRef]
48. Eringen, A.C. On differential equations of nonlocal elasticity and solutions of screw dislocation and surface waves. *J. Appl. Phys.* **1983**, *54*, 4703–4710. [CrossRef]
49. Mindlin, R.D. *An Introduction to the Mathematical Theory of Vibrations of Elastic Plates*; Yang, J., Ed.; World Scientific: Singapore, 2006.
50. Challamel, N.; Nguyen, H.P.; Ruta, G.; Wang, C.M.; Zhang, Y.P. Meta-truss and generalized elastic lattices with long-range interactions. In *Dynamics of Discrete and Continuum Structures and Media*; Altenbach, H., Eremeyev, V., Eds.; Springer: Berlin/Heidelberg, Germany, 2024; in Press.
51. Karampour, S.; Ghavanloo, E.; Fazelzadeh, S.A. In-plane vibration characteristics of metamaterial plates with locally resonant microstructures. *Acta Mech.* **2024**, *235*, 819–831. [CrossRef]

mathematics

Article

Modal Analysis of a Multi-Supported Beam: Macroscopic Models and Boundary Conditions

Antoine Rallu * and Claude Boutin

Ecole Nationale des Travaux Publics de l'Etat, Université de Lyon, LTDS UMR CNRS 5513, Rue Maurice Audin, 69518 Vaulx-en-Velin, France; claude.boutin@entpe.fr
* Correspondence: antoine.rallu@entpe.fr

Abstract: This paper deals with the long-wavelength behaviour of a Euler beam periodically supported by co-located rotation and compression springs. An asymptotic homogenization method is applied to derive the several macroscopic models according to the stiffness contrasts between the elastic supports and the beam. Effective models of differential order two or four are obtained, which can be merged into a single unified model whose dispersion relations at long and medium wavelengths fit those derived by Floquet-Bloch. Moreover, the essential role of rotation supports is clearly evidenced. A mixed "discrete/continuous" approach to the boundary conditions is proposed, which allows the boundary conditions actually applied at the local scale to be expressed in terms of Robin-type boundary conditions on macroscopic variables. This approach can be applied to both dominant-order and higher-order models. The modal analysis performed with these boundary conditions and the homogenised models gives results in good agreement with a full finite element calculation, with great economy of numerical resources.

Keywords: homogenisation; dynamics; boundary conditions; asymptotic method; multi-supported beams

MSC: 35B27;74H10;74Q10;74Q15

1. Introduction

Multi-supported beams are widely used in a variety of engineering fields, including long multi-supported bridges, multi-span steel structures, rails connected to sleepers, pipelines supported by regularly distributed foundations, and pipe networks attached to structural elements. Understanding their vibratory behaviour is of interest for safer and more efficient design, in particular for better control of their response to moving loads or to stresses induced by external actions such as ambient vibrations, earthquakes, etc. Another field of application is the monitoring and control of the in-service condition of these structures using structural health monitoring (SHM) vibration methods.

Due to their wide range of applications, many studies have focused on dynamics of beams that rest on continuous foundations or on discrete flexible supports. For good reviews in this field one can refer to [1,2].

As for continuous foundations, the precursor works of [3–5] deal with Euler or Timoshenko beams, respectively, lying on an elastic or visco-elastic foundation of a Winkler compressional type. Let us mention the instability analysis developed by [6] and the work of [7,8] dealing with non-linear foundations and stratified foundations, respectively. The continuous approach has also been applied to discrete foundations by redistributing discrete forces in continuous form, e.g., [9].

Studies that explicitly take into account the discrete nature of supports were initiated by Mead for periodic [10] or non-periodic [11] distributions with the aim of determining their vibro-acoustic behaviour. More recently, ref. [2] investigates the steady-state dynamic response under the influence of an axial force. It should be noted that in these studies the supports are often restricted to compression springs, and their rotational stiffness is much less frequently considered [10]. However, in a bending beam system, this type of spring necessarily plays an essential role that cannot be neglected in practice.

Several methods have been used to address the dynamic behaviour of multi-supported beams. Among them, the most usual are the transfer matrix method [9,12,13], the use of Green functions e.g., [2], and the numerical approach by the finite element method e.g., [7]. Asymptotic methods have been used less frequently. Particular mention should be made of the work of [14], who is interested in the dynamics of stretched beams with concentrated masses and discrete compressional elastic supports.

The objective of this paper is twofold: first, to develop a unified model that encompasses the phenomena of rotation/compression/bending within a synthetic formulation, thereby translating the overall behaviour of multi-supported structures; and second, to formulate the relevant boundary conditions in order to perform the modal analysis.

Despite recent advances in the field, to our knowledge, no unified model is available in the literature. Given the large number of possible situations, the study focuses on an Euler beam periodically supported by co-localised compression and rotation elastic springs, as illustrated in Figure 1. The stiffnesses of these springs, in comparison to that of the beam, can vary over a wide range. From the academic point of view, this example is both simple enough to allow analytical developments and rich enough to illustrate how to overcome some theoretical issues. Nevertheless, the framework of the method would allow Timoshenko beams or viscoelastic supports to be considered.

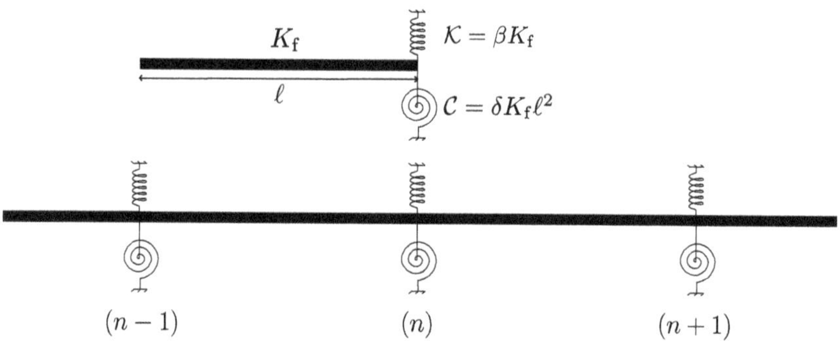

Figure 1. Above, irreducible period Ω. Bottom, beam periodically supported by the spring system $(\mathcal{C}, \mathcal{K})$.

To establish this model, we will use an asymptotic method (homogenization of discrete periodic media; HPDM) whose advantages are as follows: (i) the determination, within a rigorous framework, of explicit macroscopic models that describe the long-wavelength dynamics on the basis of period parameters and their contrasts; (ii) the identification of local and global efforts; (iii) the delimitation of the domains of validity of each obtained model; (iv) the significant savings in computational resources (compared with a full calculation) thanks to the model reduction thus achieved; and (v) the physical understanding of the various possible situations by retaining only the dominant effects on a macroscopic scale.

The modal analysis in the long-wave range requires, in addition to the identification of the equivalent macroscopic model, the determination of the appropriate boundary conditions. Let us briefly comment on these two aspects.

The homogenization method is based on the assumption of a strong separation of scale between the macroscopic phenomenon and the microscopic structure. It has been first developed for continuous media [15–17], then extended to discrete systems, such as reticulated structures or continuous networks by [14,18–20]. The homogenization of periodic discrete media (HPDM) has in particular been used to establish generalised beam models [21–23]. Furthermore, homogenization analysis has also shown its interest in exploring effects linked to imperfect scale separation through higher-order terms [24]. In quasi-static regimes, this work has made it possible to establish a link between phenomenological theories [25–27] of higher-order or micromorphic media and homogenised higher-order descriptions e.g., [28–30]. In the field of elastodynamics, these higher-order terms make it possible to tackle long wavelength dispersion, whether in 3D composite or stratified media, cf. [31–33].

The homogenised models are to be applied within the domain and must be completed at the edges of the domain. Nevertheless, the boundary conditions expressed on the variables of the homogenised model only imperfectly reflect the actual boundary conditions. This difficulty has been identified since the work of [34]. A way to overcome this drawback is to introduce a boundary layer that connects the true boundary condition to the homogenised condition inside the domain. The work of [35–37] provides a mathematical framework for these concepts. Despite these advances, this question is still topical, as shown by the recent work [38] in static regime, [39] in dynamic regime, and [40] for defining boundary layers in higher-order continuous media. As far as discrete media are concerned, to our knowledge, no specific work on this subject has been developed.

The study presented in this article deals with the various aspects mentioned above. The article is structured as follows: The first part is devoted to the determination of macroscopic models. These are constructed at the leading order using the HPDM method for each order of magnitude of stiffness contrasts in rotation and compression. The role of the contrasts on the order of the differential operator governing the overall dynamic behaviour is clearly highlighted. A so-called generic model, which encompasses all the descriptions, is established. It is shown by comparison with a Floquet-Bloch analysis that its range of validity extends well beyond the theoretical hypothesis of very large scale separation. The second part focuses on the formulation of relevant boundary conditions. A mixed "discrete/continuous" approach is proposed, which gives the effective boundary conditions to be applied to the macroscopic variables in order to reproduce the real conditions. The Robin-type conditions thus established associated with the macroscopic models can be used directly to carry out the modal analysis of these systems. The results obtained are very close to those obtained from a complete finite element calculation.

2. Method

In order to determine the eigenmodes of the periodic structure from the macroscopic model, the general procedure proposed is as follows:

1. Reduce the continuous structure studied to an equivalent discrete model. This is done by condensing the forces at the period nodes.
2. Derive the exact finite difference equation for the equilibrium between adjacent nodes in the discrete system.
3. Define the small scale parameter ε, which is used (i) to define the orders of magnitude of the spring stiffnesses as a function of the bending stiffness of the beams and (ii) to develop the kinematic variables asymptotically.
4. Solve the problems in increasing order of ε until you get a macroscopic model where the order of the differential operator allows all physical boundary conditions to be considered.
5. A boundary condition relating to a kinematic variable is applied directly to the corresponding macroscopic variable. A force boundary condition, on the other hand, is applied by re-expressing the local equilibrium at the node under consideration using the macroscopic kinematic variables.

Note that the Floquet-Bloch theory provides the dispersion relation for a periodic material at any wavelength. However, knowledge of these properties is not sufficient to carry out a modal analysis of a domain of finite size, which requires the boundary conditions to be taken into account.

2.1. Euler Beam Dynamics

Let us consider a Euler beam (width a, length ℓ, depth h, inertia $I = a^3h/12$, cross-section $A = ah$, Young modulus E, volumic mass ρ) in a harmonic regime at angular frequency ω that denotes $(\mathrm{v}(s), \theta(s))$ the transversal displacement and rotation field, where s is the abscissa along the beam (the time dependency $e^{i\omega t}$ is omitted) and linked by the relationship:

$$\theta(s) = \frac{d\mathrm{v}(s)}{ds} \tag{1}$$

The shear force $\mathrm{T}_{\mathrm{beam}}(s)$ and the momentum $\mathrm{M}_{\mathrm{beam}}(s)$ follow the local balances:

$$\frac{d\mathrm{T}_{\mathrm{beam}}(s)}{ds} = \rho A\omega^2 \mathrm{v}(s) \tag{2a}$$

$$\frac{d\mathrm{M}_{\mathrm{beam}}(s)}{ds} + \mathrm{T}_{\mathrm{beam}}(s) = 0 \tag{2b}$$

The bending behaviour law for small deformations reads:

$$\mathrm{M}_{\mathrm{beam}}(s) = -EI\frac{d\theta(s)}{ds} \tag{3}$$

Introducing (1)–(3) in the derivative of (2b) yields:

$$\frac{d^4\mathrm{v}(s)}{ds^4} - \left(\frac{\gamma}{\ell}\right)^4 \mathrm{v}(s) = 0 \qquad \text{with } \gamma = \sqrt[4]{\frac{\rho A\omega^2}{EI}}\,\ell = \frac{2\pi\ell}{\lambda_b} \tag{4}$$

where γ and λ_b are, respectively, the dimensionless bending wavenumber and the bending wavelength. Denoting by m and K_{f} the mass and bending stiffness of the beam of length ℓ, respectively, we define the reference angular frequency ω_{b} as

$$\mathrm{m} = \rho A\ell;\ K_{\mathrm{f}} = \frac{12EI}{\ell^3};\ \omega_{\mathrm{b}} = \sqrt{\frac{EI}{\rho A\ell^4}} = \sqrt{\frac{K_{\mathrm{f}}}{12\mathrm{m}}};\ \gamma^4 = \left(\frac{\omega}{\omega_{\mathrm{b}}}\right)^2 \tag{5}$$

Integrating (4) with the displacement/rotations conditions (v^B, θ^B) and (v^E, θ^E) at the extremities B and E of the beam (see Figure 2) provides:

$$\mathrm{v}(s) = \mathbb{C}_1\cos(\gamma\frac{s}{\ell}) + \mathbb{C}_2\sin(\gamma\frac{s}{\ell}) + \mathbb{C}_3\cosh(\gamma\frac{s}{\ell}) + \mathbb{C}_4\sinh(\gamma\frac{s}{\ell}) \tag{6}$$

with

$$\begin{cases} 2\mathbb{C}_1 = \left(1 + \frac{6}{\gamma^2}f_3(\gamma)\right)\mathrm{v}^B - \frac{6}{\gamma^2}f_6(\gamma)\mathrm{v}^E + \frac{4\ell}{\gamma^2}f_2(\gamma)\theta^B + \frac{2\ell}{\gamma^2}f_5(\gamma)\theta^E \\ 2\mathbb{C}_2 = -\frac{12}{\gamma^3}f_1(\gamma)\mathrm{v}^B + \frac{12}{\gamma^3}f_4(\gamma)\mathrm{v}^E + \frac{\ell}{\gamma}\left(1 - \frac{6}{\gamma^2}f_3(\gamma)\right)\theta^B - \frac{6\ell}{\gamma^3}f_6(\gamma)\theta^E \\ 2\mathbb{C}_3 = \left(1 - \frac{6}{\gamma^2}f_3(\gamma)\right)\mathrm{v}^B + \frac{6}{\gamma^2}f_6(\gamma)\mathrm{v}^E - \frac{4\ell}{\gamma^2}f_2(\gamma)\theta^B - \frac{2\ell}{\gamma^2}f_5(\gamma)\theta^E \\ 2\mathbb{C}_4 = \frac{12}{\gamma^3}f_1(\gamma)\mathrm{v}^B - \frac{12}{\gamma^3}f_4(\gamma)\mathrm{v}^E + \frac{\ell}{\gamma}\left(1 + \frac{6}{\gamma^2}f_3(\gamma)\right)\theta^B + \frac{6\ell}{\gamma^3}f_6(\gamma)\theta^E \end{cases}$$

and:

$$
\begin{aligned}
f_1(\gamma) &= \frac{\sin(\gamma)\cosh(\gamma)+\cos(\gamma)\sinh(\gamma)}{1-\cos(\gamma)\cosh(\gamma)}\frac{\gamma^3}{12} & \quad f_4(\gamma) &= \frac{\sinh(\gamma)+\sin(\gamma)}{1-\cos(\gamma)\cosh(\gamma)}\frac{\gamma^3}{12} \\
f_2(\gamma) &= \frac{\sin(\gamma)\cosh(\gamma)-\cos(\gamma)\sinh(\gamma)}{1-\cos(\gamma)\cosh(\gamma)}\frac{\gamma}{4} & \quad f_5(\gamma) &= \frac{\sinh(\gamma)-\sin(\gamma)}{1-\cos(\gamma)\cosh(\gamma)}\frac{\gamma}{2} \\
f_3(\gamma) &= \frac{\sinh(\gamma)\sin(\gamma)}{1-\cos(\gamma)\cosh(\gamma)}\frac{\gamma^2}{6} & \quad f_6(\gamma) &= \frac{\cosh(\gamma)-\cos(\gamma)}{1-\cos(\gamma)\cosh(\gamma)}\frac{\gamma^2}{6}
\end{aligned}
\tag{7}
$$

The functions the functions $\{f_i(\gamma)\}, i = 1,\ldots,6$ have the following properties:

$$
\lim_{\gamma\to 0} f_i(\gamma) = 1 \quad ; \quad \forall\gamma, \forall i \in [\![4;6]\!], \quad f_i(\gamma) \neq 0 \tag{8}
$$

Introducing (6) into (3) provides the expressions of the forces $\underline{F} = \begin{pmatrix} T_{beam} \\ M_{beam}/\ell \end{pmatrix}$ exerted by the beam at its extremities ($\underline{F}^B$ at the beginning; $\underline{F}^E$ at the end) as a function of the displacement $\underline{U} = \begin{pmatrix} v \\ \ell\theta \end{pmatrix}$ at its extremities ($\underline{U}^B$ at the beginning; $\underline{U}^E$at the end). They are written in matrix form as:

$$
\begin{aligned}
\underline{F}^B(\underline{U}^B,\underline{U}^E) &= K_f\left(\underline{\underline{H}}^{BB}.\underline{U}^B + \underline{\underline{H}}^{BE}.\underline{U}^E\right) \\
\underline{F}^E(\underline{U}^B,\underline{U}^E) &= K_f\left(\underline{\underline{H}}^{EB}.\underline{U}^B + \underline{\underline{H}}^{EE}.\underline{U}^E\right)
\end{aligned}
\tag{9}
$$

where the following elasto-inertial matrices have been introduced:

$$
\underline{\underline{H}}^{BB} = \begin{bmatrix} f_1(\gamma) & \frac{1}{2}f_3(\gamma) \\ \frac{1}{2}f_3(\gamma) & \frac{1}{3}f_2(\gamma) \end{bmatrix} \quad \underline{\underline{H}}^{EE} = \begin{bmatrix} -f_1(\gamma) & \frac{1}{2}f_3(\gamma) \\ \frac{1}{2}f_3(\gamma) & -\frac{1}{3}f_2(\gamma) \end{bmatrix}
$$

$$
\underline{\underline{H}}^{BE} = \begin{bmatrix} -f_4(\gamma) & \frac{1}{2}f_6(\gamma) \\ -\frac{1}{2}f_6(\gamma) & \frac{1}{6}f_5(\gamma) \end{bmatrix} \quad \underline{\underline{H}}^{EB} = -{}^{T}\underline{\underline{H}}^{BE} = \begin{bmatrix} f_4(\gamma) & \frac{1}{2}f_6(\gamma) \\ -\frac{1}{2}f_6(\gamma) & -\frac{1}{6}f_5(\gamma) \end{bmatrix}
$$

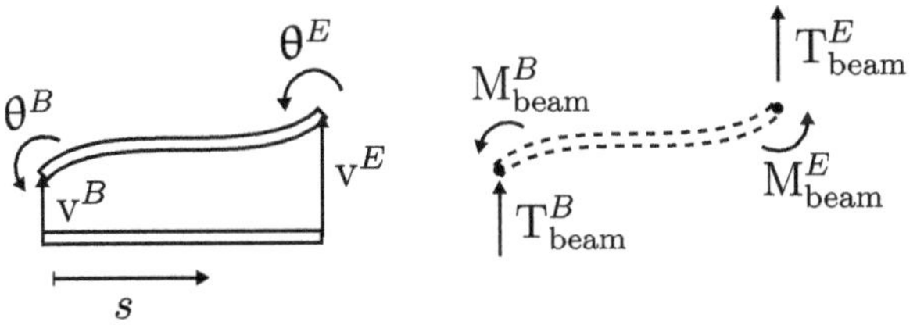

Figure 2. Kinematic variables and forces at the nodes defined in the local frame of each beam element.

2.2. Dynamics of Multi-Supported Beam—Finite Difference Formulation

The period Ω of the system is composed by a beam of size ℓ and a system of springs $(\mathcal{K},\mathcal{C})$ of positive stiffnesses. At each node $x_n = n\ell$ of the system, the springs exert a restoring force $\mathcal{F}_n^{\mathcal{K}} = -\mathcal{K}v_n$ and a restoring moment $\mathcal{M}_n^{\mathcal{C}} = -\mathcal{C}\theta_n$. Hence, we define the stiffness matrix $\underline{\underline{R}}$ and the restoring forces $\underline{f}_n$ as:

$$
\underline{f}_n = -K_f\underline{\underline{R}}.\underline{U}_n \quad ; \quad K_f\underline{\underline{R}} = \begin{bmatrix} \mathcal{K} & 0 \\ 0 & \mathcal{C}/\ell^2 \end{bmatrix} = K_f\begin{bmatrix} \beta & 0 \\ 0 & \delta \end{bmatrix}
$$

Expressing, at each node n the balance T_n of the transverse force, and the balance M_n/ℓ of the ℓ-reduced momentum yields

$$
\begin{pmatrix} T_n \\ M_n/\ell \end{pmatrix} = \underline{F}^E(\underline{U}_{n-1},\underline{U}_n) - \underline{F}^B(\underline{U}_n,\underline{U}_{n+1}) + \underline{f}_n = \underline{0} \tag{10}
$$

which can be rewritten:

$$\underline{\underline{H}}^{EB}.\underline{U}_{n-1} + \left(\underline{\underline{H}}^{EE} - \underline{\underline{H}}^{BB} - \underline{\underline{R}}\right).\underline{U}_n + ^T\underline{\underline{H}}^{EB}.\underline{U}_{n+1} = \underline{0} \tag{11}$$

It is convenient to introduce the diagonal matrices $\underline{\underline{F}}(\omega) = \underline{\underline{H}}^{BB} - \underline{\underline{H}}^{EE}$ and $\underline{\underline{D}}(\omega) = \frac{1}{2}\left(\underline{\underline{H}}^{EB} + ^T\underline{\underline{H}}^{EB}\right)$ then the antisymmetric matrix $\underline{\underline{C}}(\omega) = \frac{1}{2}\left(\underline{\underline{H}}^{EB} - ^T\underline{\underline{H}}^{EB}\right)$, i.e., explicitly:

$$\underline{\underline{D}}(\omega) = \begin{bmatrix} f_4(\gamma) & 0 \\ 0 & -\frac{1}{6}f_5(\gamma) \end{bmatrix}; \underline{\underline{C}}(\omega) = \begin{bmatrix} 0 & \frac{1}{2}f_6(\gamma) \\ -\frac{1}{2}f_6(\gamma) & 0 \end{bmatrix}; \underline{\underline{F}}(\omega) = \begin{bmatrix} f_1(\gamma) & 0 \\ 0 & \frac{1}{3}f_2(\gamma) \end{bmatrix}$$

to rearrange (10) in a finite difference set. The two balance equations are coupled through the antisymmetric matrix $\underline{\underline{C}}(\omega)$:

$$\begin{aligned} \frac{1}{K_f}\begin{pmatrix} T_n \\ M_n/\ell \end{pmatrix} &= \underline{\underline{D}}(\omega).(\underline{U}_{n-1} - 2\underline{U}_n + \underline{U}_{n+1}) \\ &\quad + \underline{\underline{C}}(\omega).(\underline{U}_{n-1} - \underline{U}_{n+1}) + (2\underline{\underline{D}}(\omega) - 2\underline{\underline{F}}(\omega) - \underline{\underline{R}})\underline{U}_n = \underline{0} \end{aligned} \tag{12}$$

2.3. *Asymptotic Method*

The switch from the finite difference description to a continuous description is made using the homogenization method for discrete media. This requires a net scale separation between the characteristic vibration length L and the period length ℓ, i.e., $\ell = \varepsilon L$, with $\varepsilon \ll 1$. Consequently, each local beam undergoes a quasi-static regime with a small dimensionless bending number $\gamma \ll 1$.

Notation: In the following, comparisons of orders of magnitude are made. The following notations are used:

- If an adimensional variable (e.g., γ) is of order ε^m, we note that

$$\gamma = \gamma^\star \varepsilon^m = O(\varepsilon^m) \text{ with } \gamma^\star = O(1)$$

- an expression $A = \mathcal{O}(\varepsilon) \Leftrightarrow A(1 + O(\varepsilon^1)) = 0$ means that all terms smaller than ε with respect to A are neglected. With this notation, it is not necessary to specify the dimensions of the terms to be compared.

The steps of the method are as follows:

Continuation of discrete variables: introduction of a continuous displacement $\underline{U}(x) = \begin{pmatrix} v(x) \\ \ell\theta(x) \end{pmatrix}$ as in the nodes of the system:

$$\underline{U}(x_n = n\ell) = \underline{U}_n = \begin{pmatrix} v_n \\ \ell\theta_n \end{pmatrix} \tag{13}$$

Expansion of kinematic variables: Under the assumption of scale separation, the variation in displacements between two consecutive nodes is small. It is then possible to expand the displacements at the neighbouring nodes of the nth node into Taylor series, giving rise to macroscopic derivatives denoted by ∂_x:

$$\underline{U}_{n\pm1} = \underline{U}(x_n \pm \varepsilon L) = \underline{U}(x_n) \pm \varepsilon L \partial_x \underline{U}(x_n) + \frac{(\varepsilon L)^2}{2}\partial_x^2\underline{U}(x_n) \pm \frac{(\varepsilon L)^3}{6}\partial_x^3\underline{U}(x_n) + \frac{(\varepsilon L)^4}{24}\partial_x^4\underline{U}(x_n) + O(\varepsilon^5)$$

Thus, the two finite difference terms of (12) become in the continuous representation:

$$\begin{aligned} \underline{U}_{n-1} - 2\underline{U}_n + \underline{U}_{n+1} &= (\varepsilon L)^2\partial_x^2\underline{U}(x_n) + \frac{(\varepsilon L)^4}{12}\partial_x^4\underline{U}(x_n) + O(\varepsilon^6) \\ \underline{U}_{n-1} - \underline{U}_{n+1} &= -2\varepsilon L\partial_x\underline{U}(x_n) - \frac{(\varepsilon L)^3}{3}\partial_x^3\underline{U}(x_n) + O(\varepsilon^5) \end{aligned}$$

Moreover, the continuous kinematic variables are also asymptotically expanded into powers of ε:

$$\underline{U}(x) = \sum_{j\geq 0} \varepsilon^j \underline{U}^{(j)}(x) \qquad \text{with } \underline{U}^{(j)}(x) = O(\varepsilon^0 \underline{U}^{(0)}(x))$$

Expansion of frequencies: as $\gamma \ll 1$, functions $\{f_i\}_{1\leq i\leq 6}$ (see (7)) are expanded in a Taylor series:

$$f_i(\gamma) = 1 + \sum_{j\geq 1} \gamma^j \partial^j_\gamma f_i(0) \tag{14}$$

Finally, the angular frequency is small compared to the reference frequency ω_b (see (5)), thus is expanded asymptotically around the zero pulsation:

$$\omega = \gamma^2 \omega_b = 0 + \sum_{j\geq 1} \varepsilon^j \omega^{(j)} \qquad \text{with } \omega^{(j)} = O(\varepsilon^0 \omega_b) \tag{15}$$

By introducing (15) into (14), the matrices $\underline{\underline{D}}$, $\underline{\underline{C}}$ and $\underline{\underline{F}}$ can be asymptotically expanded into powers of ε.

Normalization: In this step, the geometric and mechanical contrasts are defined in order of magnitude of ε. In accordance with the Euler-Bernoulli beam hypothesis, each beam element must be sufficiently slender (width $a \ll \ell$). This geometric contrast is expressed by considering (without loss of generality) that:

$$a/\ell = e_0 \varepsilon = O(\varepsilon)$$

In addition, the order of magnitude of the spring stiffnesses in translation and rotation is weighed in power of ε compared to the bending stiffness of a local beam:

$$\beta = \frac{\mathcal{K}}{K_f} = O(\varepsilon^p) \quad ; \quad \delta = \frac{\mathcal{C}}{K_f \ell^2} = O(\varepsilon^q)$$

Playing with these two orders of magnitude can generate different behaviours in the medium on a macroscopic scale. This will be the subject of Section 3.

Resolution: All the previous expansions are introduced in (12), so that:

$$\frac{1}{K_f}\begin{pmatrix} \mathrm{T}_n \\ \mathrm{M}_n/\ell \end{pmatrix} = -\sum_{j\geq 0} \varepsilon^j \begin{pmatrix} \mathbb{T}^{(j)} \\ \mathbb{M}^{(j)}/\ell \end{pmatrix} = 0 \quad ; \quad \forall \varepsilon \to 0 \tag{16}$$

Considering the equilibria for each order separately, the following successive equations arise for the first four orders:

$$\begin{pmatrix} \mathbb{T}^{(-1)} \\ \mathbb{M}^{(-1)} \end{pmatrix} = -\underline{\underline{R}}^{(-1)}.\underline{U}^{(0)}(x) = \underline{0}$$

$$\begin{pmatrix} \mathbb{T}^{(0)} \\ \mathbb{M}^{(0)} \end{pmatrix} = -\underline{\underline{R}}^{(-1)}.\underline{U}^{(1)}(x) + \left(2\left(\underline{\underline{D}}^{(0)} - \underline{\underline{F}}^{(0)}\right) - \underline{\underline{R}}^{(0)}\right).\underline{U}^{(0)}(x) = \underline{0}$$

$$\begin{pmatrix} \mathbb{T}^{(1)} \\ \mathbb{M}^{(1)} \end{pmatrix} = -\underline{\underline{R}}^{(-1)}.\underline{U}^{(2)}(x) + \sum_{i=0}^{1}\left(2\left(\underline{\underline{D}}^{(i)} - \underline{\underline{F}}^{(i)}\right) - \underline{\underline{R}}^{(i)}\right).\underline{U}^{(1-i)}(x) - 2L\underline{\underline{C}}^{(0)}.\partial_x \underline{U}^{(0)}(x) = \underline{0}$$

$$\begin{pmatrix} \mathbb{T}^{(2)} \\ \mathbb{M}^{(2)} \end{pmatrix} = -\underline{\underline{R}}^{(-1)}.\underline{U}^{(3)}(x) + \sum_{i=0}^{2}\left(2\left(\underline{\underline{D}}^{(i)} - \underline{\underline{F}}^{(i)}\right) - \underline{\underline{R}}^{(i)}\right).\underline{U}^{(2-i)}(x) - 2L\sum_{i=0}^{1}\underline{\underline{C}}^{(i)}.\partial_x \underline{U}^{(1-i)}(x) + L^2 \underline{\underline{D}}^{(0)}.\partial_x^2 \underline{U}^{(0)}(x) = \underline{0}$$

...

The explicit formulation of these equations can be found in Appendix A. In order to encompass in a single framework the different possible levels of stiffness contrast expressed by β and δ, these parameters are also developed in the form of $\beta = \sum_{j=-1}^{4} \varepsilon^j \beta^{(j)}$ and $\delta = \sum_{j=-1}^{4} \varepsilon^j \delta^{(j)}$. Thus, a ε^p contrast for β will result in $\beta^{(p)} \neq 0$ and $\beta^{(j \neq p)} = 0$ (and similarly for a ε^q contrast for δ)

3. Multi-Supported Beam Models

The aim of this section is to establish the equivalent macroscopic models for multi-supported beams corresponding to different pairs of spring stiffnesses. First, two basic models are presented that show the crucial influence of stiffness contrasts. It is then shown that by combining the characteristics of these two models, it is possible to build a generic model that covers all possible contrast situations. Finally, the validity of these macroscopic descriptions is analysed by comparing the dispersion curves obtained for these models with those obtained using the classical Floquet-Bloch approach.

3.1. On the Level of Beam/Support Contrasts

Obviously, the stiffness contrast between the beam and the supports plays a crucial role. Indeed, in the case of vanishing support stiffness, the multi-beam system tends to behave as a single long beam, while at very high support stiffness, the multi-beam system tends to behave like a chain of clamped (and therefore independent) beams. The situations of interest lie between these two extreme cases.

In order for the supports to play an effective role, it is necessary for the restoring forces to act as a source in the force and moment balance. With a view to an assessment, the localised restoring forces can be redistributed along the length ℓ of the beam. This gives the following relationships:

$$O(\frac{d\mathrm{T}_{\mathrm{beam}}}{ds}) \leq O(\frac{\mathcal{K}\mathrm{v}}{\ell}) \quad ; \quad O(\frac{d\mathrm{M}_{\mathrm{beam}}}{ds}) \leq O(\frac{\mathcal{C}\theta}{\ell})$$

In the scale separation assumption, assuming that both v and θ vary with the macroscopic length, one derives from the beam constitutive laws that $O(\frac{d\mathrm{T}_{\mathrm{beam}}}{ds}) = O(EI\frac{\mathrm{v}}{L^4}) = O(\frac{K_f \ell^3}{L^4}\mathrm{v})$ and $O(\frac{d\mathrm{M}_{\mathrm{beam}}}{ds}) = O(EI\frac{\theta}{L^2}) = O(\frac{K_f \ell^3}{L^2}\theta)$. These assessments reported in the previous relations provide the lower bound of the contrasts below that support the effect vanishes, i.e.,

$$O(K_f \frac{\ell^3}{L^4}) \leq O(\frac{\mathcal{K}}{\ell}) \implies O(\beta) \geq \varepsilon^4 \quad ; \quad O(K_f \frac{\ell^3}{L^2}) \leq O(\frac{\mathcal{C}}{\ell}) \implies O(\delta) \geq \varepsilon^2$$

On the other hand, the contrast β is also limited by an upper bound. In fact, a strong support imposes that the rotation and the curvatures vary at the local scale so that $O(\theta) = O(\mathrm{v}/\ell)$ and $O(\frac{d\theta}{ds}) = O(\mathrm{v}/\ell^2)$, while, according to the scale separation assumption, the transverse force and moment vary at the large scale. This gives the following maximum estimate of $\frac{d\mathrm{T}_{\mathrm{beam}}}{ds}$ that the restoring force can balance

$$O(\frac{d\mathrm{T}_{\mathrm{beam}}}{ds}) = O(EI\frac{\mathrm{v}}{\ell^2 L^2}) = O(\frac{K_f \ell}{L^2}\mathrm{v})$$

Hence

$$O(K_f \frac{\ell}{L^2}) \geq O(\frac{\mathcal{K}}{\ell}) \implies O(\beta) \leq \varepsilon^2$$

Thus, the contrasts of interest are in the ranges $\varepsilon^4 \leq O(\beta) \leq \varepsilon^2$ and $\varepsilon^2 \leq O(\delta)$.

3.2. Basic Models

The two basic models are established by considering the cases where the compressional supports are as stiff as possible, i.e., $O(\beta) = \varepsilon^2$ and the rotational supports are either stiff or soft.

3.2.1. Multi-Supported Beam with Stiff Rotational Support

Let us consider multi-beam supported by stiff supports such that

$$\beta = \frac{\mathcal{K}}{K_f} = \beta^{(2)}\varepsilon^2 \quad ; \quad \delta = \frac{\mathcal{C}}{K_f\ell^2} = \delta^{(0)}\varepsilon^0$$

The macroscopic description at leading order is obtained by solving problems (A2) to (A4) in the Appendix A) in the growing order of powers of ε. Problems (A1) and (A2) are trivially verified with $\theta^{(-1)} = 0$. At the order ε^1, (A3) gives:

$$\begin{cases} \mathbb{T}^{(1)} = 0 \\ \mathbb{M}^{(1)} = \left(\delta^{(0)} + 1\right)L\theta^{(0)} - L\partial_x v^{(0)} = 0 \end{cases}$$

from which we can deduce the kinematic relationship to the dominant order:

$$\theta^{(0)} = \frac{1}{\left(\delta^{(0)} + 1\right)}\partial_x v^{(0)} \tag{17}$$

The equilibrium equation for the shear force at order 2 (A4) gives:

$$\mathbb{T}^{(2)} = L^2\partial_x\theta^{(0)} + \left[\beta^{(2)} - \frac{\rho L^2}{Ee_0^2}(\omega^{(1)})^2\right]v^{(0)} - L^2\partial_x^2 v^{(0)} = 0$$

By introducing (17) into $\mathbb{T}^{(2)}$, we obtain the macroscopic description at leading order, with $\omega = \varepsilon\omega^{(1)} + o(\varepsilon)$:

$$\left(\mathcal{K} - \mathrm{m}\omega^2\right)v(x) - \mathcal{C}_{\mathrm{eq}}\partial_x^2 v(x) = \mathcal{O}(\varepsilon) \tag{18a}$$

$$\theta(x) = \frac{\mathcal{C}_{\mathrm{eq}}}{\mathcal{C}}\partial_x v(x) + \mathcal{O}(\varepsilon) \tag{18b}$$

where $\mathcal{C}_{\mathrm{eq}}$ represents the effective rotational stiffness, which combines in series the stiffness $\mathcal{C}$ and stiffness $K_f\ell^2$ of the beam, i.e.,

$$\frac{1}{\mathcal{C}_{\mathrm{eq}}} = \frac{1}{\mathcal{C}} + \frac{1}{K_f\ell^2} \tag{19}$$

The rotational spring $\mathcal{C}$ plays an essential role in the macroscopic balance equation (18a), which differs fundamentally from that of a bending beam. $\mathcal{C}$ also affects the kinematic relationship (18b) between the rotation θ and the displacement v. In addition, the compression spring imposes a cut-off pulsation ω_c defined by:

$$\omega_c = \sqrt{\frac{\mathcal{K}}{\mathrm{m}}} \tag{20}$$

below which the differential operator (18a) driving v is not propagative.

The macroscopic description (18a) can be interpreted in terms of the balance of a macroscopic transverse force $-\mathcal{T}(x)$ linked to the displacement gradient by an apparent modulus $\mu = \frac{\mathcal{C}_{\mathrm{eq}}}{\ell}$.

$$\partial_x\mathcal{T}(x) = \rho A\omega^2 v(x) - \frac{\mathcal{K}}{\ell}v(x) \tag{21a}$$

$$\mathcal{T}(x) = -\frac{\mathcal{C}_{\mathrm{eq}}}{\ell}\partial_x v(x) \tag{21b}$$

Hence this model, denoted MSB-Stiff in the sequel behaves macroscopically as a kind of shear-beam for which (i) the effects of the local bending and of the rotational spring are encapsulated in the effective shear modulus μ and (ii) the effect of the compressional spring results in an effective lineic force $-\frac{\mathcal{K}}{\ell}v(x)$.

3.2.2. Multi-Supported Beam with Soft Rotational Support

This model is obtained by reducing the stiffness of the rotational spring (compared to the previous case), so that:

$$\beta = \frac{\mathcal{K}}{K_f} = \beta^{(2)}\varepsilon^2 \quad ; \quad \delta = \frac{\mathcal{C}}{K_f \ell^2} = \delta^{(2)}\varepsilon^2$$

As before, the problems (A1) to (A2) are trivially verified with $\theta^{(-1)} = 0$. For the next order, (A3) gives:

$$\begin{cases} \mathbb{T}^{(1)} = 0 \\ \mathbb{M}^{(1)} = L\theta^{(0)} - L\partial_x v^{(0)} = 0 \end{cases}$$

which leads to the classical kinematic relationship of a Euler beam:

$$\theta^{(0)} = \partial_x v^{(0)}$$

At order ε^2, the equilibrium equations (A4) are:

$$\begin{cases} \mathbb{T}^{(2)} = \left[\beta^{(2)} - \frac{\rho L^2}{E e_0^2}(\omega^{(1)})^2\right] v^{(0)} = 0 \\ \mathbb{M}^{(2)} = L\theta^{(1)} - L\partial_x v^{(1)} = 0 \end{cases}$$

$\mathbb{T}^{(2)}$ gives the expression for the first angular frequency term $\omega^{(1)}$, and $\mathbb{M}^{(2)}$ gives the same kinematic relationship as in the previous order. Thus:

$$\omega^{(1)} = \sqrt{\beta^{(2)}}\omega_b > 0 \quad ; \quad \theta^{(1)} = \partial_x v^{(1)}$$

At the order ε^3, (A5) gives:

$$\begin{cases} \mathbb{T}^{(3)} = -\frac{\rho L^2}{E e_0^2} 2\omega^{(1)}\omega^{(2)} v^{(0)} = 0 \\ \mathbb{M}^{(3)} = L\theta^{(2)} + \left[\delta^{(2)} - \frac{\rho L^2}{210 E e_0^2}(\omega^{(1)})^2\right] L\theta^{(0)} - L\partial_x v^{(2)} - \frac{13\rho L^2}{210 E e_0^2}(\omega^{(1)})^2 L\partial_x v^{(0)} = 0 \end{cases}$$

As $\omega^{(1)} \neq 0$, $\mathbb{T}^{(3)}$ imposes $\omega^{(2)} = 0$, and $\mathbb{M}^{(3)}$ provides a second-order correction to the kinematic relationship:

$$\theta^{(2)} = \partial_x v^{(2)} + \left(\frac{\beta^{(2)}}{15} - \delta^{(2)}\right)\theta^{(0)}$$

Finally, the equilibrium (A6) in shear force at order ε^4 is written:

$$\mathbb{T}^{(4)} = -L^2\delta^{(2)}\partial_x^2 v^{(0)} + \frac{L^4}{12}\partial_x^4 v^{(0)} - \left[\frac{(\beta^{(2)})^2}{60} + \frac{\rho L^2}{E e_0^2} 2\omega^{(1)}\omega^{(3)}\right] v^{(0)} = 0$$

Since $\omega^{(2)} = 0$, the angular frequency expands to: $\omega = \varepsilon\omega^{(1)} + \varepsilon^3\omega^{(3)} + o(\varepsilon^3)$ and consequently $\varepsilon^4 2\omega^{(1)}\omega^{(3)} = \omega^2 - (\varepsilon\omega^{(1)})^2 + o(\varepsilon^4)$. By introducing this expression into $\mathbb{T}^{(4)}$, as well as all those resulting from the solutions to the previous problems, we obtain the following macroscopic description:

$$\left(-\mathrm{m}\omega^2 + \mathcal{K}_{\mathrm{eq}}\right)v(x) - \mathcal{C}\partial_x^2 v(x) + EI\ell\partial_x^4 v(x) = \mathcal{O}(\varepsilon) \tag{22a}$$

$$\theta(x) = \partial_x v(x) + \mathcal{O}(\varepsilon) \tag{22b}$$

$\mathcal{K}_{\text{eq}}$ is an effective stiffness that takes into account the stiffness $60K_{\text{f}}$ due to the beam in bending and $\mathcal{K}$ of the support in compression. Note that as $\mathcal{K} = \beta K_{\text{f}} = \mathcal{O}(\varepsilon^2 K_{\text{f}})$, $\mathcal{K}_{\text{eq}}$ corresponds at this order to a series association between these two elements.

$$\mathcal{K}_{\text{eq}} = \mathcal{K}(1 - \frac{\mathcal{K}}{60K_{\text{f}}}) = \frac{1}{\frac{1}{\mathcal{K}} + \frac{1}{60K_{\text{f}}}} + o(\varepsilon^2) \tag{23}$$

As before, the presence of $\mathcal{K}_{\text{eq}}$ introduces a cutoff angular frequency below which the differential operator (22a) is not propagative.

$$\omega_{\text{c,eq}} = \sqrt{\frac{\mathcal{K}_{\text{eq}}}{\text{m}}} < \omega_{\text{c}} \tag{24}$$

We note that in the model described by (22a) and (22b), the kinematic relationship of the elementary beam is transferred to the macroscopic scale as well as the bending term, which is translated by the same fourth-order differential operator. This leads to the introduction of the macroscopic transverse force $\mathcal{T}(x)$ and moment $\mathcal{M}(x)$, which allow the model to be rewritten in terms of the transverse force and moment balances as shown below.

$$\partial_x \mathcal{T}(x) = \rho A \omega^2 v(x) - \frac{\mathcal{K}_{\text{eq}}}{\ell} v(x) \tag{25a}$$

$$\partial_x \mathcal{M}(x) + \mathcal{T}(x) = -\frac{\mathcal{C}}{\ell}\theta(x) \tag{25b}$$

$$\mathcal{M}(x) = -EI\partial_x^2 v(x) \tag{25c}$$

$$\theta(x) = \partial_x v(x) \tag{25d}$$

This formulation emphasises that the presence of the springs $(\mathcal{K}, \mathcal{C})$ simply results in forces of volume $-\mathcal{K}_{\text{eq}} v(x)$ and a moment of volume $-\mathcal{C}\partial_x v(x)$, without modifying the inherent behaviour of the beams.

3.3. Multi-Supported Generic Beam Model

The basic models make it possible to build a single generic model that incorporates the characteristics of each. To do this, we keep the equilibrium Equations (25a) and (25b), and we replace the kinematic relationship between the macroscopic variables (25d) by (18b). This leads to the following set, where $(\mathcal{K}_{\text{eq}}, \mathcal{C}_{\text{eq}})$ are the equivalent stiffnesses (19) and (23):

$$\partial_x \mathcal{T}(x) = \rho A \omega^2 v(x) - \frac{\mathcal{K}_{\text{eq}}}{\ell} v(x)$$

$$\partial_x \mathcal{M}(x) + \mathcal{T}(x) = -\frac{\mathcal{C}}{\ell}\theta(x)$$

$$\mathcal{M}(x) = -EI\partial_x^2 v(x)$$

$$\theta(x) = \frac{\mathcal{C}_{\text{eq}}}{\mathcal{C}}\partial_x v(x)$$

Combining these equations leads to the leading-order macroscopic behaviour equation, where the $\omega_{\text{c,eq}}$ cut-off angular frequency is defined as (24):

$$\text{m}\left(\omega^2 - \omega_{\text{c,eq}}^2\right)v(x) + \mathcal{C}_{\text{eq}}\partial_x^2 v(x) - EI\ell\partial_x^4 v(x) = 0 \tag{26}$$

3.4. Dimensional Analysis

A dimensional analysis of (26) divided by $m\omega_b^2 = K_f/12$ highlights the dominant effects according to the order of magnitude of the stiffnesses ($\mathcal{K} = \beta K_f$ and $\mathcal{C} = \delta K_f \ell^2$). To do this, the displacement is rewritten in the form

$$v = v^r V^\star \tag{27}$$

where v^r is the reference displacement and $V^\star$ the adimensional displacement, $V^\star = O(\varepsilon^0)$. In addition, we perform the change of variables $\hat{x} = x/L$ (where $L = \ell/\varepsilon \gg \ell$ is related to the wavelength λ by the relation $L = \lambda/2\pi$). By defining the contrasts ($\beta = \beta^\star \varepsilon^p, \delta = \delta^\star \varepsilon^q$) (26) becomes:

$$\underbrace{\left(\frac{\omega}{\omega_b}\right)^2 V^\star}_{\mathcal{T}_\omega} \underbrace{-12\frac{\beta^\star \varepsilon^p}{1+\frac{\beta^\star}{60}\varepsilon^p} V^\star}_{\mathcal{T}_\beta} \underbrace{+12\frac{\delta^\star \varepsilon^{q+2}}{1+\delta^\star \varepsilon^q}\partial_{\hat{x}}^2 V^\star}_{\mathcal{T}_\delta} \underbrace{-\varepsilon^4 \partial_{\hat{x}}^4 V^\star}_{\mathcal{T}_b} = 0 \tag{28}$$

The terms $\mathcal{T}_\beta, \mathcal{T}_\delta, \mathcal{T}_b$ represent respectively the effect of the elastic restoring of the compression springs, the rotation springs, and the bending of the beam elements. The term $\mathcal{T}_\omega$ represents the force of inertia of the system represented adimensionally by $(\omega/\omega_b)^2 = \gamma^4 \ll 1$. The dynamic balance (28) requires the $\mathcal{T}_\omega - \mathcal{T}_\beta$ term linked to $V^\star$ to be balanced by the elastic terms $\mathcal{T}_\delta$ and/or $\mathcal{T}_b$. This makes it possible to determine, according to the order of magnitude $(\mathcal{K}, \mathcal{C})$, the associated model, which retains only the dominant terms of the same order and the order of magnitude of the angular frequency.

3.4.1. Family of Models

The generic model (26) degenerates into less rich models depending on the order of magnitude of the different terms. Three main types of degenerate models (29)–(31) occur in the case of strong compression springs ($\beta = O(\varepsilon^2)$). These expressions are general and, depending on the contrast, the $(\mathcal{C}_{eq}, \mathcal{K}_{eq})$ parameters themselves may degenerate and the cut-off angular frequency $\omega_{c,eq}$ may disappear.

Multi-supported beam with stiff rotational support (MSB-Stiff) ($q \leq 1$):
In that case $\mathcal{T}_\beta = O(\mathcal{T}_\delta) = O(\mathcal{T}_\omega) \gg \mathcal{T}_b$, thus the action of the rotational springs and internal bending, included in $\mathcal{C}_{eq}$, balances the inertia and elastic restoring forces of the compression springs.

$$m\left(\omega^2 - \omega_{c,eq}^2\right)v(x) + \mathcal{C}_{eq}\partial_x^2 v(x) = \mathcal{O}(\varepsilon) \tag{29}$$

Multi-supported beam with medium rotational support (MSB-Medium) ($q = 2$):
In that case, all the terms are of the same order of magnitude. The action of the rotational springs and the bending of the beam elements are differentiated and balance the inertia and elastic restoring forces of the compression springs.

$$m\left(\omega^2 - \omega_{c,eq}^2\right)v(x) + \mathcal{C}\partial_x^2 v(x) - EI\ell\partial_x^4 v(x) = \mathcal{O}(\varepsilon) \tag{30}$$

Multi-supported beam with soft rotational support (MSB-Soft) ($q \geq 3$):
In that case, $\mathcal{T}_\omega - \mathcal{T}_\beta = O(\mathcal{T}_b) \gg O(\mathcal{T}_\delta)$ This means that the action of the rotational springs disappears at the dominant order, and only the global bending balances the inertia forces of the restoring elastic forces of the compression springs.

$$m\left(\omega^2 - \omega_{c,eq}^2\right)v(x) - EI\ell\partial_x^4 v(x) = \mathcal{O}(\varepsilon) \tag{31}$$

Table 1 shows all macroscopic descriptions depending on stiffness contrasts.

Table 1. Set of macroscopic descriptions of multi-supported beams as a function of the order of magnitude of the stiffness constraints (β, δ). The first column shows the three models (i) MSB-Stiff (in blue), (ii) MSB-Medium (in orange), and (iii) MSB-Soft (in green).

	$\beta^{(p\leq 1)}$	$\beta^{(2)}$	$\beta^{(3)}$	$\beta^{(4)}$	$\beta^{(p\geq 5)}$
$\delta^{(q\leq -1)}$	No scale separation	$\mathrm{m}(\omega^2-\omega_c^2)v(x)+K_f\ell^2\partial_x^2 v(x)=0$	$\mathrm{m}\omega^2 v(x)+K_f\ell^2\partial_x^2 v(x)=0$		
$\delta^{(0)}$		$\mathrm{m}(\omega^2-\omega_c^2)v(x)+C_{eq}\partial_x^2 v(x)=0$	$\mathrm{m}\omega^2 v(x)+C_{eq}\partial_x^2 v(x)=0$		
$\delta^{(1)}$		$\mathrm{m}(\omega^2-\omega_c^2)v(x)+C\partial_x^2 v(x)=0$		$\mathrm{m}\omega^2 v(x)+C\partial_x^2 v(x)=0$	
$\delta^{(2)}$		$\mathrm{m}\left(\omega^2-\omega_{c,eq}^2\right)v(x)+C\partial_x^2 v(x)+EI\ell\partial_x^4 v(x)=0$	$\mathrm{m}(\omega^2-\omega_c^2)v(x)+C\partial_x^2 v(x)+EI\ell\partial_x^4 v(x)=0$		$\mathrm{m}\omega^2 v(x)+C\partial_x^2 v(x)+EI\ell\partial_x^4 v(x)=0$
$\delta^{(q\geq 3)}$		$\mathrm{m}\left(\omega^2-\omega_{c,eq}^2\right)v(x)-EI\ell\partial_x^4 v(x)=0$	$\mathrm{m}(\omega^2-\omega_c^2)v(x)-EI\ell\partial_x^4 v(x)=0$		$\mathrm{m}\omega^2 v(x)-EI\ell\partial_x^4 v(x)=0$

Some of the results presented in [14] are recovered, in particular the role of compression springs, which reveal a cut-off frequency. Our study emphasises the essential role of rotation springs, which was not considered in [14].

3.4.2. Order of Magnitude of Frequencies

Denoting $\max(\mathcal{T}_\delta, \mathcal{T}_b) = O(\varepsilon^m)$, (with $m \geq 2$ by construction) there are three distinct situations for the order of magnitude of the frequency:

① If $\mathcal{T}_\beta \ll \max(\mathcal{T}_\delta, \mathcal{T}_b)$, then $\mathcal{T}_\omega = O(\max(\mathcal{T}_\delta, \mathcal{T}_b))$. In these cases, the macroscopic descriptions are always propagative, and the angular frequency expands as follows:

$$\omega^2 = O(\omega_b^2 \varepsilon^m) \tag{32}$$

② If $\mathcal{T}_\beta = O(\max(\mathcal{T}_\delta, \mathcal{T}_b))$, then $\mathcal{T}_\omega = O(\varepsilon^m)$ and the macroscopic behaviour equations are propagative only for angular frequencies (32) greater than ω_c.

③ If $\mathcal{T}_\beta \gg \max(\mathcal{T}_\delta, \mathcal{T}_b)$, then it is necessary that the difference is $\mathcal{T}_\omega - \mathcal{T}_\beta = O(\varepsilon^m)$. In other words, the expansion of the angular frequency is of the form:

$$\omega^2 - \omega_c^2 = O(\varepsilon^m \omega_b^2) \qquad \text{where} \qquad \left(\frac{\omega_c}{\omega_b}\right)^2 = O(\varepsilon^p) \tag{33}$$

Macroscopic descriptions are only propagative for angular frequencies greater than a cut-off frequency $\omega_{c,eq}$.

3.5. *Dispersion Relationships at Long Wavelengths*

Macroscopic models assume that $\varepsilon \to 0$. To determine the extent of the validity domain of these models for finite values of ε, the dispersion relations of the various degenerate homogenised models (cf. Section 3.5.1) and of the generic model are compared with those obtained by the exact Floquet-Bloch analysis (cf. Section 3.5.2), which is not restricted by the scale separation condition.

3.5.1. Dispersion Relationships for the Different Models

By inserting v of the form $v(x) = V^\star e^{ik_\omega x}$ in each of the three models (29)–(31) and (26), the following dispersion relationships are obtained:

MSB-Stiff $(q \leq 1)$:

$$\left(\frac{\omega}{\omega_b}\right)^2 = \left(\frac{\omega_{c,eq}}{\omega_b}\right)^2 + \frac{12\delta}{1+\delta}(k_\omega \ell)^2 \tag{34}$$

MSB-Medium $(q = 2)$:

$$\left(\frac{\omega}{\omega_b}\right)^2 = \left(\frac{\omega_{c,eq}}{\omega_b}\right)^2 + 12\delta(k_\omega \ell)^2 + (k_\omega \ell)^4 \tag{35}$$

MSB-Soft ($q \geq 3$):

$$\left(\frac{\omega}{\omega_b}\right)^2 = \left(\frac{\omega_{c,eq}}{\omega_b}\right)^2 + (k_\omega \ell)^4 \tag{36}$$

MSB-Generic ($\forall q$):

$$\left(\frac{\omega}{\omega_b}\right)^2 = \left(\frac{\omega_{c,eq}}{\omega_b}\right)^2 + \frac{12\delta}{1+\delta}(k_\omega \ell)^2 + (k_\omega \ell)^4 \tag{37}$$

The wavenumber k_ω is real only above the cut-off angular frequency $\omega_{c,eq}$; otherwise, it is purely imaginary. These relationships are presented in Figure 3 for three multi-supported systems with β and δ stiffness contrasts corresponding to each of the models, which correspond to a typical value of $\varepsilon = 0.1$:

Case $(\beta = 10^{-2}, \delta = 1)$ **(in blue):** model MSB-Stiff (34);
Case $(\beta = 10^{-2}, \delta = 10^{-2})$ **(in orange):** model MSB-Medium (35);
Case $(\beta = 10^{-5}, \delta = 10^{-3})$ **(in green):** model MSB-Stiff (36) in which $\omega_{c,eq} = 0$.

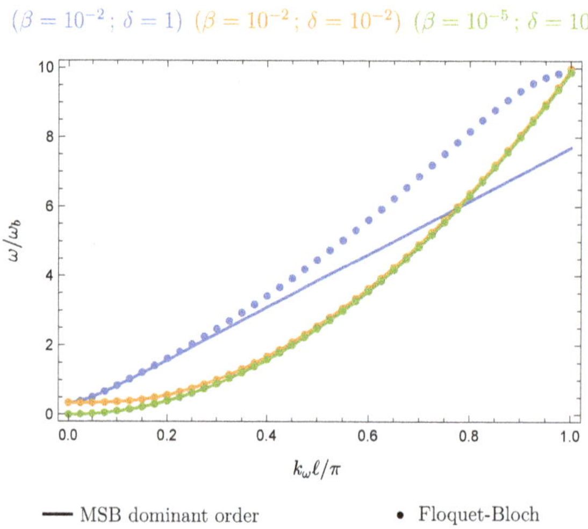

Figure 3. Dispersion relationships ω/ω_b versus the adimensional wavenumber $k_\omega \ell/\pi$, $k_\omega \in [0, \frac{\pi}{\ell}[$. For the three types of contrast, the dispersion relationships derived from the models are plotted as solid lines. The relationships calculated by Floquet-Bloch for the same contrasts are in dots.

Figure 3 also shows the dispersion curves obtained by the Floquet-Bloch analysis described below for the same contrast values.

3.5.2. Dispersion Relations by Floquet-Bloch

The exact dispersion relation for the system described by (12) is obtained by searching for $\underline{U}_n$ in the form:

$$\underline{U}_n = \underline{U}_0 e^{-ik_\omega n\ell} \qquad k_\omega \in [0, \frac{\pi}{\ell}[\tag{38}$$

By introducing (38) into (12), we get:

$$\underline{\underline{D}}.\underline{U}_0 = \left[2\cos(k\ell)\underline{\underline{D}}(\omega) + 2I\sin(k\ell)\underline{\underline{C}}(\omega) - 2\underline{\underline{F}}(\omega) - \underline{\underline{R}}\right].\underline{U}_0 = \underline{0}$$

The determinant of $\underline{\underline{\mathbb{D}}}$ must vanish so that $\underline{U}_0$ is not identically zero. The dispersion relation is then written as $\det(\underline{\underline{\mathbb{D}}}) = 0$, i.e.,:

$$\begin{aligned}\mathcal{E}(\omega, k_\omega) = 0 = &\frac{1}{3}(\beta + 2f_1(\gamma) - 2f_4(\gamma)\cos(k_\omega \ell))(3\delta + 2f_2(\gamma) + f_5(\gamma)\cos(k_\omega \ell)) \\ &- f_6(\gamma)^2 \sin^2(k_\omega \ell)\end{aligned} \tag{39}$$

This non-linear relationship is solved numerically for the contrasting stiffnesses (β, δ) indicated in Section 3.5.1. The results reported in Figure 3 show that the models MSB-Soft and MSB-Medium coincide with the exact calculations for wavenumbers $k_\omega \in [0, \frac{\pi}{\ell}[$. The MSB-Stiff model is tangent to the exact solution and is a good approximation of the latter up to $k_\omega \ell = \varepsilon \approx 0.3\pi$, i.e., wavelengths greater than three periods. These homogenised models at the dominant order are therefore valid well beyond the scale separation condition $\varepsilon \ll 1$.

This conclusion is supported by the use of the dispersion relation of the generic model. This is shown in Figure 4, where the dispersion from the generic model and that from the exact calculation are compared for the three types of contrast. For $k_\omega \in [0, \frac{\pi}{\ell}[$ the dispersion of the generic model coincides with that given by the exact calculations for the models MSB-Soft and MSB-Medium. Dispersion relationships are therefore not affected by model enrichment. Applying the generic model to the MSB-Stiff case makes it possible to follow not only the tangency but also the curvature of the exact solution, which leads to a good approximation up to $\varepsilon \approx 0.5\pi$.

In conclusion, the macroscopic models make it possible not only to recover (well beyond the limit of the scale separation hypothesis) the dispersion curves derived from the Floquet-Bloch approach but also to access the macroscopic forces (in a generalized sense), which are not accessible by Floquet-Bloch.

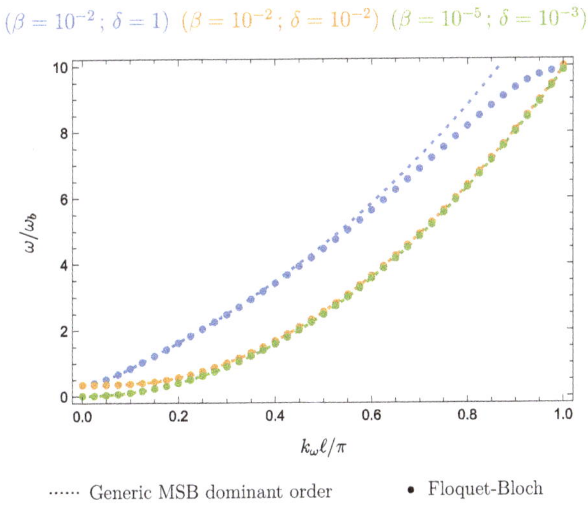

Figure 4. Dispersion relationships for $k_\omega \in [0, \frac{\pi}{\ell}[$ from the generic model (dashed line) for the three types of contrast, compared with the dispersions from the Floquet-Bloch calculation (in dots).

4. Boundary Conditions

The macroscopic dynamic models established in the previous section have shown their validity in infinite media within the limit of wavelengths greater than three periods.

The modal analysis of such finite-length systems requires the specification of the boundary conditions. This question involves two difficulties:

- Is it possible (and if so, how?) to use the macroscopic force and/or displacement variables to express the boundary conditions actually imposed on the real system that condition the microscopic force and/or displacement variables? This is the subject of Section 4.1;

- If the order of the differential operator of the macroscopic model is less than that of the 4th order operator of the bending beams (4), the number of macroscopic conditions is less than that required to express the real local conditions. A higher-order model is constructed in Section 4.4 in order to obtain a well-posed problem that is compatible with the real boundary conditions.

4.1. Energy Approach

Let us consider a multi-supported beam consisting of N periods Ω (see Figure 1) with a total length of $H = N\ell$. The energy balance of the generic beam is established by multiplying (26) by $\frac{1}{2}v(x)$ and integrating over $[0, H]$. After two integrations by parts, we obtain:

$$\begin{gathered}\frac{1}{2}\int_0^H \mathrm{m}\omega^2 v(x)^2\,dx = \frac{1}{2}\int_0^H \left(\ell\frac{\mathcal{M}(x)^2}{EI} + \frac{\left(\mathcal{K}_{\mathrm{eq}} v(x)\right)^2}{\mathcal{K}_{\mathrm{eq}}} + \frac{\left(\mathcal{C}\theta(x)\right)^2}{\mathcal{C}_{\mathrm{eq}}}\right) dx + \mathcal{W}_{\mathrm{BC}} + \mathcal{O}(\varepsilon) \\ \text{where} \quad \mathcal{W}_{\mathrm{BC}} = \frac{\ell}{2}\left[\mathcal{T}(x)v(x) + \mathcal{M}(x)\theta(x)\frac{\mathcal{C}}{\mathcal{C}_{\mathrm{eq}}}\right]_0^H \quad \text{and} \quad \theta(x) = \frac{\mathcal{C}_{\mathrm{eq}}}{\mathcal{C}}\partial_x v(x)\end{gathered} \tag{40}$$

Equation (40) gives, at the leading order, the kinetic energy, the deformation energy, and the work $\mathcal{W}_{\mathrm{BC}}$ of the boundary conditions expressed in macroscopic variables.

4.2. Effective Boundary Conditions

4.2.1. Displacement/Rotation Boundary Conditions

By construction in Section 2.3, the kinematic variables at the nodes are the macroscopic kinematic variables, see (13). Consequently, the kinematic boundary conditions imposed locally on the real system are directly related to the macroscopic kinematic variables. If at $x = 0$ (node 1) or $x = H$ (node $N+1$), the imposed kinematic (denoted $\underline{\mathrm{U}}^{\mathrm{imp}}$) conditions give the following macroscopic boundary conditions:

$$\underline{U}(0) = \underline{\mathrm{U}}_1^{\mathrm{imp}} = \begin{pmatrix} \mathrm{v}^{\mathrm{imp}} \\ \ell\theta^{\mathrm{imp}} \end{pmatrix}_{|x=0} \quad ; \quad \underline{U}(H) = \underline{\mathrm{U}}_{N+1}^{\mathrm{imp}} = \begin{pmatrix} \mathrm{v}^{\mathrm{imp}} \\ \ell\theta^{\mathrm{imp}} \end{pmatrix}_{|x=H} \tag{41}$$

4.2.2. Force/Moment Boundary Conditions

These forces, which are correctly described within the structure, are disturbed at the ends due to the absence of the adjacent node. In fact, at the extreme nodes, the equilibrium is no longer governed by (11) but by an expression that can be derived by (i) removing the contributions of elements outside the domain and adding the imposed forces; (ii) replacing the discrete variables at the nodes by continuous variables at these points according to (13). We therefore write at the extremities $x = 0$ and $x = H$:

$$\begin{gathered}-\underline{\underline{\mathrm{H}}}^{BB}.\underline{U}(0) + {}^{T}\underline{\underline{\mathrm{H}}}^{EB}.\underline{U}(\ell) = \underline{\mathrm{F}}^{B}\left(\begin{pmatrix} v(0) \\ \ell\theta(0) \end{pmatrix}, \begin{pmatrix} v(\ell) \\ \ell\theta(\ell) \end{pmatrix}\right) = \begin{pmatrix} T^{\mathrm{imp}} \\ M^{\mathrm{imp}}/\ell \end{pmatrix}_{|x=0} \\ \underline{\underline{\mathrm{H}}}^{EB}.\underline{U}(H-\ell) + \left(\underline{\underline{\mathrm{H}}}^{EE} - \underline{\underline{\mathrm{R}}}\right).\underline{U}(H) = \underline{\mathrm{F}}^{E}\left(\begin{pmatrix} v(H-\ell) \\ \ell\theta(H-\ell) \end{pmatrix}, \begin{pmatrix} v(H) \\ \ell\theta(H) \end{pmatrix}\right) - \underline{\underline{\mathrm{R}}}.\begin{pmatrix} v(H) \\ \ell\theta(H) \end{pmatrix} = \begin{pmatrix} T^{\mathrm{imp}} \\ M^{\mathrm{imp}}/\ell \end{pmatrix}_{|x=H}\end{gathered} \tag{42}$$

whose explicit formulation in terms of macroscopic kinematic variables is as follows at $x = 0$:

$$v(0)f_1(\gamma) - v(\ell)f_4(\gamma) + \frac{\ell}{2}[\theta(0)f_3(\gamma) + \theta(\ell)f_6(\gamma)] = \left(T^{\mathrm{imp}}\right)_1/K_{\mathrm{f}} \tag{43a}$$

$$\frac{1}{2}[v(0)f_3(\gamma) - v(\ell)f_6(\gamma)] + \frac{\ell}{3}\left[f_2(\gamma)\theta(0) + \frac{1}{2}\theta(\ell)f_5(\gamma)\right] = \left(M^{\mathrm{imp}}\right)_1/(\ell K_{\mathrm{f}}) \tag{43b}$$

and at $x = H$

$$-v(H)(f_1(\gamma)+\beta)+v(H-\ell)f_4(\gamma)+\frac{\ell}{2}[\theta(H)f_3(\gamma)+\theta(H-\ell)f_6(\gamma)]=\left(T^{\text{imp}}\right)_{N+1}/K_{\text{f}} \tag{44a}$$

$$\frac{1}{2}[v(H)f_3(\gamma)-v(H-\ell)f_6(\gamma)]-\frac{\ell}{3}\left[\theta(H)(f_2(\gamma)+\delta)+\frac{1}{2}\theta(H-\ell)f_5(\gamma)\right]=\left(M^{\text{imp}}\right)_{N+1}/(K_{\text{f}}\ell) \tag{44b}$$

Note that this type of condition is similar to the Robin conditions coupled in v and θ. In particular, note that the nullity of the macroscopic forces $\mathcal{T}(H)$ and $\mathcal{M}(x)$ or of $\mathcal{W}_{\text{BC}}$ does not reflect the free end condition. To put it in a nutshell:

1. A real condition on a kinematic variable is directly transposed into the homogenised model for the corresponding macroscopic kinematic variable.
2. A real condition on force or moment does not apply to the corresponding macroscopic effort of the homogenized model. In this case, it is necessary to express the force/moment equilibrium at the node using the macroscopic kinematic variables of this node and the adjacent internal node.

In all cases, this approach gives four boundary conditions expressed in macroscopic variables, which, for models of differential order four, make it possible to respect the conditions actually applied to a multi-supported beam of finite length. However, for models of differential order two, these four conditions are overabundant. This question is addressed in the next section.

4.3. *Modal Analysis of a Differential Model of the Second Order*

In this section, we focus on structures with stiff rotational springs that behave at the leading order as the model MSB-Stiff described by (29) (see Section 3.2.1). This situation needs to address the discrepancy between the number of boundary conditions required (two conditions at each edge beam) and the order of the differential operator in the macroscopic description (one condition at each extremity). In this view, three strategies are studied:

- Use the macroscopic model (29) whose differential operator is of order two and where only two boundary conditions can be expressed. The driving variable is the displacement, while the rotation is merely a hidden variable derived from it. This forces us to consider the boundary condition in displacement v^{imp} and forces T^{imp} (and disregard rotations θ^{imp} and moments M^{imp}) in (41) and (42).
- Use the extension of the model by the generic model (26) with a differential operator of the fourth order. In this case, four conditions (41) and (42) can be effectively applied.
- Use the model (55) developed in Section 4.4.2, which is enriched at higher orders and results in a fourth-order differential operator that allows the application of four boundary conditions (41) and (42).

These options have been explored in the context of the modal analysis of a multi-supported beam of kind MSB-Stiff ($\beta = 10^{-2}; \delta = 1$) with fixed-fixed and fixed-free boundary conditions.

4.3.1. "Reduced" Macroscopic Modal Analysis

As an example, let us consider a fixed-fixed MSB-Stiff beam consisting of $N = 10$ periods. The "reduced" macroscopic modal analysis is based on (i) the MSB-Stiff model and (ii) the two zero displacement conditions imposed at the edges (and therefore ignores the zero rotation conditions imposed). This situation leads directly to the macroscopic modal deformation of the jth mode:

$$v(x)=A\sin\big((k_\omega)_j x\big)\,;\,\theta(x)=\frac{C_{\text{eq}}}{C}\partial_x v(x)=A(k_\omega)_j\cos\big((k_\omega)_j x\big),\ \text{with}\ (k_\omega)_j=\frac{j\pi}{H},\ j\in\mathbb{N}^\star \tag{45}$$

where the angular frequency ω_j, which is associated with $(k_\omega)_j$ is defined by the dispersion relation (34). Figure 5 shows the modal deformations of the first three modes thus obtained. For comparison, the numerical results from a finite element calculation (taking into account the zero rotation conditions) of the real fixed-fixed structure are also shown (after mesh study, eight bending beam elements are considered per beam).

In the model MSB-Stiff, the stiffness of the rotation springs is of the same order of magnitude as the bending stiffness of each beam element. The modal deformation of the system can be decomposed into (i) the amplitude of the displacements at the nodes and (ii) the deformation of the beams between the nodes. It can be seen that in the case of MSB-Stiff beams, the displacements (and rotations) of the nodes are fixed by the total length of the system, regardless of the stiffness of the supports and beams. On the other hand, the rotational supports play a significant role in the inter-node deformation of the beams. This effect is clearly visible both for rotational deformations (angular points at the nodes of the system caused by the discontinuity of moments between two elements) and for displacement deformations (curvature of each beam element linked to the rotation differential between two nodes). The compression springs do not influence the modal deformation, but they do influence the modal frequency; see (34).

There is good general agreement between the macroscopic model and the numerical model. The omission of the zero rotation condition in the model MSB-Stiff is clearly visible near the edges, mainly in the first period on each side. It should be noted that the differences between the homogenised model and the calculation increase from the fundamental mode to the third mode. In the latter case, however, they remain relatively limited due to the small-scale separation.

Remark 1. *The choice of zero rotation would have led to* $\theta(x) = A\sin((k_\omega)_j x)$ *and therefore a deformation* $v(x) = B_1\cos((k_\omega)_j x) + B_2$. *This expression is not admissible because it does not respect the fixed-fixed conditions on the driving variable. This is confirmed by the modal deformations obtained from the numerical simulations (see Figure 5), which show that* v *is indeed of the form* (45).

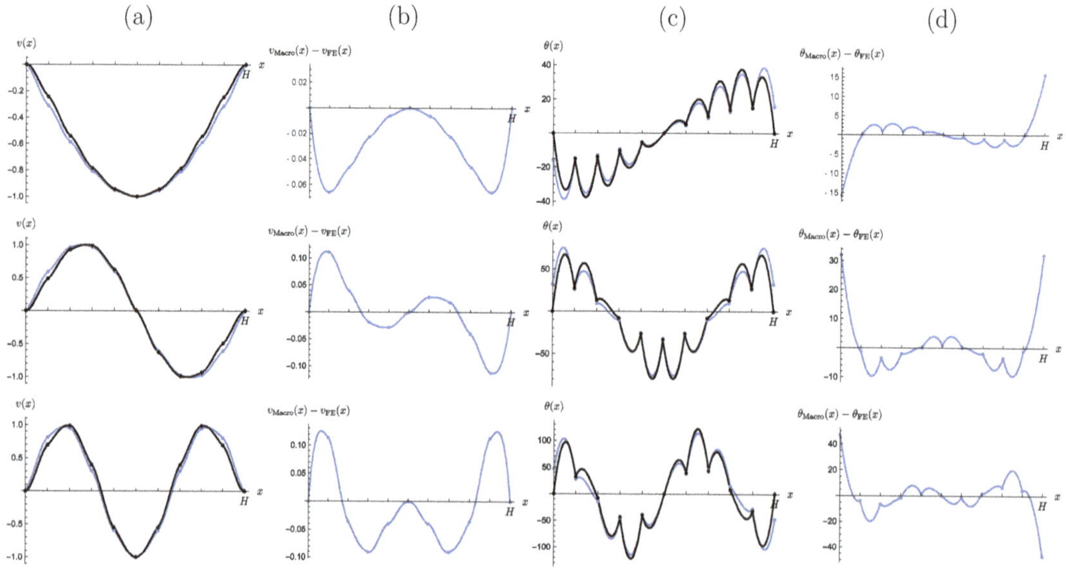

Figure 5. First three modal deformations of a 10−period fixed−fixed structure ($\beta = 10^{-2}; \delta = 1$) of the MSB-Stiff model (in blue) compared with the finite element model (in black). From left to right: (**a**) modal deformation in displacement $v(x)$, (**b**) difference between homogenised model and finite element model $v_{\text{Macro}}(x) - v_{\text{FE}}$, (**c**) modal deformation in rotation $\theta(x)$, (**d**) difference between homogenised model, and finite element model $\theta_{\text{Macro}}(x) - \theta_{\text{FE}}(x)$. The dots represent the nodes between the beams where the springs are located.

4.3.2. "Full" Macroscopic Modal Analysis

The full macroscopic modal analysis was carried out on a structure consisting of $N = 30$ periods. The two models used are that of the generic beam and that of the beam MSB-Stiff enriched with the higher orders; see Section 4.4. In each case, the four boundary conditions are those indicated in Section 4.2.

Figure 6 shows the fundamental modes determined by these two models with boundary conditions either fixed-fixed (Figure 6-top) or fixed-free (Figure 6-bottom). For comparison, the results obtained using finite elements and those obtained using "reduced" modal analysis are also shown.

Compared to the 10-period fixed-fixed structure, we can see that tripling the number of periods, as assumed, results in better scale separation and a gain in accuracy of the reduced modal analysis approach. Moreover, the concentration of the error for rotations on the periods at the edges is even more obvious.

For both types of boundary conditions, it can be seen that "full" modal analyses performed with four boundary conditions significantly improve the results. Furthermore, it appears that the higher-order model leads to a better approximation than the generic model. This is explained by the fact that the additional term in the higher-order model is inherited from effects specific to the contrast under study. This improves the description as it provides a model of an accuracy up to $O(\varepsilon^3)$. In contrast, the generic model is built up in such a way that it gives an accuracy up to $O(\varepsilon)$ regardless of the contrasts. It is always valid at the dominant order in the sense that the terms naturally vanish when they are not relevant. Nevertheless, even if the differential operator of the generic model is of fourth order, one cannot expect that it will reach the same level of accuracy as the higher-order model.

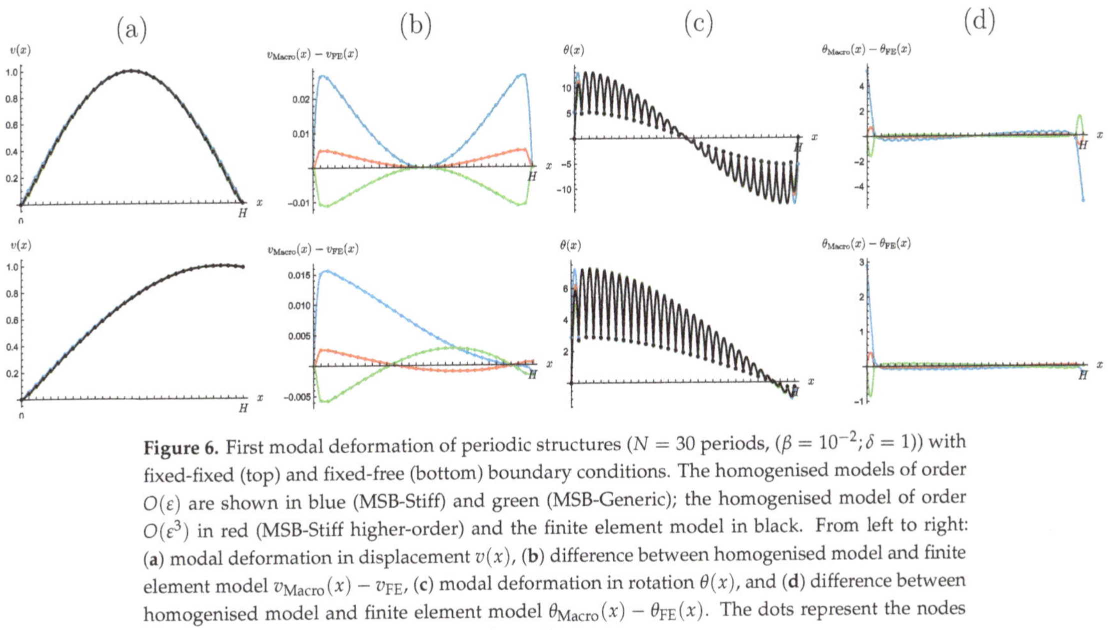

Figure 6. First modal deformation of periodic structures ($N = 30$ periods, ($\beta = 10^{-2}; \delta = 1$)) with fixed-fixed (top) and fixed-free (bottom) boundary conditions. The homogenised models of order $O(\varepsilon)$ are shown in blue (MSB-Stiff) and green (MSB-Generic); the homogenised model of order $O(\varepsilon^3)$ in red (MSB-Stiff higher-order) and the finite element model in black. From left to right: (**a**) modal deformation in displacement $v(x)$, (**b**) difference between homogenised model and finite element model $v_{\mathrm{Macro}}(x) - v_{\mathrm{FE}}$, (**c**) modal deformation in rotation $\theta(x)$, and (**d**) difference between homogenised model and finite element model $\theta_{\mathrm{Macro}}(x) - \theta_{\mathrm{FE}}(x)$. The dots represent the nodes between the beams where the springs are located.

4.3.3. Synthesis

In order to quantify the differences between the modes calculated according to a homogenised model and those calculated according to a finite element analysis, two indicators have been defined: (i) following (46), the modal deformations in displacement or in rotation are compared only at the nodes, such as the more $\widetilde{\mathrm{MAC}}_{X=v \text{ or } \theta}$ is weak and the more the compared modal shapes are close; (ii) the modal angular frequencies are compared according to (47), where γ represents the local bending wavenumber; see (5).

$$\widetilde{\text{MAC}}_X = 1 - \text{MAC}_X = 1 - \left(\frac{\underline{X}^{\text{homo}}.\underline{X}^{\text{FE}}}{\|\underline{X}^{\text{homo}}\| \; \|\underline{X}^{\text{FE}}\|} \right)^2 \qquad \underline{X} \in \{\underline{v}, \underline{\theta}\} \tag{46}$$

$$\%\gamma = 100 \frac{|\gamma^{\text{homo}} - \gamma^{\text{FE}}|}{\gamma^{\text{FE}}} \quad ; \quad \gamma^4 = \left(\frac{\omega}{\omega_b}\right)^2 \tag{47}$$

Figure 7 shows the three indicators $\left(\widetilde{\text{MAC}}_v, \widetilde{\text{MAC}}_\theta, \%\gamma\right)$ for the three first modes of the structure constituted by $N = 30$ periods. Systematically, it can be noted that (i) the "full" boundary conditions lead to more accurate descriptions, the best being obtained for the higher-order model, and (ii) the higher modes are described with less precision due to the smaller scale separation. It should be noted, however, that the dynamics of the system are relatively undisturbed by the "reduced" boundary conditions, as indicated by the good estimation of the eigenfrequencies ($\%\gamma \leq 0.8\%$ for the third mode); see Figure 7-right.

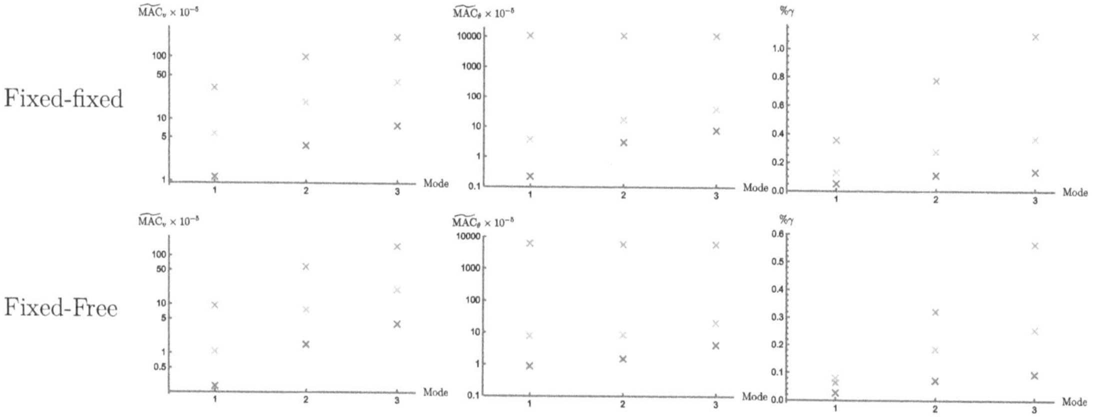

Figure 7. Comparisons of the modes of the structures with $N = 30$ periods according to the indicators (from left to right) $\left(\widetilde{\text{MAC}}_v, \widetilde{\text{MAC}}_\theta, \%\gamma\right)$ for two types of boundary conditions: fixed-fixed (**top**), fixed-free (**bottom**). The homogenised models of order $O(\varepsilon)$ are shown in blue (MSB-Stiff) and green (MSB-Generic); the homogenised model of order $O(\varepsilon^3)$ in red (MSB-Stiff higher-order).

4.4. Higher-Order Macroscopic Model

The aim is to enrich the MSB-Stiff model with higher order terms so that the order of the macroscopic differential operator corresponds to the order of the differential operator of the local problem (4) and allows all local conditions to be translated. In the enriched representation of discrete media, e.g., [39,41], and in higher-order continuous media, e.g., [25,26,31,42], it is necessary to construct generalised boundary conditions that are consistent with the order of the differential operator. On the other hand, for MSB-Stiff beams, the macroscopic model must be enriched to take into account all the real boundary conditions.

Returning to the asymptotic expansion of Section 3.2.1, solving it to second-order leads to the macroscopic kinematic relation (18b). Moreover, the transverse force equilibrium at order 2 gave the macroscopic balance at the dominant order. This equilibrium is rewritten in the form (48), which is systematically used in higher-order expressions.

$$\frac{\rho L^2}{E e_0^2} (\omega^{(1)})^2 v^{(0)} = \beta^{(2)} v^{(0)} - L^2 \frac{\delta^{(0)}}{\delta^{(0)} + 1} \partial_x^2 v^{(0)} \tag{48}$$

4.4.1. First-Order Macroscopic Model

The second-order moment (A4) gives a kinematic relation at the first order identical to that of the dominant order:

$$\theta^{(1)} = \frac{1}{\delta^{(0)}+1}\partial_x v^{(1)} \tag{49}$$

The third-order shear equilibrium (A5), in which the relation (49) has been introduced, provides the first-order macroscopic equation:

$$-L^2\frac{\delta^{(0)}}{\delta^{(0)}+1}\partial_x^2 v^{(1)} + \left[\beta^{(2)} - \frac{\rho L^2}{Ee_0^2}(\omega^{(1)})^2\right]v^{(1)} - \frac{\rho L^2}{Ee_0^2}2\omega^{(1)}\omega^{(2)}v^{(0)} = 0 \tag{50}$$

Denoting $\Theta^{(1)} = \theta^{(0)} + \varepsilon\theta^{(1)}$ and $V^{(1)} = v^{(0)} + \varepsilon v^{(1)}$ the first-order approximations of θ and v, respectively, the kinematic relation is written as (17) + ε(49):

$$\Theta^{(1)} = \frac{1}{\delta+1}\partial_x V^{(1)}$$

The asymptotic expansion of the angular frequency is $\omega = \varepsilon\omega^{(1)} + \varepsilon^2\omega^{(2)} + o(\varepsilon^2)$. This involves that $\omega^2 = \varepsilon^2(\omega^{(1)})^2 + 2\varepsilon^3\omega^{(1)}\omega^{(2)} + o(\varepsilon^3)$. Consequently, the first-order macroscopic equation is written as:

$$\mathrm{m}\left(\omega^2 - \omega_c^2\right)\mathrm{V}^{(1)} + C_{\mathrm{eq}}\partial_x^2\mathrm{V}^{(1)} = O(\varepsilon^2) \tag{51}$$

The Equation (51) is identical to (18a); only the precision of the expansion differs. Since the derivation order (two) of the operator is unchanged, it is not possible to apply all the boundary conditions. It is therefore necessary to expand asymptotically to the next order.

4.4.2. Second-Order Macroscopic Model

The second-order moment (A5) gives the second-order kinematic relation, rewritten using (48):

$$\theta^{(2)} = \frac{1}{\delta^{(0)}+1}\left(\partial_x v^{(2)} + \frac{\beta^{(2)}}{210}\left(13 + \frac{1}{1+\delta^{(0)}}\right)\partial_x v^{(0)} + \frac{L^2}{210}\frac{\delta^{(0)}}{\delta^{(0)}+1}\left(22 - \frac{\delta^{(0)}}{\delta^{(0)}+1}\right)\partial_x^3 v^{(0)}\right) \tag{52}$$

The second-order approximations of θ and v are denoted by

$$\Theta^{(2)} = \theta^{(0)} + \varepsilon\theta^{(1)} + \varepsilon^2\theta^{(2)} \quad ; \quad \mathrm{V}^{(2)} = v^{(0)} + \varepsilon v^{(1)} + \varepsilon^2 v^{(2)}$$

By summing (17) + ε(49) + ε^2(52), the kinematic relation enriched to the second order is:

$$\Theta^{(2)} = \frac{C_{\mathrm{eq}}}{C}\left(\partial_x \mathrm{V}^{(2)}\left[1 + \frac{\beta}{210}\left(13 + \frac{1}{1+\delta}\right)\right] + \frac{\ell^2}{210}\frac{\delta}{\delta+1}\left(22 - \frac{\delta}{\delta+1}\right)\partial_x^3\mathrm{V}^{(2)}\right) + O(\varepsilon^3) \tag{53}$$

The fourth-order shear equilibrium (A6) provides the following second-order equilibrium equation, in which (48) has been used to swap the time and space derivatives:

$$\begin{aligned}&\beta^{(2)}v^{(2)} - \frac{1}{60}(\beta^{(2)})^2 v^{(0)} - \frac{\rho L^2}{Ee_0^2}\left((\omega^{(1)})^2 v^{(2)} + 2\omega^{(1)}\omega^{(2)}v^{(1)} + \left((\omega^{(2)})^2 + 2\omega^{(1)}\omega^{(3)}\right)v^{(0)}\right)\\ &- L^2\frac{\delta^{(0)}}{\delta^{(0)}+1}\left(\partial_x^2 v^{(2)} + \frac{\beta^{(2)}}{210}\left(20 + \frac{1}{\delta^{(0)}+1}\right)\partial_x^2 v^{(0)}\right) + \frac{L^4}{12}\left(1 - \frac{2}{35}\left(\frac{\delta^{(0)}}{\delta^{(0)}+1}\right)^2\left(\frac{23}{2} - \frac{\delta^{(0)}}{\delta^{(0)}+1}\right)\right)\partial_x^4 v^{(0)} = 0\end{aligned} \tag{54}$$

The sum of the three Equations (18a) + ε(50) + ε^2(54) provides the second-order macroscopic behaviour equation in ε called MSB-Stiff higher-order:

$$\mathrm{m}\left(\omega^2 - \omega_{\mathrm{c,eq}}^2\right)\mathrm{V}^{(2)}(x) + \mathrm{C}^{(2)}\partial_x^2\mathrm{V}^{(2)}(x) - \mathrm{E}^{(2)}I\ell\partial_x^4\mathrm{V}^{(2)}(x) = O(\varepsilon^3) \tag{55}$$

with $(\mathrm{C}^{(2)}, \mathrm{E}^{(2)})$ the elastic coefficients $(\mathcal{C}, E)$ enriched by taking into account the stiffnesses (β, δ).

$$\mathrm{C}^{(2)} = \mathcal{C}_{\mathrm{eq}}\left[1 + \frac{\beta}{210}\left(20 + \frac{1}{1+\delta}\right)\right] \quad ; \quad \mathrm{E}^{(2)} = E\left[1 - \frac{1}{35}\left(\frac{\delta}{1+\delta}\right)^2\left(23 - 2\frac{\delta}{1+\delta}\right)\right] \tag{56}$$

The model (55) is enriched by two orders compared to MSB-Stiff (18a). This can be interpreted mechanically in terms of transverse forces and moments. This leads to the definition of beam model forces, generalised to the second order $\mathcal{T}^{(2)}(x)$, $\mathcal{M}^{(2)}(x)$, which satisfy the following system of equations:

$$\partial_x\mathcal{T}^{(2)}(x) = \rho A\omega^2\mathrm{V}^{(2)}(x) - \frac{\mathcal{K}_{\mathrm{eq}}}{\ell}\mathrm{V}^{(2)}(x) \tag{57a}$$

$$\partial_x\mathcal{M}^{(2)}(x) + \mathcal{T}^{(2)}(x) = -\frac{\mathrm{C}^{(2)}}{\ell}\partial_x\mathrm{V}^{(2)}(x) \tag{57b}$$

$$\mathcal{M}^{(2)}(x) = -\mathrm{E}^{(2)}I\partial_x\mathrm{V}^{(2)}(x) \tag{57c}$$

$$\frac{\mathrm{C}^{(2)}}{\ell}\partial_x\mathrm{V}^{(2)}(x) = \frac{\mathcal{C}_{\mathrm{eq}}}{\ell}\frac{1 + \frac{\beta}{210}\left(20 + \frac{1}{1+\delta}\right)}{1 + \frac{\beta}{210}\left(13 + \frac{1}{1+\delta}\right)}\left(\frac{\mathcal{C}}{\mathcal{C}_{\mathrm{eq}}}\Theta^{(2)}(x) - \frac{\ell^2}{210}\frac{\delta}{\delta+1}\left(22 - \frac{\delta}{\delta+1}\right)\partial_x^3\mathrm{V}^{(2)}\right) \tag{57d}$$

Indeed, it takes into account a bending moment $\mathcal{M}^{(2)}(x)$, which raises the order of the differential operator to the fourth order. Note that the effective elastic modulus $\mathrm{E}^{(2)}$ is modified by the presence of the rotation springs. The force balance (57a) involves a distributed force source whose elastic parameter $\mathcal{K}_{\mathrm{eq}}$ takes into account both the compression spring support and the bending stiffness of the beam. Finally, the momentum balance (57b) is subjected to a source of distributed momentum associated with the gradient of the deflection. The elastic parameter $\mathrm{C}^{(2)}$ involves both the compression and rotation spring support as well as the bending stiffness of the beam. The enriched kinematic relation shows that the deflection depends on the rotation $\Theta^{(2)}$ and on the variation of curvature $\partial_x^3\mathrm{V}^{(2)}$.

4.4.3. Dispersion Curve Based on the Model MSB-Stiff Higher-Order

The dispersion equation of the model MSB-Stiff higher-order is obtained by inserting $\mathrm{V}^{(2)}$ in the form $\mathrm{V}^{(2)} = V^{\star}e^{ik_\omega x}$ into (55).

$$\left(\frac{\omega}{\omega_{\mathrm{b}}}\right)^2 = \left(\frac{\omega_{\mathrm{c,eq}}}{\omega_{\mathrm{b}}}\right)^2 + \frac{12\delta}{1+\delta}\frac{\mathrm{C}^{(2)}}{\mathcal{C}_{\mathrm{eq}}}(k_\omega\ell)^2 + \frac{\mathrm{E}^{(2)}}{E}(k_\omega\ell)^4 \tag{58}$$

Figure 8 compares the dispersion relationships corresponding to the following models: (i) MSB-Stiff (34), (ii) MSB-Generic (37), (iii) MSB-Stiff higher-order (58), and (iv) Floquet-Bloch (39).

The MSB-Stiff higher-order model, like MSB-Generic, takes into account the bending moment and therefore the $(k_\omega\ell)^4$ term, which makes it possible to follow the variation in the dispersion branch more precisely than MSB-Stiff. For MSB-Stiff higher-order, the improvement over MSB-Generic lies in the introduction of correctors $(\mathrm{C}^{(2)}, \mathrm{E}^{(2)})$, which are specific to the magnitudes of the contrasts $(\beta^{(2)}, \delta^{(0)})$. Improving the dispersion by integrating higher orders in the homogenised description is a well-known method; see, for example, [31] for composite elastic media, [43] for mass/spring networks, and [39] for 1D stratified media.

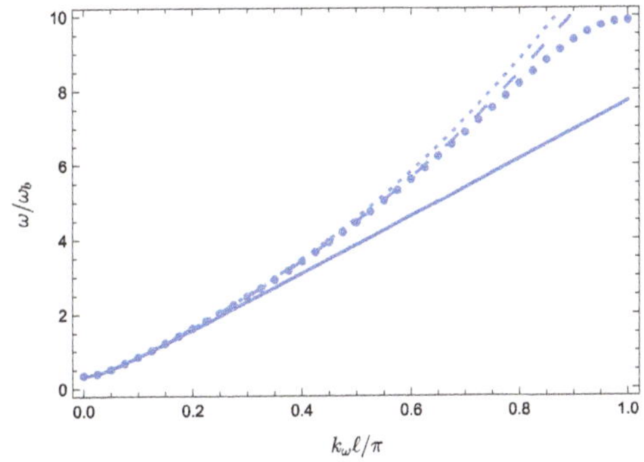

—— MSB dominant order ······ Generic MSB dominant order

– – MSB higher order • Floquet-Bloch

Figure 8. Dispersion relationships (large wavelength) for the system of contrasts ($p = 2, q = 0$) with $\varepsilon = 0.1$ according to four models: MSB-Stiff (solid line), MSB-Generic (dotted line), MSB-Stiff higher-order (dashed line), and Floquet-Bloch (dots).

5. Conclusions

The discrete medium homogenization method (HPDM) applied to multi-supported beams yields three distinct models (29)–(31), which correspond to stiff, medium, and soft rotation supports, respectively. Independently of the rotation springs, the stiffness of the compression springs gives rise to the emergence of a cut-off angular frequency below which the macroscopic dynamics is not propagative. The different models are synthesised into a single generic model that fits the Floquet-Bloch dispersion curve up to wavelengths of the order of three periods. Furthermore, the homogenization also provides an interpretation in terms of macroscopic forces/moments.

A mixed "discrete/continuous" approach to the boundary conditions is proposed, which allows the boundary conditions (natural or essential) actually applied at the local scale to be expressed in terms of Robin-type boundary conditions on macroscopic variables. The modal analysis performed with these boundary conditions and the generic model gives results in good agreement with a full finite element calculation, with great economy of numerical resources. Note that this condition can be used to define interface conditions that allow different models to be connected.

Nevertheless, for sufficiently rigid supports whose homogenised model at the dominant order is of differential order two, it is shown that the results provided by the generic model and the "discrete/continuous" type boundary conditions can be further improved by the dedicated homogenised model of higher order.

This work will be expanded in several directions. The findings of this study can be extended to Timoshenko beams. The developments will be of a similar nature but will require the introduction of another dimensionless parameter concerning the relative effects of shear versus bending in the beams. This additional parameter will influence the macroscopic model without modifying the general framework of the models or the treatment of the boundary conditions.

Furthermore, it is straightforward to integrate a linear dissipative contribution to the purely conservative supports in order to address, for instance, visco-elastic polymer fixation systems or to deal with support foundations on soils expressed by impedance conditions. In order to consider such supports, it is sufficient to consider springs with complex frequency-dependent values. In this case, the dissipation introduced is transferred to the macroscopic model, whose apparent dissipation and its frequency dependence will depend on the beam/support system.

It is also worth noting the possibility of dealing with high frequencies, which results in the modulation of local beam modes following the approach implemented in [44,45]. The transposition of the mixed "discrete/continuous" boundary conditions to these high-frequency dynamic regimes makes it possible to envisage a better consideration of real boundary conditions.

The same method could also be used to analyse the instability of multi-supported beams by considering the static equilibrium of bended beams in their deformed state under the action of the normal force instead of the dynamic equilibrium expressed on the undeformed state as assumed in this study.

Finally, these diverse perspectives also extend to the analysis of the mechanical behaviour of multi-supported plates or two-dimensional networks of multi-supported beams at their nodes.

Author Contributions: Conceptualization, A.R. and C.B.; methodology, A.R. and C.B.; formal analysis, A.R. and C.B.; writing—original draft preparation, A.R. and C.B.; writing—review and editing, A.R. and C.B. All authors have read and agreed to the published version of the manuscript.

Funding: This research received no external funding.

Data Availability Statement: Data are contained within the article.

Conflicts of Interest: The authors declare no conflicts of interest.

Abbreviations

The following abbreviations are used in this manuscript:

FE	Finite Element
HPDM	Homogenization of Periodic and Discrete Media
MSB	Multi-Supported Beam

Appendix A. Expansion of Equations of Equilibrium by Order of ε

In this appendix are presented the full expansions of the transverse and momentum equilibria (16) to successive first orders of ε:

Order -1:

$$\begin{cases} \mathbb{T}^{(-1)} = 0 \\ \mathbb{M}^{(-1)} = \delta^{(-1)} L\theta^{(-1)} = 0 \end{cases} \tag{A1}$$

Order 0:

$$\begin{cases} \mathbb{T}^{(0)} = 0 \\ \mathbb{M}^{(0)} = \delta^{(-1)} L\theta^{(0)} + \left(\delta^{(0)} + 1\right) L\theta^{(-1)} = 0 \end{cases} \tag{A2}$$

Order 1:

$$\begin{cases} \mathbb{T}^{(1)} = L^2 \partial_x \theta^{(-1)} = 0 \\ \mathbb{M}^{(1)} = \delta^{(-1)} L\theta^{(1)} + \left(\delta^{(0)} + 1\right) L\theta^{(0)} + \delta^{(1)} L\theta^{(-1)} - L\partial_x v^{(0)} = 0 \end{cases} \tag{A3}$$

Order 2:

$$\begin{cases} \mathbb{T}^{(2)} = L^2 \partial_x \theta^{(0)} + \left[\beta^{(2)} - \dfrac{\rho L^2}{E e_0^2} (\omega^{(1)})^2\right] v^{(0)} - L^2 \partial_x^2 v^{(0)} = 0 \\ \mathbb{M}^{(2)} = \delta^{(-1)} L\theta^{(2)} + \left(\delta^{(0)} + 1\right) L\theta^{(1)} + \delta^{(1)} L\theta^{(0)} + \left[\delta^{(2)} - \dfrac{\rho L^2}{210 E e_0^2} (\omega^{(1)})^2\right] L\theta^{(-1)} - L\partial_x v^{(1)} + \dfrac{L^3}{6} \partial_x^2 \theta^{(-1)} \end{cases} \tag{A4}$$

Order 3:

$$\begin{cases} \mathbb{T}^{(3)} = L^2\partial_x\theta^{(1)} + \left[\beta^{(2)} - \frac{\rho L^2}{Ee_0^2}(\omega^{(1)})^2\right]v^{(1)} - L^2\partial_x^2 v^{(1)} \\ \quad + \left[\beta^{(3)} - \frac{\rho L^2}{Ee_0^2}2\omega^{(1)}\omega^{(2)}\right]v^{(0)} + \frac{13\rho L^2}{210Ee_0^2}(\omega^{(1)})^2 L^2\partial_x\theta^{(-1)} + \frac{L^4}{6}\partial_x^3\theta^{(-1)} = 0 \\ \mathbb{M}^{(3)} = \delta^{(-1)}L\theta^{(3)} + \left(\delta^{(0)}+1\right)L\theta^{(2)} + \delta^{(1)}L\theta^{(1)} + \left[\delta^{(2)} - \frac{\rho L^2}{210Ee_0^2}(\omega^{(1)})^2\right]L\theta^{(0)} - L\partial_x v^{(2)} + \frac{L^3}{6}\partial_x^2\theta^{(0)} \\ \quad + \left[\delta^{(3)} - \frac{\rho L^2}{210Ee_0^2}2\omega^{(1)}\omega^{(2)}\right]L\theta^{(-1)} - \frac{13\rho L^2}{210Ee_0^2}(\omega^{(1)})^2 L\partial_x v^{(0)} - \frac{L^3}{6}\partial_x^3 v^{(0)} \end{cases} \tag{A5}$$

Order 4:

$$\begin{cases} \mathbb{T}^{(4)} = L^2\partial_x\theta^{(2)} + \left[\beta^{(2)} - \frac{\rho L^2}{Ee_0^2}(\omega^{(1)})^2\right]v^{(2)} - L^2\partial_x^2 v^{(2)} + \left[\beta^{(3)} - \frac{\rho L^2}{Ee_0^2}2\omega^{(1)}\omega^{(2)}\right]v^{(1)} \\ \quad + \frac{13\rho L^2}{210Ee_0^2}(\omega^{(1)})^2 L^2\partial_x\theta^{(0)} + \frac{L^4}{6}\partial_x^3\theta^{(0)} + \left[\beta^{(4)} - \frac{\rho L^2}{Ee_0^2}(\omega^{(2)})^2 - \frac{\rho L^2}{Ee_0^2}2\omega^{(1)}\omega^{(3)} - \left(\frac{\rho L^2}{Ee_0^2}\right)^2\frac{\left(\omega^{(1)}\right)^4}{60}\right]v^{(0)} \\ \quad - \frac{9\rho L^2}{70Ee_0^2}(\omega^{(1)})^2 L^2\partial_x^2 v^{(0)} - \frac{L^4}{12}\partial_x^4 v^{(0)} + \frac{13\rho L^2}{210Ee_0^2}2\omega^{(1)}\omega^{(2)}L^2\partial_x^2\theta^{(-1)} = 0 \\ \mathbb{M}^{(4)} = \delta^{(-1)}L\theta^{(4)} + \left(\delta^{(0)}+1\right)L\theta^{(3)} + \delta^{(1)}L\theta^{(2)} + \left[\delta^{(2)} - \frac{\rho L^2}{210Ee_0^2}(\omega^{(1)})^2\right]L\theta^{(1)} - L\partial_x v^{(3)} + \frac{L^3}{6}\partial_x^2\theta^{(1)} \\ \quad + \left[\delta^{(3)} - \frac{\rho L^2}{210Ee_0^2}2\omega^{(1)}\omega^{(2)}\right]L\theta^{(0)} - \frac{13\rho L^2}{210Ee_0^2}(\omega^{(1)})^2 L\partial_x v^{(1)} - \frac{L^3}{6}\partial_x^3 v^{(1)} \\ \quad + \left[\delta^{(4)} - \frac{\rho L^2}{210Ee_0^2}(\omega^{(2)})^2 - \frac{\rho L^2}{210Ee_0^2}2\omega^{(1)}\omega^{(3)} - \frac{13}{970200}\left(\frac{\rho L^2}{Ee_0^2}\right)^2\left(\omega^{(1)}\right)^4\right]L\theta^{(-1)} \\ \quad + \frac{\rho L^2}{140Ee_0^2}(\omega^{(1)})^2 L^3\partial_x^2\theta^{(-1)} - \frac{13\rho L^2}{210Ee_0^2}2\omega^{(1)}\omega^{(2)}L\partial_x v^{(0)} + \frac{L^5}{72}\partial_x^4\theta^{(-1)} = 0 \end{cases} \tag{A6}$$

References

1. Colmenares, D.; Andersson, A.; Karoumi, R. Closed-form solution for mode superposition analysis of continuous beams on flexible supports under moving harmonic loads. *J. Sound Vib.* **2022**, *520*, 116587. [CrossRef]
2. Li, J.; Chen, B.; Mao, H. Exact closed-form solution for vibration characteristics of multi-span beams on an elastic foundation subjected to axial force. *Structures* **2024**, *60*, 105884. [CrossRef]
3. Mathews, P.M. Vibrations of a beam on elastic foundation. *ZAMM J. Appl. Math. Mech. Z. Angew. Math. Mech.* **1958**, *38*, 105–115. [CrossRef]
4. Achenbach, J.D.; Sun, C.T. Moving load on a flexibly supported Timoshenko beam. *Int. J. Solids Struct.* **1965**, *1*, 353–370. [CrossRef]
5. Chonan, S. Moving Harmonic Load on an Elastically Supported Timoshenko Beam. *ZAMM J. Appl. Math. Mech. Z. Angew. Math. Mech.* **1978**, *58*, 9–15. [CrossRef]
6. Metrikine, A.V.; Verichev, S.N. Instability of vibrations of a moving two-mass oscillator on a flexibly supported Timoshenko beam. *Arch. Appl. Mech.* **2001**, *71*, 613–624. [CrossRef]
7. Rodrigues, C.; Simões, F.M.; Pinto da Costa, A.; Froio, D.; Rizzi, E. Finite element dynamic analysis of beams on nonlinear elastic foundations under a moving oscillator. *Eur. J. Mech.-A/Solids* **2018**, *68*, 9–24. [CrossRef]
8. Morfidis, K. Vibration of Timoshenko beams on three-parameter elastic foundation. *Comput. Struct.* **2010**, *88*, 294–308. [CrossRef]
9. Ding, L.; Wu, L.; Zhu, H.P. Propagation and localization of wave in multi-span Timoshenko beams on elastic foundations under moving harmonic loads. *Int. J. Numer. Anal. Methods Geomech.* **2017**, *41*, 1687–1710. [CrossRef]
10. Mead, D.J. A new method of analyzing wave propagation in periodic structures; Applications to periodic timoshenko beams and stiffened plates. *J. Sound Vib.* **1986**, *104*, 9–27. [CrossRef]

11. Mead, D.J.; Yaman, Y. The harmonic response of uniform beams on multiple linear supports: A flexural wave analysis. *J. Sound Vib.* **1990**, *141*, 465–484. [CrossRef]
12. Lin, Y.K.; McDaniel, T.J. Dynamics of Beam-Type Periodic Structures. *J. Eng. Ind.* **1969**, *91* , 1133–1141.
13. Lin, H.P.; Chang, S.C. Free vibration analysis of multi-span beams with intermediate flexible constraints. *J. Sound Vib.* **2005**, *281*, 155–169. [CrossRef]
14. Manevitch, L.I.; Oshmyan, V.G. An Asymptotic Study of the Linear Vibrations of a Stretched Beam with Concentrated Masses and Discrete Elastic Supports. *J. Sound Vib.* **1999**, *223*, 679–691. [CrossRef]
15. Sanchez-Palencia, E. *Non-Homogeneous Media and Vibration Theory*; Springer: Berlin/Heidelberg, Geramny, 1980. [CrossRef]
16. Auriault, J.L.; Boutin, C.; Geindreau, C. *Homogenization of Coupled Phenomena in Heterogenous Media*; Wiley-ISTE: London, UK; Hoboken, NJ, USA, 2009. [CrossRef]
17. Allaire, G. Homogenization and Two-Scale Convergence. *SIAM J. Math. Anal.* **1992**, *23*, 1482–1518. [CrossRef]
18. Caillerie, D.; Trompette, P.; Verna, P. Homogenization of periodic trusses. In Proceedings of the Congress IASS Madrid, 1989; pp. 7139–7180.
19. Moreau, G.; Caillerie, D. Continuum modeling of lattice structures in large displacement applications to buckling analysis. *Comput. Struct.* **1998**, *68*, 181–189. [CrossRef]
20. Tollenaere, H.; Caillerie, D. Continuous modeling of lattice structures by homogenization. *Adv. Eng. Softw.* **1998**, *29*, 699–705. [CrossRef]
21. Boutin, C.; Hans, S.; Chesnais, C. Generalized Beams and Continua. Dynamics of Reticulated Structures. In *Advances in Mechanics and Mathematics*; Springer: New York, NY, USA, 2010; Volume 21, pp. 131–141. [CrossRef]
22. Hans, S.; Boutin, C. Dynamics of discrete framed structures: A unified homogenized description. *J. Mech. Mater. Struct* **2008**, *3*, 1709–1739. [CrossRef]
23. Kolpakov, A.G.; Andrianov, I.V. Asymptotic decomposition in the problem of joined elastic beams. *ZAMM Z. Angew. Math. Mech.* **2014**, *94*, 818–836. [CrossRef]
24. Gambin, B.; Kröner, E. Higher-Order Terms in the Homogenized Stress-Strain Relation of Periodic Elastic Media. *Phys. Status Solidi (b)* **1989**, *151*, 513–519. [CrossRef]
25. Germain, P. The Method of Virtual Power in Continuum Mechanics. Part 2: Microstructure. *SIAM J. Appl. Math.* **1973**, *25*, 556–575. [CrossRef]
26. Mindlin, R.D. Second gradient of strain and surface-tension in linear elasticity. *Int. J. Solids Struct.* **1965**, *1*, 417–438. [CrossRef]
27. Eringen, A.C. Linear theory of nonlocal elasticity and dispersion of plane waves. *Int. J. Eng. Sci.* **1972**, *10*, 425–435. [CrossRef]
28. Boutin, C. Microstructural effects in elastic composites. *Int. J. Solids Struct.* **1996**, *33*, 1023–1051. [CrossRef]
29. Smyshlyaev, V.P.; Cherednichenko, K.D. On rigorous derivation of strain gradient effects in the overall behaviour of periodic heterogeneous media. *J. Mech. Phys. Solids* **2000**, *48*, 1325–1357. [CrossRef]
30. Cherednichenko, K.; Smyshlyaev, V.P. On full two-scale expansion of the solutions of nonlinear periodic rapidly oscillating problems and higher-order homogenised variational problems. *Arch. Ration. Mech. Anal.* **2004**, *174*, 385–442. [CrossRef]
31. Boutin, C.; Auriault, J. Rayleigh scattering in elastic composite materials. *Int. J. Eng. Sci.* **1993**, *31*, 1669–1689. [CrossRef]
32. Fish, J.; Chen, W. Higher-Order Homogenization of Initial/Boundary-Value Problem. *J. Eng. Mech.* **2001**, *127*, 1223–1230. [CrossRef]
33. Andrianov, I.V.; Bolshakov, V.I.; Danishevs'kyy, V.V.; Weichert, D. Higher order asymptotic homogenization and wave propagation in periodic composite materials. *Proc. R. Soc. A Math. Phys. Eng. Sci.* **2008**, *464*, 1181–1201. [CrossRef]
34. Dumontet, H. Study of a boundary layer problem in elastic composite materials. *ESAIM Math. Model. Numer. Anal.* **1986**, *20*, 265–286. [CrossRef]
35. Moskow, S.; Vogelius, M. First-order corrections to the homogenised eigenvalues of a periodic composite medium. A convergence proof. *Proc. R. Soc. Edinb. Sect. A Math.* **1997**, *127*, 1263–1299. [CrossRef]
36. Allaire, G.; Amar, M. Boundary layer tails in periodic homogenization. *ESAIM Control Optim. Calc. Var.* **1999**, *4*, 209–243. [CrossRef]
37. Cioranescu, D.; Donato, P. *An Introduction to Homogenization*; Oxford University Press: Oxford, UK, 1999. [CrossRef]
38. Armstrong, S.; Kuusi, T.; Mourrat, J.C.; Prange, C. Quantitative Analysis of Boundary Layers in Periodic Homogenization. *Arch. Ration. Mech. Anal.* **2017**, *226*, 695–741. [CrossRef]
39. Cornaggia, R.; Guzina, B.B. Second-order homogenization of boundary and transmission conditions for one-dimensional waves in periodic media. *Int. J. Solids Struct.* **2020**, *188–189*, 88–102. [CrossRef]
40. Fergoug, M.; Parret-Fréaud, A.; Feld, N.; Marchand, B.; Forest, S. A general boundary layer corrector for the asymptotic homogenization of elastic linear composite structures. *Compos. Struct.* **2022**, *285*, 115091. [CrossRef]
41. Kaplunov, J.D.; Pichugin, A.V. *On Rational Boundary Conditions for Higher-Order Long-Wave Models*; IUTAM Bookseries; Springer: Dordrecht, The Netherlands, 2009, Volume 10, pp. 81–90. [CrossRef] [PubMed]
42. Andrianov, I.V.; Bulanova, N.S.; Sedin, V.L. Vibrations of ribbed plates on elastic bases. *Int. Appl. Mech.* **1999**, *35*, 64–68. [CrossRef]
43. Pichugin, A.V.; Askes, H.; Tyas, A. Asymptotic equivalence of homogenisation procedures and fine-tuning of continuum theories. *J. Sound Vib.* **2008**, *313*, 858–874. [CrossRef]

44. Craster, R.V.; Kaplunov, J.; Pichugin, A.V. High-frequency homogenization for periodic media. *Proc. R. Soc. A Math. Phys. Eng. Sci.* **2010**, *466*, 2341–2362. [CrossRef]
45. Rallu, A.; Hans, S.; Boutin, C. Asymptotic analysis of high-frequency modulation in periodic systems. Analytical study of discrete and continuous structures. *J. Mech. Phys. Solids* **2018**, *117*, 123–156. [CrossRef]

mathematics

MDPI

Article

A Multiscale Statistical Analysis of Rough Surfaces and Applications to Tribology

Feodor M. Borodich [1,*,†], Andrey Pepelyshev [2,†] and Xiaoqing Jin [1,†]

[1] College of Aerospace Engineering, Chongqing University, Chongqing 400044, China; jinxq@cqu.edu.cn
[2] School of Mathematics, Cardiff University, Senghennydd Road, Cardiff CF24 4AG, UK; pepelyshevan@cardiff.ac.uk
* Correspondence: borodichfm@cqu.edu.cn
† These authors contributed equally to this work.

Abstract: Mathematical modelling of surface roughness is of significant interest for a variety of modern applications, including, but not limited to, tribology and optics. The most popular approaches to modelling rough surfaces are reviewed and critically examined. By providing counterexamples, it is shown that approaches based solely on the use of the fractal geometry or power spectral density have many drawbacks. It is recommended to avoid these approaches. It is argued that the surfaces that cannot be distinguished from the original rough surfaces can be synthesised by employing the concept of the representative elementary pattern of roughness (REPR), i.e., the smallest interval (or area) of a rough surface that statistically represents the whole surface. The REPR may be extracted from surface measurement data by the use of the "moving window" technique in combination with the Kolmogorov–Smirnov statistic.

Keywords: roughness; statistics; fractal; multiscale; surface; tribology; friction; multilevel

MSC: 74A55; 62G05; 28A80

1. Introduction

Tribology is the science devoted to studies of the interactions between rubbing surfaces. It includes contact mechanics and research on friction, wear, and lubrication. Studies of contact interactions between elastic solids started from the classic works of Hertz [1,2] and Boussinesq [3]. They considered problems for solids having ideal canonical shapes, e.g., elliptical paraboloids, spheres, cones, and flat-ended cylinders. However, it was realised soon that the topography of surfaces may have a great influence on the resulting stress–strain fields, while all engineering surfaces possesses some roughness, regardless of their preparation method. The surface roughness and asperity deformations have very significant influences on friction, wear, energy dissipation during the relative sliding of surfaces, and other processes of tribology. For example, it was shown that the fatigue lives of gears with superfinished teeth, i.e., whose average roughness was reduced by a factor of 5, were about four-times greater than the lives of gears without the improved surface finish [4].

The competition "Contact Mechanics Challenge" [5] was dedicated to celebrate 50 years of the Greenwood and Williamson (GW) model [6]. This research competition was designed to test various approximate methods for solving the problem of an elastic contact between the mathematically defined fractal rough surfaces under small adhesion [5]. Although 12 international groups took part in this "contact sport of rough surfaces", the results were not very impressive. The formulation of the task and the outcomes of the competition raise doubts about their correctness: "this is disappointing as friction has been studied since the times of Leonardo and in 500 years, no predictive model has emerged, nor any significant improvement from rough contact models" [7]. Thus, although no convincing models so far have been presented, the international scientific community realises that proper statistical approaches to surface roughness and modelling of contact between the surfaces are of great practical interest, not only for the dry friction phenomenon, but also for almost all areas of modern tribology. Here, we discuss various attempts to characterise real surfaces by some statistical parameters or by various statistical models of rough surfaces. We also discuss some standard parameters of roughness, surfaces having a Gaussian distribution of asperity heights, along with models that are based on the fractal assumptions or attempts to model rough surfaces by their power spectral density (PSD). As Carpick has correctly noted, "For more realistic rough surfaces, the problem becomes a multiscale exercise in surface-height statistics, even before including complex phenomena such as adhesion, plasticity, and fracture" [8].

The term multiscale analysis has many rather different meanings. Usually, multiscale asymptotic analysis is applied to systems that may be separated to subsystems whose components have specific features at multiple temporary and/or spatial scales. If the components or variables of the systems vary on different time or space scales, then it is often convenient to introduce the so-called fast–slow or large–small variables. The introduction of such variables enables the researchers to reduce the problem formulation to problems for two or more subsystems. The multiscale analysis in application to some physical problems is often treated as the renormalisation group method, which reduces a difficult problem for a system to a sequence of simpler problems for separated scales. Multiscale approaches are used in many branches of science ranging from Kolmogorov's multiscale approach to turbulent eddy cascade [9] to asymptotic analysis of space-periodic steady convective magnetic dynamos in a rotating layer of incompressible electrically conducting fluid [10]. As A. Chorin wrote in his Foreword to Barenblatt's book [9], "We are at the beginning of the age of multiscale science and multiscale computation, with a growing need to understand not only phenomena on each of many scales but also the interaction between phenomena at very different scales".

Here, the term "scale" is related to the capabilities of the system to involve a different physical/chemical mechanism of interaction between surfaces at each specific scale. The model scale depends on the characteristic length under investigation, and it depends also on the processes involved in interactions between surfaces. For example, the atomic scale is responsible for chemical interactions (if any), the nanoscale asperities are mainly responsible for molecular adhesion, and the microscale asperities are related to the mechanical interlocking of surface protuberances and their plastic deformations. If two or more mechanisms of interaction are involved in the model, then the model is a multiscale one. This review does not intend to describe only recent results in the field of the analysis of rough surfaces, but rather, it intends to provide a critical evaluation of key papers having a crucial influence on engineers working in the area of tribology. We present a new look at the papers that are customarily considered as the classic ones. We critically analysed these papers and analysed some of the common misinterpretations of these paper that may be considered as myths. Therefore, we refer to the original papers and not to recent papers, which just repeat the common mistakes.

2. Preliminaries, Scales, and Levels of Surface Roughness

Questions related to the modelling of rough surfaces have been intensively studied; see, e.g., the fundamental monograph by Whitehouse [11] and the references therein. In spite of many attempts to involve surface topography in the problems of tribology, up to now, there is no clear understanding of how roughness affects contact and dry friction phenomena.

The scale of consideration of any tribological problem is very important. For example, further development of micro-/nanotechnologies requires studies of surfaces whose asperities are at the nano-/molecular scale. Roughness has a crucial influence on the dry friction phenomenon when liquid lubricants are not used, e.g., for nano-/microscale devices and devices used in a vacuum environment. The tribological properties of macro-mechanical objects may differ by many orders of magnitude from the properties of nanoscale objects.

The following distinctions between the scales l may be introduced [12]: (i) the atomic scale reflects the phenomena specific to the characteristic length $l \leq 1$ nm, in particular chemical interactions between surfaces; (ii) the nanoscale reflects phenomena specific to the characteristic length 1 nm $\leq l \leq 1$ µm, mainly the van der Waals interaction between surfaces; (iii) the microscale is specific to the characteristic length 1 µm $\leq l \leq 1$ mm, namely the mechanical interlocking of asperities or their plastic deformations.

The nanoscale has several specific features. For example, studying the asperities of crystalline materials, one has to take into account the Polonsky–Keer effect: "when asperity size decreases and becomes comparable to the characteristic microstructural length, contact plastic deformation becomes increasingly difficult, and finally impossible". Hence, "below a certain threshold asperity size, the microcontact response becomes purely elastic" [13,14].

Another very importance feature of real surface roughness is its multilevel nature. In 1957, Archard argued that real roughness has a hierarchical multilevel structure and presented an iterative procedure of the model construction [15]. In fact, he considered a spherical punch of radius R_0 that was covered by small spherical asperities of radius R_1 superimposed on the initial sphere. In turn, the latter set of asperities was covered by the next generation of spherical protuberances of radius R_2, and so on. Note that the Archard construction leads to a number of overlapping regions, and hence, it was noted [16] that the Archard profile is not a self-similar fractal, contrary to the statement of Greenwood and Wu [17].

Let us specify our terms. The model is hierarchical if its vertical construction consists of several generations of asperities. The question of whether a surface topography may be specified as multilevel is related to the surface structure and to whether the asperities of the m-th generation are located at the same height or not. If all asperities of the same generation are located at the same height, then it is a single-level model, even if it was a multiscale model, i.e., the configuration of the model would allow the single-level asperities to be engaged in the mechanisms of the interactions between surfaces specific to at least two length scales. If the location heights of the same-generation asperities vary, then it is a multilevel model. For example, the C_B profile introduced by one of the authors (F.B.) [16,18] and some modifications of the profile (see, e.g., [19–21]) are single-level models, while the profile introduced by Borodich and Onishchenko [22,23] is a multilevel model.

It is interesting to note that, although the statement that the roughness or attachment devices of animals exhibit a multilevel hierarchical structure is currently widely accepted, and it is attributed to the findings by S. Gorb (Kiel University) and his colleagues [24], along with other experts in the biology of animals [25], the Archard concept of a hierarchical multilevel "bump on bump" or "protuberances on protuberances on protuberances" structure of surface roughness remained largely undeveloped until 1993 when Borodich and Onishchenko [22] introduced the multilevel hierarchical model of roughness.

The Archard concept was employed in numerical simulations of friction between a rough surface and a nominally flat slider that was represented as a multiscale, hierarchical system of connected deformable components [12]. The friction force has been calculated as the ratio between dissipated energy $U_{dissipated}$ and sliding distance x. The assumption of the vacuum environment allowed reducing the energy dissipation to three sources: (i) breaking of metallic bonds, which corresponds to the atomic scale ($U_{chemical}$); (ii) breaking of van der Waals bonds, which corresponds to the nanoscale (U_{vdW}); and (iii) mechanical interlocking of asperities at the microscale ($U_{mechanical}$), i.e.,

$$F_f = \frac{U_{chemical} + U_{vdW} + U_{mechanical}}{x}.$$

In this case, the friction coefficient μ can then be expressed as $\mu = U_{dissipated}/(P + A)$, and it can be split into several components attributed to the mechanism of energy dissipation [26]. Here, P is the normal force and A is the force of adhesion.

Because the elastic components of the slider of different scales are coupled, a deformation of any asperity of the model causes deformations of the whole system, i.e., the assumptions that the asperities act independently was not used. In this model, the Polonsky–Keer effect [13,14] was taken into account in application to nanoscale asperities for the first time. This effect is the reason that the atomic-/nanoscale asperities do not have plastic deformations even under very high pressure.

3. Early Methods of Roughness Characterisation and Contact Problems for Rough Surfaces

Let us consider a solid that nominally occupies a negative half-space $z \leq 0$. We use the Cartesian x, y, z coordinates. The solid is bounded by a nominally flat rough surface. Take a plane $y = 0$ perpendicular to such a surface, then the cross-section between the surface and the plane is referred to as a surface profile or a roughness profile $z(x)$. The origin (O) of the Cartesian coordinates x, y is at the point of initial contact between the Atomic Force Microscope (AFM) or the stylus probe and the surface. We will consider profiles of nominal length L or $2L$.

Let $\bar{z}$ denote the mean profile line, i.e.,

$$\frac{1}{2L}\int_{-L}^{L}[z(x)-\bar{z}]dx = 0,$$

where $2L$ is the nominal length of the profile, i.e., the length of projection of the profile on the boundary plane.

In practice, we measure the roughness at equidistant points $x_1, x_2, \ldots, x_n$, where $x_1 = 0$. Evidently, $\bar{z} = 0$ due to the choice of the origin of the coordinate system.

3.1. *Some Popular Statistical Parameters of Roughness*

Let us introduce the density probability function $\phi(z)$, which shows the probability that the height $z(x)$ at a surface point x is between z and $z + dz$. The right-tailed cumulative distribution function of the surface heights $\Phi(z)$ is defined as

$$\Phi(z) = \int_{z}^{\infty}\phi(t)\,dt.$$

In 1933, Abbott and Firestone [27] suggested to calculate the total length of the slice of the profile above the level $z = h$ and equating this length to $\Phi(h)$. Apparently, this was the first use of statistical tools in application to surface roughness. In tribology, this function is called the Abbott–Firestone, Abbott, or bearing area curve.

After the Abbott–Firestone curve was introduced, there was a period that was characterised by Whitehouse [28] as, "the parameter rash", because a huge number of statistical parameters of roughness were introduced. These characteristics were related to the vertical distribution of heights, the horizontal distribution of the rough profiles, and the shapes of asperities [29]. Apparently, the most popular height parameter is the maximum height R_{max} of the profile $z(x)$ defined on an interval $[-L, L]$, which is defined as

$$R_{max} = \max_{x\in[-L,L]} z(x).$$

The arithmetical mean deviation of the surface R_a and the root-mean-squared (rms) height R_q are also very popular parameters of surface roughness:

$$R_a = \frac{1}{2L}\int_{-L}^{L}|z(x)|\,dx \approx \frac{\sum_{i=1}^{n}|z(x_i)|}{n}, \quad R_q = \left[\frac{1}{2L}\int_{-L}^{L}[z(x)]^2\,dx\right]^{1/2},$$

where n is the number of points of measurements on the interval and $z(x_i)$ is the measured height at the interval point x_i. Note that R_q is the square root of the mean-squared deviation with respect to the mean profile line $\bar{z} = 0$.

More than 30 statistical parameters of roughness are currently in use [29]. In spite of these attempts, there is no clear understanding of how these parameters affect contact and friction. This situation may confuse practical engineers because the engineers should follow the standards, while the European and British standard [30] contains over 20 surface and profile parameters. The situation is worse in the American standard [31] because the local engineers followed the fashion to use the so-called fractal terminology and they included in the standard not only all parameters of the European standard, but also many additional parameters, including procedures for the evaluation of surface topography using the fractal parameters.

Although some statistical parameters of roughness, e.g., the Abbott–Firestone curve, may be quite useful for specific engineering purposes, most of them are rather useless [28].

3.2. Contact Problems for a Thin Elastic Layer

The Abbott–Firestone curve is a quite useful parameter that may be correlated with the contact properties of the rough surfaces. For example, consider a rough punch that contacts a soft thin elastic coating on a hard elastic half-space. It is known (see, e.g., [32,33] and the references therein) that, if the characteristic size of the contact region is larger than the layer thickness, then the leading term of the asymptotic solution may be represented as the contact between the punch and the Fuss–Winkler foundation, which is also known as the spring bed mattress elastic foundation [34]. If the half-space is rigid and E and ν denote, respectively, the elastic modulus and Poisson's ratio of the layer, then the spring constant k of the Winkler–Fuss spring layer is

$$k = \frac{E(1-\nu)}{h(1+\nu)(1-2\nu)}.$$

In this case, the compressing force is

$$P(\delta) = kV(\delta),$$

where $V(\delta)$ is the volume of the body under the cross-section of height $z = \delta$.

3.3. Zhuravlev and Kragelsky Models of Contact between Rough Elastic Surfaces

In 1940, Zhuravlev presented a statistical model of contact between rough surfaces [35,36]. It was assumed that spherical asperities of the same radius R have various heights, and therefore, their number at a specific height ξ increases as the level of consideration goes deeper into the rough surface.

Assuming the distribution of the summits of the asperities at various levels ξ of a surface element are characterised by a function $\phi(\xi)$ (the same for both surfaces) and employing the Hertz contact model for each asperity, he wrote the general relation between the contact force P and the area A of true contact

$$P = \frac{\sqrt{2R}}{3\pi k_1 N}\int_0^{x/2}\int_0^{x/2}[x-(\xi_1+\xi_2)]^{3/2}\phi(\xi_1)\phi(\xi_2)d\xi_1 d\xi_2,$$
$$A = \frac{\pi R}{2N}\int_0^{x/2}\int_0^{x/2}[x-(\xi_1+\xi_2)]\phi(\xi_1)\phi(\xi_2)d\xi_1 d\xi_2, \quad (1)$$

where x is the relative compressing displacement of contacting surfaces, N is the total number of asperity summits situated at various levels of the nominal unit element of contacting surfaces from the highest summit to the level of depth x, $k_1 = (1-\nu^2)/\pi E$, ν is the Poisson ratio, and E is the Young modulus of the material. It was shown that, if $\phi(\xi)$ is linear, then analytical integration of the expressions (1) is quite easy and $A \sim P^{10/11}$.

In 1948, Kragelsky [37] developed the statistical approach of Zhuravlev and modelled contact and friction between two surfaces having different roughness. Instead of spherical asperities, he considered asperities as a number of rods at various heights based on a rigid plane. He argued that his model gives qualitatively the same results as the Zhuravlev one.

He noted that roughness may have a Gaussian distribution of heights; however, the use of this distribution would necessarily lead to numerical integration of general expressions. Hence, for the demonstration of the model applications, he used only linear function $\phi(z)$, as Zhuravlev did.

3.4. The Greenwood and Williamson Contact Model

The Greenwood and Williamson (GW) model [6] is based on the same assumptions as the Zhuravlev one; however, they developed the model considerably by studying also plastic deformations of asperities. In addition, as Greenwood noted [38], the Greenwood and Williamson model, "improves on Zhuravlev by assuming a Gaussian height distribution (still of identical spherical caps)". In fact, the calculations were performed not for a Gaussian height distribution, but rather, for an exponential distribution of the sphere summits.

We need to comment that, for processes at the nanoscale, the improvements of the Zhuravlev model by Greenwood and Williamson are rather imaginary. Indeed, the nanoscale asperities do not deform plastically due to the Polonsky–Keer effect, while the height distribution of asperities is often non-Gaussian.

The use of the Zhuravlev–Greenwood–Williamson-type models assumes that the radius of the summits of surface asperities is known. If the roughness is isotropic, then the surface roughness $z(x,y)$ is characterised by just a profile $z(x)$. If the profile heights $z_m = z(x_m)$ are measured with a regular stylus or AFM step τ, i.e., one observes z_m, then the curvature (κ) of a protuberance z_m can be defined as

$$\kappa = -(z_{m-1} + z_{m+1} - 2z_m)/\tau^2$$

where $z_{m-1}, z_{m+1} < z_m$ [39]. Whitehouse and Archard [40] found that different sampling intervals gave different mean curvature, i.e., the mean curvature is a scale-dependent parameter.

4. Surface Topography as a Realisation of a Gaussian Process

In addition to attempts to characterise the roughness of nominally flat surfaces by the selection of several roughness parameters, there are very popular models of surface topography as realisations of random processes. A core assumption underlying most of these models is the normality (Gaussian distribution) of asperity heights, either explicitly stated or implied.

Models considering roughness as realisations of Gaussian (normal) processes are well established. The vast majority of research employing statistical methods for surface roughness description relies on the assumption that asperity heights follow a Gaussian distribution. A Gaussian surface is completely defined by two parameters, a mean and a covariance function $K(\delta)$ [40,41]:

$$K(\delta) = \lim_{T\to\infty} \frac{1}{2T} \int_{-T}^{T} [z(x+\delta) - \bar{z}][z(x) - \bar{z}]\, dx = \langle [z(x+\delta) - \bar{z}][z(x) - \bar{z}] \rangle.$$

The function $K(\delta)$ measures the statistical dependence of the height values of roughness at two points separated by the distance δ and, therefore, influences the horizontal distribution of the asperities of a rough surface profile. Indeed, one needs to specify both the vertical and horizontal distributions because, as was clearly formulated by Maugis [42], two profiles may have the same height and peak height (local extrema) distributions, but they may differ in the horizontal behaviour. Instead of $K(\delta)$, one can use its Fourier transform, the power spectral density (PSD) $G(\omega)$ of the signal frequency ω:

$$G(\omega) = \frac{2}{\pi} \int_0^{\infty} K(\delta) \cos \omega\delta \, d\delta.$$

The idea to model rough surfaces as such surfaces may be traced back to Linnik and Khusu [43,44] and Whitehouse and Archard [40]. In fact, Linnik and Khusu suggested to describe surface roughness using graphs of a stationary Gaussian random process having a covariance function [43]:

$$K(x) = K(0) \cdot e^{-\alpha|x|},$$

where $K(0)$ and α are the parameters of the roughness.

Various mathematical properties of Gaussian (normal) surfaces were intensively studied (see, e.g., [41,45] and the references therein). Many researchers studied non-adhesive contact problems for surfaces having Gaussian roughness (see, e.g., [42,46,47]), while several models of adhesive contact between such surfaces were developed by Fuller, Tabor, and Galanov [48–51]. However, to use these models, one needs to demonstrate that the surface roughness follows a Gaussian distribution. Unfortunately, the tools developed for Gaussian surfaces are often used without proper justification in application to surfaces whose height distribution is non-normal.

There exist various tests of normality of experimental data [52]. The most popular tests of normality are the Kolmogorov–Smirnov (KS), Lilliefors (LF), Shapiro–Wilk (SW), Anderson–Darling (AD), Cramer–von Mises (CVM), Pearson, and Shapiro–Francia (SF) ones. Our statistical analysis of grinding surface roughness showed that the distribution is not normal, either at the micro- or nanoscale [53]. The height distribution is normal for polishing papers with nominal asperity sizes of 0.3, 1, 3, 9, and 12 μm [54]. The roughness of carbon-based coatings deposited by direct current pulsed magnetron sputtering was normal at the microscale and at the nanometre scale, when AFM measurements were taken with 117 nm steps, while the same surface measured with the 10 nm AFM steps showed that the roughness of the non-biased sample was not normal [55]. Thus, there are many cases when the models developed for Gaussian surfaces are not applicable.

5. Models of Roughness Based on the Use of Fractal or PSD Concepts

The concept of "fractals" is often used in various applications. Many researchers believe that they understand the concept very well. However, their belief is usually based on the consideration of self-similar sets such as the von Koch curve, the Sierpinski carpet, or the Cantor discontinuum, i.e., the sets that are described in the book for Russian teenagers published by Vilenkin in 1965 [56]. The difficulty of studying fractals is caused by the rather vague description of the subject introduced by Mandelbrot [57]. As Greenwood commented on the situation: "Mandelbrot is somewhat reluctant to define 'fractals' or 'fractal dimension' preferring to offer examples" [39]. This is true (see, e.g., [58]). The lack of established definitions has led to the situation that the papers dedicated to fractal analysis of applied problems are often based on the repetition of common myths about fractals, which are uncritically reported in a number of papers. As we have mentioned earlier [59], the state-of-the-art approach is replaced by the state of-the-street one or, in another words, by the vulgar fractal approach.

There are many papers trying to model roughness using fractal language. It is argued here that the researchers should avoid the fractal approaches at least in application to problems of tribology because these approaches do not have rigorous mathematical support and they suffer from the lack of clear definitions of the main concepts. In addition, during the last couple of decades, many papers have been published that tried to model rough surfaces by their power spectral density (PSD). Unfortunately, the overwhelming majority of these papers are just some interpretations of the fractal approaches. Hence, papers based on the employment of fractal or PSD approaches are considered together here because they share the same drawbacks.

5.1. *The Weierstrass–Mandelbrot Function*

The use of fractal concepts in tribology is due to the comment of Berry and Hannay [60] on the results by Sayles and Thomas [61], who observed the power-law behaviour of the scaled PSD of many real surfaces. Later, Berry and Lewis published a very popular paper [62], where they studied in detail the Weierstrass function modified by Mandelbrot [57]. They called this function W of complex variable t:

$$W(t;p) = \sum_{n=-\infty}^{\infty} p^{(D-2)n}(1 - e^{ip^n t})e^{i\phi_n}, \quad 1 < D < 2, \quad p > 1,$$

the Weierstrass–Mandelbrot (W-M) function. Here, p is a parameter and ϕ_n are arbitrary phases that can be chosen so that W exhibits either deterministic or stochastic behaviour. D is equal to the box-counting (fractal) dimension of the graph of $W(t)$. They also considered the real part of the W-M function with $\phi_n = 0$:

$$W_1(x;p) = \sum_{n=-\infty}^{\infty} p^{(D-2)n}(1 - \cos p^n)x, \tag{2}$$

A paper by Majumdar and Bhushan [63], which had a big impact on the tribology community, suggested to measure the PSD of a real rough surface and replace studying the surface by studying a truncated W-M function:

$$\tilde{W_1}(x;p) = \Lambda^{(D-1)} \sum_{n=n_1}^{\infty} p^{(D-2)n} \cos 2\pi p^n x, \quad 1 < D < 2, \quad p > 1, \tag{3}$$

having the same PSD. The integer number n_1 corresponds to the low cut-off frequency of the profile, and Λ is the so-called characteristic length scale of the profile.

The first papers where contact problems were studied for fractal-shaped indenters were published in 1991 [18,64]. In [18], the C_B profile was introduced, and the model has been discussed above. The model of elastic–plastic contact presented by Majumdar and Bhushan (M-B) employed the fractal W-M function (3). Borodich argued that the M-B model's assumptions are controversial and the M-B model exhibits inherent flaws. Indeed, the model employs of superposition of the solutions of the nonlinear Hertz-type contact problems with non-fixed contact regions. Evidently, one cannot decompose the W-M function and use such superposition; nevertheless, it was used by many authors (see, e.g., [64,65]). In addition, the Hertz approach is not applicable to their model because the ratio of the contact radius to the radius of the asperity goes to infinity as the number of generations of the prefractal increases [66]. Although this M-B model was withdrawn by Bhushan, who wrote that, he, "now finds that any fractal characterisation model including M-B model does not give unique values of fractal parameters" for roughness [67] and "the fractal approach is not practical for engineering surfaces" [68], it is still very popular and it is discussed by many researchers.

5.2. *Mathematical and Physical Fractals*

Studying fractal approaches to solid mechanics, one of the authors (F.B.) realised that one should clearly distinguish mathematical and physical fractals [69]. Although both kinds of fractals use the concept of a cover of the object (surface) by cubes or spheres of some size δ, they are fundamentally different. Indeed, mathematical fractals are studied considering limits for $\delta \to 0$, while physical fractals have to demonstrate a kind of power-law behaviour in some interval between the lower δ_* and upper Δ_* cut-offs.

5.2.1. Some Features of Mathematical Fractals

Graphs of mathematical fractal curves are nowhere differentiable. Any part of a rough profile modelled as a mathematical fractal has infinite length. Hence, one cannot use such concepts as the surface energy or work of adhesion, and one can only use the concept of the energy attributed to a unit of fractal measure [70].

One cannot use the concepts of tangential and normal vectors for a solid having a fractal boundary. Even to formulate and prove the Gauss–Ostrogradsky theorem, one needs to use quite sophisticated mathematical tools [71].

It has been proven that the box-counting (Minkowski) fractal dimension of the W-M function graph (2) is equal, D, while nobody could calculate the Hausdorff (Hausdorff–Besicovitch) dimension of the graph. Hence, it is difficult to understand the statement [72] that' "for the Gaussian-based models that we have in mind, all the common definitions are applicable, and all produce the same numerical value for the dimension".

Thus, modelling of surface roughness by a mathematical fractal translates the difficult problem of tribology to the more difficult problem of mathematical physics.

5.2.2. Limits of Physical Fractal Description of Roughness

Although it was often claimed that the fractal dimension is a scale-independent or scale-invariant parameter of roughness (see, e.g., [18,64]), this is not true. A research group from the Hebrew University of Jerusalem analysed data published in *Physical Reviews* journals. They showed that the so-called fractal behaviour spreads usually for just 1.5 orders of magnitude [73,74]. Mandelbrot disagreed with the conclusion of the Jerusalem group; he said that the prevalence of published examples demonstrating power-law correlations in tribology research might be attributable, in part, to an over-enthusiasm for the PSD approach, potentially exacerbated by limitations in the peer-review process [75]. However, the group insisted that limited-range relations for physical fractals dominate observations in Nature [76]. If the FD value remains stable for less than two or three orders of magnitude, the fractal approach becomes impractical, as Whitehouse pointed out [77]. Indeed, in this case, if the lower cut-off of the observed power-law behaviour of roughness is, say, $\delta_* = 10$ nm, then the upper cut-off is just $\Delta_* = 310$ nm, and one needs to use another model for microscale roughness.

5.3. *Parametric–Homogeneous Rough Surfaces*

One can understand better the properties of the W-M function employing the theory of parametric–homogeneous (PH) functions introduced and developed by Borodich [78–80]. In fact, applications of the concept to contact problems was discussed at ICTAM in Haifa in 1992 [81]. In particular, the concept includes PH and parametric–quasi-homogeneous (PQH) functions. They are defined in the following way. The function $B_d : \mathbb{R}^n \to \mathbb{R}$ is called a PQH function of degree d and parameter p with weights $\boldsymbol{\alpha} = (\alpha_1, \ldots, \alpha_n)$ if there exists a positive parameter p, $p \neq 1$ such that it satisfies the following identity:

$$B_d(p^{m\alpha_1}x_1, \ldots, p^{m\alpha_n}x_n; p) = p^{md}B_d(\mathbf{x}; p), \quad m \in \mathbb{Z}.$$

The PH functions B_d are the particular case of PQH functions when $\alpha_1 = \ldots = \alpha_n$.

From the definition of the PQH function, one can obtain the following properties. Any PQH function of zero degree $B_0 : \mathbb{R}^n_+ \to \mathbb{R}$ is an automorphic function with respect to a discrete group of quasi-homogeneous dilations $\Gamma_{p^{\boldsymbol{\alpha} m}}$, where $\Gamma_{p^{\boldsymbol{\alpha} m}}\mathbf{x} = (p^{m\alpha_1}x_1, \ldots, p^{m\alpha_n}x_n)$.

For each $B_d : \mathbb{R}^n_+ \to \mathbb{R}$, there exists a homogeneous function of degree d $H_d : \mathbb{R}^n_+ \to \mathbb{R}$ and a PQH function of zero degree $B_0 : \mathbb{R}^n_+ \to \mathbb{R}$ such that

$$B_d(\mathbf{x}; p) = H_d(\mathbf{x})B_0(\mathbf{x}; p).$$

In fractal geometry, the quasi-homogeneous dilations of coordinates are usually called the self-affine dilations. The graphs of PH functions can be smooth, e.g., the log-periodic sinus functions, or they can be nowhere differentiable, in particular fractal. It can be shown that such base sets used in fractal geometry as the von Koch curve, the W-M function, and the Cantor staircase are examples of PH sets. However, in general, the fractal properties of a curve and the PH property are independent of each other.

The W-M function is a PH function of degree $d = (2 - D)$ and parameter p; hence, the graph of $W_1(x; p)$ has a specific trend that can be described as x^{2-D}. The function is a particular example of a fractal PH function, and it cannot be considered as the general fractal model of rough surfaces.

5.4. PSD Approach to Rough Surfaces

Over the past two decades, the PSD approach has gained significant traction within the tribology community (see, e.g., [82,83]). In fact, the authors of the papers have focused solely on the PSD to characterise rough surfaces in tribology.

It was noted that quantities varying with time have often fractal graphs, and one way in which their fractal nature is manifested is by a power-law behaviour of the covariance function $K(x)$ [84]. This note was misinterpreted in many papers, which assumed that, if $K(x)$ and, respectively, $G(\omega)$ have power-law behaviour, then the graph or profile is fractal. For example, Persson [83] states that the power spectrum of a self-affine fractal surface is $G(\omega) \sim \omega^{-2(H+1)}$, where H is the Hurst exponent and H is related to the fractal dimension as $D = 3 - H$.

Actually, one can find the last statement in many other papers, i.e., it is often assumed that the scaling properties of rough surfaces are defined by the exponent H (the Hurst exponent), and the fractal dimension D of the surface or a profile can be calculated as $H = E - D$. Here, E is the Euclidean dimension of the space: $E = 3$ for a surface and $E = 2$ for a curve. This is usually explained by an example saying that, under quasi-homogeneous scaling of coordinates $x \to \lambda_1 x$ and $y \to \lambda_2 y$, the fractal surfaces appear as approximately the same or

$$\mathbf{x} \to \lambda \mathbf{x}; \quad z(\mathbf{x}) \to \lambda^H z(\mathbf{x}).$$

Here, $\lambda_1 = \lambda > 0$ and $\lambda_2 = \lambda^H$ are scaling factors. Using the above-mentioned properties of PQH and PH functions, one can construct a function having an arbitrarily given box-counting fractal dimension D and an arbitrarily given trend x^H, $H > 0$ (see the examples in [78,79]). Hence, the term Hurst exponent is ill-defined and the statement that the fractal dimension $D = E - H$ is not correct.

5.5. Some Counterexamples to Popular Statements of Fractal and PSD Approaches

5.5.1. PSD of Some Smooth PH Functions

Let us show that the connections of the fractal dimension and the slopes of PSD functions in logarithmic coordinates are meaningless. Let us show that some functions having smooth graphs may have a power-law PSD. Consider a smooth PH unction, namely a sin log-periodic function. Figure 1 shows the power spectra for $z(x) = \sin(2\pi \ln(x)/\ln(p))$ with $x \in [0.0001, 1]$.

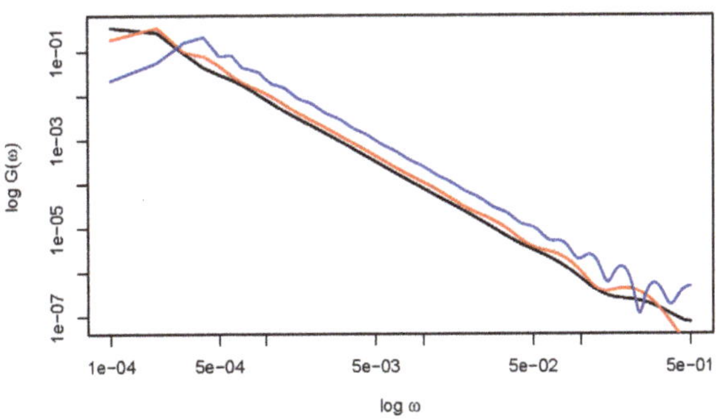

Figure 1. The power spectra of the profile $z(x) = \sin(2\pi \ln(x)/\ln(p))$ for $x \in [0.0001, 1]$ with $p = 3.5$ (black), $p = 2.5$ (red), and $p = 1.5$ (blue).

Although the function itself is not fractal, its power spectrum exhibits a linear trend on a logarithmic scale within a specific range.

5.5.2. PSD Does Not Describe the Contact Properties of Surfaces

Using rather vague arguments, Persson represented the diffusion of stress fields in a solid contacting with a rough counterpart using the diffusion equation [85]. Criticising the Persson approach [85], Borodich [86] wrote that two punches with identical fractal surface textures can exhibit distinct asymptotic behaviours in their load–displacement relationships, depending on whether they are positioned above or below the surface. Then, it was shown [53] that any surface and its replica have the same PSD; therefore, it is not sufficient on its own to fully characterise the contact properties of rough surfaces. In spite of this critique, this erroneous approach has been multiplied in a number of papers; see, e.g., [87].

Let us demonstrate this by simple examples. A rough surface has asperities and valleys that can be schematically considered as pins whose tips are directed up and down (Figure 2a). If the surface was polished, then the asperities (the pins whose tips are directed up) were removed (Figure 2b).

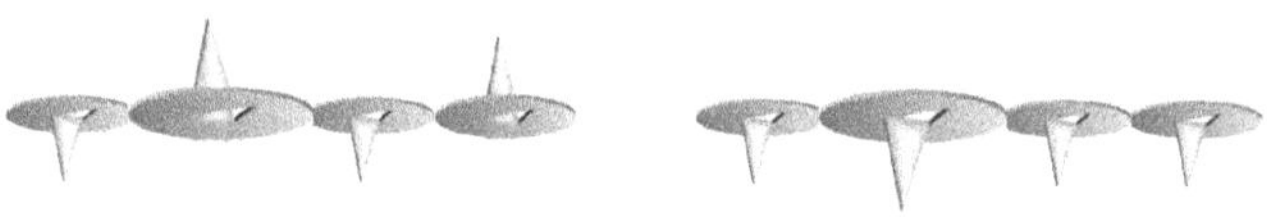

(a) (b)

Figure 2. Schematic representations of surfaces: (**a**) up (asperities) and down (valleys) pins; (**b**) down pins (asperities were removed).

Now, let us consider a solid that is described by the fractal Mandelbrot function. It is evident that, for the tribological effect of contact between the solid and the intact rough surface having sharp asperities, the pins whose tips are directed up (Figure 3a) and the polished surface that is the replica of the former one (Figure 3b) will be rather different. However, the PSD of the former and the latter surfaces are the same. As a Russian poet S. Marshak wrote ironically about a philosopher [88], "The world," he taught, "is my representation!". When his son stuck a pin under the seat of his chair, he cried out, "Horror! How terrible is my representation!" Thus, the rough surfaces cannot be represented solely by the PSD.

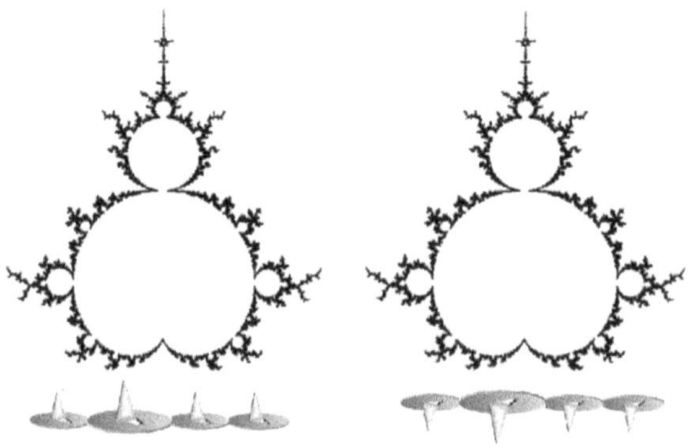

(a) (b)

Figure 3. Schematic representations of contact between a fractal solid and two surfaces: (**a**) up and down pins; (**b**) down pins (a polished surface).

5.5.3. PSD of Truncated Weierstrass–Mandelbrot Functions

Now, we demonstrate that the slopes of the PSD for different truncated Weierstrass–Mandelbrot functions are almost the same.

Let us consider two truncated Weierstrass–Mandelbrot functions:

$$C_t(x;p) = \sum_{n=-N}^{N} p^{-(2-D)n}(1-\cos(p^n x)), \quad A(x;p) = \sum_{n=-N}^{N} (-1)^n p^{-(2-D)n} \sin(p^n x),$$

where N is a large natural number, $p > 1$, and $D \in (1,2)$. In addition, let us consider a function $x^{D-2}C_t(x;p)$. If $N \to \infty$, then $C_t(x;p) \to C(x;p)$ and D is their box dimension; however, they have different trends.

Let us fix the values $p = 1.6$ and $D = 1.6$ and vary N. The PSDs for these functions can be calculated numerically and are presented in logarithmic coordinates in Figure 4.

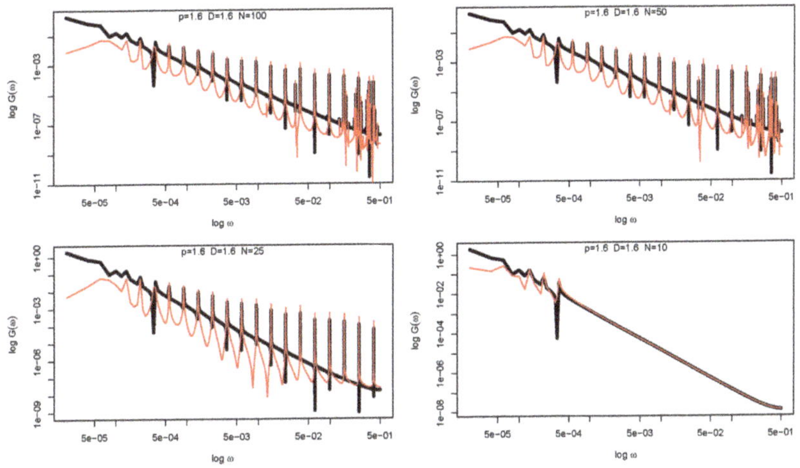

Figure 4. The power spectra of the profile $C_t(x;p)$ (black) and the profile $x^{D-2}C_t(x;p)$ (red) for $x \in [0,1]$ with $p = 1.6$, $D = 1.6$, and $N = 10, 25, 50, 100$.

One can see that the average PSDs are approximately the same, while the functions are rather different.

Now, let us fix the values $p = 1.6$ and $N = 100$, vary D, and numerically calculate the PSD for the obtained functions. The results are presented in logarithmic coordinates in Figure 5.

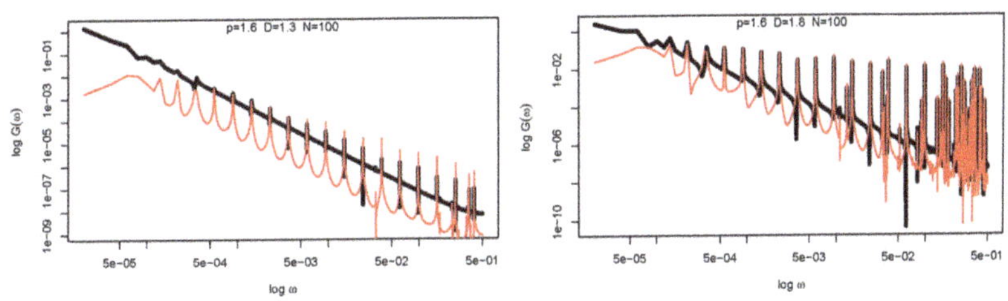

Figure 5. The power spectra of the profile $C_t(x;p)$ (black) and the profile $x^{D-2}C_t(x;p)$ (red) for $x \in [0,1]$ with $p = 1.6$ and $N = 100$ for $D = 1.3$ (**left**) and $D = 1.8$ (**right**).

Note that, for $N = 100$, the graphs of $C_t(x;p)$ and $x^{D-2}C_t(x;p)$ are very close to the graphs of the PH functions of degrees $d = 2 - D$ and $d = 0$, respectively. We know that these functions have the same D and different trends (see the discussion above), while the slopes of their PSDs are almost equal.

The same procedure we have outlined can be directly applied to analyse $A(x;p)$ and $x^{D-2}A(x;p)$. The PSDs for these profiles are shown in Figure 6.

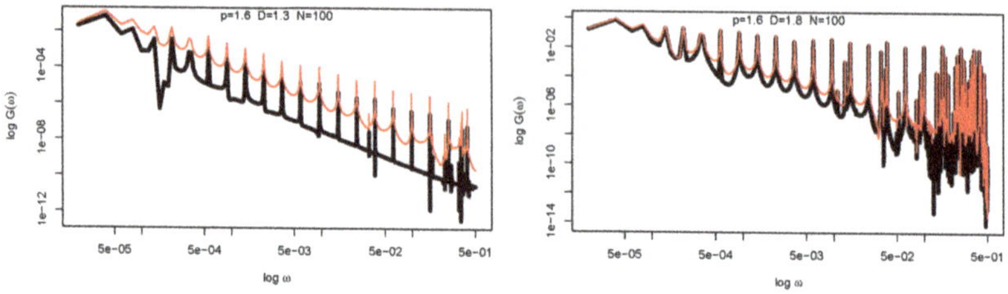

Figure 6. The power spectra of the profile $A(x;p)$ (black) and the profile $x^{D-2}A(x;p)$ (red) for $x \in [0,1]$ for $x \in [0,1]$ with $p = 1.6$ and $N = 100$ for $D = 1.3$ (**left**) and $D = 1.8$ (**right**).

We see that to the same conclusion holds as above.

5.5.4. PSD of Polished Surfaces

As has been mentioned, enhancing the surface finish can significantly extend the lifespan of gears [4]. Considering the roughness of polished parts, Chen et al. [89] argued that, after vibratory polishing, some elements of the topography of the initial surface and the topography of the polished parts coexist in the final topography. To model the topography obtained, they discussed the Pawlus superposition method [90] and the Tang method [91]. Let us use some ideas of the Pawlus superposition method. Let an intact surface be described by a realisation $f_1(x)$ of a Gaussian process. Let us assume that the topography of polished parts is described by $0.1f_2(x)$, where the profile $f_2(x)$ is a realisation of another Gaussian process. Hence, the fully polished surface is described by $f_3(x) = \min[f_1(x), 0.1f_2(x)]$, shown in Figure 7.

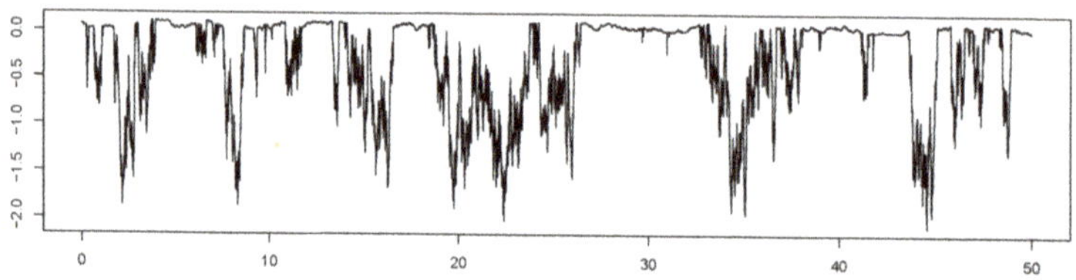

Figure 7. The "polished" profile $f_3(x) = \min[f_1(x), 0.1f_2(x)]$, where the profiles $f_1(x)$ and $f_2(x)$ are realisations of independent Gaussian processes.

Figure 8 shows the power spectra for the profile $f_1(x)$ (black), the profile $f_2(x)$ (red), and the "polished" profile $f_3(x) = \min[f_1(x), 0.1f_2(x)]$ (blue).

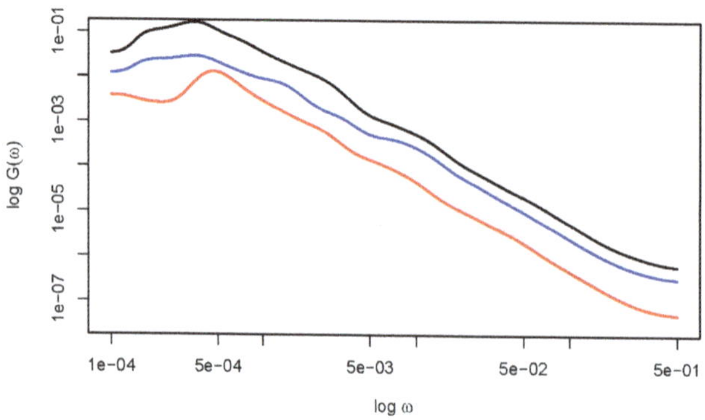

Figure 8. The power spectra for the profile $f_1(x)$ (black), the profile $f_2(x)$ (red), and the "polished" profile $f_3(x) = \min[f_1(x), 0.1f_2(x)]$ (blue).

These three surface profiles, despite exhibiting distinct appearances and tribological properties, surprisingly share the same exponent of 1.95 in their power spectra, i.e., $G(\omega){\sim}1/\omega^{1.95}$. Thus, the slope of the power spectral density alone is not sufficient to characterise the tribological properties of polished surfaces.

6. Statistically Justified Descriptions of Surface Roughness

The process of mimicking a real surface in order to create a synthetic surface is called the surface synthesis. Synthesised surfaces offer tremendous value to tribology researchers. By enabling the creation of surfaces with controlled properties, they allow for the modelling and investigation of various tribological phenomena, such as friction, wear, etc. [92]. It is assumed usually that some characteristics of synthetic roughness are the same as the characteristics of the original surface. The chosen characteristics may be taken as some characteristics from the international standards [30,31], the fractal dimension of the roughness, or its PSD function. It can be also assumed that the roughness may be described by some random process, e.g., the Gaussian process. However, one has to justify that a new surface has the same statistical properties as the original one. As has been demonstrated above, these methods of surface synthesis have various drawbacks, and usually, they do give a solid statistical support. Therefore, we will discuss a statistical approach to problems related to the synthesis of rough surfaces of solids introduced by two of the authors (F.B. and A.P.).

Similar to the notion of the representative elementary volume used in the mechanics of random inhomogeneous materials, we introduce the notion of the representative elementary pattern of roughness (REPR) and the statistically representative pattern of surface roughness (SRPSR). The REPR is the smallest interval (or area) of a rough surface that statistically represents the whole surface. Hence, from a statistical point of view, the REPR cannot be distinguished from the original rough surface. For patterns smaller than the REPR, a representative property cannot be obtained. Indeed, the Kolmogorov–Smirnov (K-S) statistic is a nonparametric goodness-of-fit test, which may be used to determine whether two distributions differ, i.e., the K-S statistic may test whether a given set of data (its empirical distribution) deviates significantly from a hypothesised reference distribution or another set of data. The K-S statistic provides a solid statistical support to the described method and enabled us to extract the REPR of the surfaces. This will enable us to synthesise simpler surfaces with equivalent roughness in terms of the height distribution.

The employment of some statistical techniques of time series analysis, e.g., the moving window technique, may be effectively used for the extraction of the REPR from the experimental data [93]. The moving window techniques are novel only for their use in tribology; however, these techniques are used in many methods of time series analysis, in particular in Singular Spectrum Analysis (SSA) methods [94]. It is known that the SSA methods may be effectively used in many areas of applied statistics, and they are not computationally demanding [95,96]. The presented algorithm for the determination of the REPR is even simpler than the SSA algorithm. The REPR algorithm has been described by the authors elsewhere [97]. It should be applied to representative sets of roughness profiles.

For the sake of simplicity, we describe it here just for one profile, which is described as heights $z_1, z_2, \ldots, z_N$.

Take the moving window of length N_s, where $N_s < N$, and slide it along the profile. Specifically, for the starting point i, $0 \le i \le N - N_s$, take a subset series of heights within the sliding window $(z_{i+1}, z_{i+2}, \ldots z_{i+N_s})$.

Compute the similarity between the long series $z_1, z_2, \ldots, z_N$ of the profile and the subset sample with the shift i using the Kolmogorov–Smirnov statistic:

$$D_{N,N_s:i} = \sup_x \left| F_N(z) - F_{N_s:i}(z) \right|,$$

where $F_N(z)$ and $F_{N_s:i}(z)$ are the cumulative distribution functions for two samples, respectively.

For the fixed window of length N_s, define a subset sample that minimises the Kolmogorov–Smirnov statistic with respect to the shift i. Also, we consider the KS-based distance measure defined by

$$D_N(N_s) = \min_{i=0,\ldots,N-N_s} D_{N,N_s:i}$$

as a function N_s.

Choose N_s such that the measure $D_N(N_s)$ is close to zero and N_s is not big, e.g., take N_s as the smallest integer such that $D_N(N_s) < 0.05$.

Let us consider an example. From real profilometer measurements of a cooper engineering surface of length $N = 668$, one can obtain the REPR of length $N_s = 52$, as is shown in Figure 9. As one can see, this REPR does not behave like a fractal or a realisation of a Gaussian process.

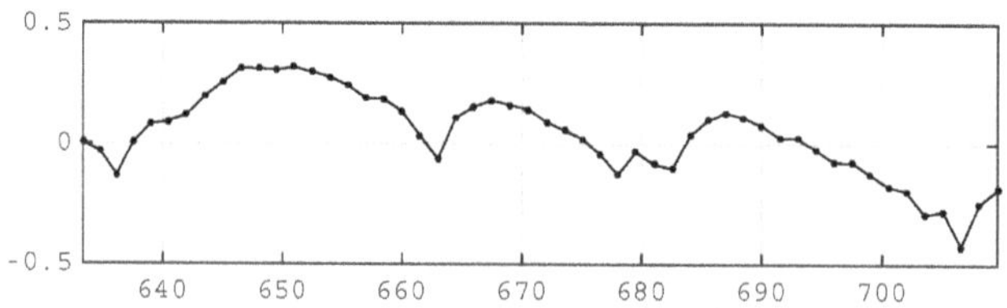

Figure 9. The REPR for the roughness data obtained by a profilometer; $N = 668$, $N_s = 52$, and $i = 423$.

However, we need to find the size of the moving window and its location such that the appropriate pattern (the SRPSR) satisfies not only the condition that it is entirely typical of the whole surface, but also it satisfies some additional conditions depending on the contact problem formulation and the numerical solver used. For example, the synthesised surface may be used by the Polonsky–Keer solver of contact problems for rough surfaces [98,99] based on the use a Fast Fourier Transform (FFT). It is known that the Fourier Transform converts a signal or a graph of a function to the frequency domain. In the case of measurements of roughness, one needs to use a discrete Fourier Transform, which is a rather slow process having complexity $O(n^2)$, where n is the number of discrete measurements. The Fast Fourier Transform (FFT) may greatly reduce the complexity of computation to $O(n \log n)$. However, the classic realisations of the FFT algorithm, e.g., the Cooley–Tukey algorithm, are usually restricted to power-of-two sizes. Hence, if the the Cooley–Tukey or a similar FFT algorithm is employed in the contact solver, then the REPR should be extended to 2^n points.

Another restriction is that the direct employment of the REPR for the synthesis of a surface may cause artificial stress singularities. For example, for a profile (i.e., in the 2D problem), to avoid stress singularities, one can request that the heights at the first point and at the last point are the same.

It is assumed that the synthetic surfaces based on the described REPR techniques will be used within the framework of various contact solvers. For example, the REPR-based method may be combined with the Polonsky–Keer algorithm, which has demonstrated its effectiveness for solving various problems of contact between rough surfaces (see, e.g., [100]).

7. Conclusions

Enhancing engineers' ability to predict real-world tribological behaviour requires modelling the key features of rough surfaces across various roughness scales. The multiscale approach is crucial for accurately capturing the complex interactions that occur between contacting surfaces. If a tribological problem under consideration is governed by both atomic/nano- and microscale phenomena, then the synthesis of surfaces should be based on data obtained by a stylus, or white light interferometry (at the microscale) and AFM, or confocal microscopy (at the atomic/nanoscale). The REPRs should be obtained for both sets of measurements.

Using examples, it was demonstrated that the many common statements about fractal or power spectral density function are wrong or just meaningless. Some terms are ill-defined, and papers based on the employment of the PSD and their applications to rough surfaces create just "new clothes" for the vulgar fractal approach.

It is argued that the use of two new concepts, namely the representative elementary pattern of roughness (REPR), i.e., the smallest interval (or area) of a rough surface that statistically represents the whole surface, and the statistically representative pattern of surface roughness (SRPSR) enabled us to synthesise surfaces that cannot be distinguished statistically from the original rough surfaces. To extract the REPR from the experimental data, one can use some modified statistical techniques of time series analysis, e.g., the moving window technique. Using the Kolmogorov–Smirnov statistic, one can prove that the surfaces consisting of the REPRs are statistically the same as the original surfaces.

Author Contributions: Conceptualisation, F.M.B. and A.P.; methodology, F.M.B. and A.P.; software, A.P.; validation, A.P. and X.J.; formal analysis, F.M.B. and A.P.; writing—original draft preparation, F.M.B. and X.J.; writing—review and editing, F.M.B. and X.J.; visualisation, F.M.B. and A.P.; funding acquisition, F.M.B. and X.J. All authors have read and agreed to the published version of the manuscript.

Funding: This work was funded by the National Natural Science Foundation of China (grant No. 11932004 and No. HWG2022001) and the Opening Fund of State Key Laboratory of Nonlinear Mechanics.

Data Availability Statement: The data will be made available by the authors on request.

Conflicts of Interest: The authors declare no conflicts of interest.

References

1. Hertz, H. Ueber die Berührung fester elastischer Körper. *J. Reine Angew. Math.* **1882**, *92*, 156–171. [CrossRef]
2. Hertz, H. On the contact of elastic solids. In *Miscellaneous Papers by H. Hertz*; Jones, D.E., Schott, G.A., Eds.; Macmillan: London, UK, 1896; pp. 146–162.
3. Boussinesq, J. *Applications des Potentiels 'a l'Étude de l'équilibre et du Mouvement des Solides Élastique*; Gauthier-Villars: Paris, France, 1885.
4. Krantz, T.L.; Alanou, M.P.; Evans, H.P.; Snidle, R.W. Surface fatigue lives of case-carburized gears with an improved surface finish. *ASME J. Tribol.* **2001**, *123*, 709–716. [CrossRef]
5. Müser, M.H.; Dapp, W.B.; Bugnicourt, R.; Sainsot, P.; Lesaffre, N.; Lubrecht, T.A.; Persson, B.N.J.; Harris, K.; Bennett, A.; Schulze, K.; et al. Meeting the contact-mechanics challenge. *Tribol. Lett.* **2017**, *65*, 118. [CrossRef]
6. Greenwood, J.A.; Williamson, J.B.P. Contact of nominally flat surfaces. *Proc. R. Soc. Lond.* **1966**, *A370*, 300–319.
7. Ciavarella, M.; Papangelo, A. The "sport" of rough contacts and the fractal paradox in wear laws. *Facta Univ.* **2018**, *16*, 65–75. [CrossRef]
8. Carpick, R.W. The contact sport of rough surfaces. *Science* **2018**, *359*, 38. [CrossRef] [PubMed]
9. Barenblatt, G.I. *Scaling*; Cambridge University Press: Cambridge, UK, 2003.
10. Chertovskih, R.; Zheligovsky, V. Large-scale weakly nonlinear perturbations of convective magnetic dynamos in a rotating layer. *Phys. D Nonlinear Phenom.* **2015**, *313*, 99–116. [CrossRef]
11. Whitehouse, D.J. *Handbook of Surface and Nanometrology*; CRC Press: Boca Raton, FL, USA, 2011.
12. Borodich, F.M.; Savencu, O. Hierarchical models of engineering rough surfaces and bioinspired adhesives. In *Bio-Inspired Structured Adhesives*; Heepe, L., Xue, L., Gorb, S.N., Eds.; Springer: Berlin/Heidelberg, Germany, 2017; pp. 179–219.
13. Polonsky, I.A.; Keer, L.M. Simulation of microscopic elastic-plastic contacts by using discrete dislocations. *Ser. A Math. Phys. Eng. Sci.* **1996**, *452*, 2173–2194.
14. Polonsky, I.A.; Keer, L.M. Scale effects of elastic-plastic behavior of microscopic asperity contacts. *J. Tribol.* **1996**, *118*, 335–340. [CrossRef]
15. Archard, J.F. Elastic deformation and the laws of friction. *Proc. R. Soc. Lond.* **1957**, *243*, 190–205.
16. Borodich, F.M.; Mosolov, A.B. Fractal roughness in contact problems. *J. Appl. Math. Mech.* **1992**, *56*, 681–690. [CrossRef]
17. Greenwood, J.A.; Wu, J.J. Surface roughness and contact: An apology. *Meccanica* **2001**, *36*, 617–630. [CrossRef]
18. Borodich, F.M.; Mosolov, A.B. Fractal contact of solids. *Sov. Phys.-Tech. Phys.* **1991**, *61*, 50–54.
19. Plesha, M.E.; Ni, D. Scaling of geological discontinuity normal load-deformation response using fractal geometry. *Int. J. Num. Anal. Meth. Geomech.* **2001**, *25*, 741–756. [CrossRef]

20. Warren, T.L.; Krajcinovic, D. Fractal models of elastic–perfectly plastic contact of rough surfaces based on the Cantor set. *Int. J. Solids Struct.* **1995**, *32*, 2907–2922. [CrossRef]
21. Warren, T.L.; Krajcinovic, D. Random Cantor set models for the elastic-perfectly plastic contact of rough surfaces. *Wear* **1996**, *196*, 1–15. [CrossRef]
22. Borodich, F.M.; Onishchenko, D.A. Fractal roughness for problem of contact and friction (the simplest models). *J. Frict. Wear* **1993**, *14*, 14–19.
23. Borodich, F.M.; Onishchenko, D.A. Similarity and fractality in the modelling of roughness by a multilevel profile with hierarchical structure. *Int. J. Solids Struct.* **1999**, *36*, 2585–2612. [CrossRef]
24. Spinner, M.; Kovalev, A.; Gorb, S.N.; Westhoff, G. Snake velvet black: Hierarchical micro- and nanostructure enhances dark colouration in Bitis rhinoceros. *Sci. Rep.* **2013**, *3*, 1846. [CrossRef]
25. Gao, H.; Wang, X.; Yao, H.; Gorb, S.; Arzt, E. Mechanics of hierarchical adhesion structures of geckos. *Mech. Mater.* **2005**, *37*, 275–285. [CrossRef]
26. Borodich, F.M.; Keer, L.M. Modeling effects of gas adsorption and removal on friction during sliding along diamond-like carbon films. *Thin Solid Films* **2005**, *476*, 108–117. [CrossRef]
27. Abbott, E.J.; Firestone, F.A. Specifying surface quality: A method based on accurate measurement and comparison. *Mech. Eng.* **1933**, *55*, 569–572.
28. Whitehouse, D.J. The parameter rash-is there a cure? *Wear* **1982**, *83*, 75–78. [CrossRef]
29. Nowicki, B. Multiparameter representation of surface roughness. *Wear* **1985**, *102*, 161–176. [CrossRef]
30. *EN ISO 4287:1998+A1:2009*; Geometrical Product Specification (GPS)—Surface Texture: Profile Method-Terms, Definitions and Surface Texture Parameters. ISO: Geneva, Switzerland, 2009.
31. *ASME B46.1-2002*; Surface Texture (Surface Roughness, Waviness and Lay). ASME: New York, NY, USA, 2002.
32. Borodich, F.M.; Galanov, B.A.; Perepelkin, N.V.; Prikazchikov, D.A. Adhesive contact problems for a thin elastic layer: Asymptotic analysis and the JKR theory. *Math. Mech. Solids* **2019**, *24*, 1405–1424. [CrossRef]
33. Kaplunov, J.; Prikazchikov, D.A.; Sultanova, L. Justification and refinement of Winkler-Fuss hypothesis. *Z. Angew. Math. Phys.* **2018**, *69*, 80. [CrossRef]
34. Johnson, K.L. *Contact Mechanics*; Cambridge University Press: Cambridge, UK, 1985.
35. Zhuravlev, V.A. On question of theoretical justification of the Amontons-Coulomb law for friction of unlubricated surfaces. *Zh. Tekh. Fiz.* **1940**, *10*, 1447–1452. [CrossRef]
36. Zhuravlev, V.A. On question of theoretical justification of the Amontons-Coulomb law for friction of unlubricated surfaces. *Proc. Inst. Mech. Eng. Part J J. Eng. Tribol.* **2007**, *221*, 893–898. [CrossRef]
37. Kragelsky, I.V. Static friction between two rough surfaces. *Bul. USSR Acad. Sci. Div. Tech. Sci.* **1948**, *10*, 1621–1625.
38. Greenwood, J.A. Surface modelling in tribology. In *Applied Surface Modelling*; Creasy, C.F.M., Craggs, C., Eds.; Ellis Horwood: New York, NY, USA, 1990; pp. 61–75.
39. Greenwood, J.A. Problems with surface roughness. In *Fundamentals of Friction*; Singer, I.L., Pollock, H.M., Eds.; Kluwer: Boston, MA, USA, 1992; pp. 57–76.
40. Whitehouse, D.J.; Archard, J.F. The properties of random surfaces of significance in their contact. *Proc. R. Soc. Lond.* **1970**, *A316*, 97–121.
41. Khusu, A.P.; Vitenberg, Y.R.; Palmov, V.A. *Roughness of Surfaces: Theoretical Probabilistic Approach*; Nauka: Moscow, Russia, 1975.
42. Maugis, D. *Contact, Adhesion and Rupture of Elastic Solids*; Springer: Berlin/Heidelberg, Germany, 2000.
43. Linnik, Y.V.; Khusu, A.P. Mathematical and statistical description of unevenness of surface profile at grinding. *Uspekhi Mat. Nauk.* **1954**, *9*, 255.
44. Linnik, Y.V.; Khusu, A.P. Mathematical and statistical description of unevenness of surface profile at grinding. *Bul. USSR Acad. Sci. Div. Techn. Sci.* **1954**, *20*, 154–159.
45. Barakat, R.; Baumann, E. Mean and variance of the arc length of a Gaussian process on a finite interval. *Int. J. Control* **1970**, *12*, 377–383. [CrossRef]
46. Nayak, P.R. Random process model of rough surfaces. *ASME J. Lub. Tech.* **1971**, *93*, 398–407. [CrossRef]
47. Nayak, P.R. Some aspects of surface roughness measurement. *Wear* **1973**, *26*, 165–174. [CrossRef]
48. Fuller, K.N.G.; Tabor, D. The effect of surface roughness on the adhesion of elastic solids. *Proc. R. Soc. Lond. A* **1975**, *345*, 327–342.
49. Fuller, K. Effect of surface roughness on the adhesion of elastomers to hard surfaces. *Mater. Sci. Forum* **2011**, *662*, 39–51. [CrossRef]
50. Galanov, B.A. Models of adhesive contact between rough elastic bodies. *Int. J. Mech. Sci.* **2011**, *53*, 968–977. [CrossRef]
51. Galanov, B.A.; Valeeva, I.K. Sliding adhesive contact of elastic solids with stochastic roughness. *Int. J. Eng. Sci.* **2016**, *101*, 64–80. [CrossRef]
52. Thode, H.C. *Testing for Normality*; Marcel Dekker: New York, NY, USA, 2002.
53. Borodich, F.M.; Pepelyshev, A.; Savencu, O. Statistical approaches to description of rough engineering surfaces at nano and microscales. *Tribol. Int.* **2016**, *103*, 197–207. [CrossRef]
54. Pepelyshev, A.; Borodich, F.M.; Galanov, B.A.; Gorb, E.V.; Gorb, S.N. Adhesion of soft materials to rough surfaces: Experimental studies, statistical analysis and modelling. *Coatings* **2018**, *8*, 350. [CrossRef]
55. Borodich, F.M.; Brousseau, E.; Clarke, A.; Pepelyshev, A.; Sanchez-Lopez, J.C. Roughness of deposited carbon-based coatings and its statistical characteristics at nano and microscales. *Front. Mech. Eng.* **2019**, *5*, 73–82. [CrossRef]

56. Vilenkin, N.Y. *Stories about Sets*; Academic Press: New York, NY, USA, 1968.
57. Mandelbrot, B.B. *Fractals: Form, Chance, and Dimension*; W.H.Freemanl: San Francisco, CA, USA, 1977.
58. Mandelbrot, B.B. Fractal geometry: What is it, and what does it do? *Proc. R. Soc. Lond.* **1989**, *423*, 3–16.
59. Borodich, F.M.; Jin, X.; Pepelyshev, A. Probabilistic, fractal, and related techniques for analysis of engineering surfaces. *Front. Mech. Eng.* **2020**, *6*, 64. [CrossRef]
60. Berry, M.V.; Hannay, J.H. Topography of random surfaces. *Nature* **1978**, *273*, 573. [CrossRef]
61. Sayles, R.S.; Thomas, T.R. Surface topography as a nonstationary random process. *Nature* **1978**, *271*, 431–434. [CrossRef]
62. Berry, M.V.; Lewis, Z.V. On the Weierstrass-Mandelbrot fractal functions. *Proc. R. Soc. Lond.* **1980**, *370*, 459–484.
63. Majumdar, A.; Bhushan, B. Role of fractal geometry in roughness characterization and contact mechanics of surfaces. *J. Tribol.* **1990**, *112*, 205–216. [CrossRef]
64. Majumdar, A.; Bhushan, B. Fractal model of elastic-plastic contact between rough surfaces. *J. Tribol.* **1991**, *113*, 1–11. [CrossRef]
65. Ciavarella, M.; Demelio, G.; Barber, J.R.; Jang, Y.H. Linear elastic contact of the Weierstrass profile. *Proc. R. Soc. Lond. A* **2000**, *456*, 387–405. [CrossRef]
66. Borodich, F.M. Fractal Contact Mechanics. In *Encyclopedia of Tribology*; Wang, Q.J., Chung, Y.-W., Eds.; Springer: Berlin/Heidelberg, Germany, 2013; Volume 2, pp. 1249–1258.
67. Bhushan, B. A fractal theory of the temperature distribution at elastic contacts of fast sliding surfaces—Discussion. *J. Tribol.* **1995**, *117*, 214–215. [CrossRef]
68. Bhushan, B. *Modern Tribology Handbook*; CRC Press: Boca Raton, FL, USA, 2001.
69. Borodich, F.M. Some fractal models of fracture. *J. Mech. Phys. Solids* **1997**, *45*, 239–259. [CrossRef]
70. Borodich, F.M.; Feng, Z. Scaling of mathematical fractals and box-counting quasi-measure. *Z. Angew. Math. Phys.* **2010**, *61*, 21–31. [CrossRef]
71. Borodich, F.M.; Volovikov, A.Y. Surface integrals for domains with fractal boundaries and some applications to elasticity. *Proc. R. Soc. Lond. Ser. A.* **2000**, *456*, 1–23. [CrossRef]
72. Davies, S.; Hall, P. Fractal analysis of surface roughness by using spatial data. *J. R. Statist. Soc. B* **1999**, *61*, 3–37. [CrossRef]
73. Avnir, D.; Biham, O.; Lidar, D.; Malcai, O. Is the geometry of Nature fractal? *Science* **1998**, *279*, 39–40. [CrossRef]
74. Malcai, O.; Lidar, D.A.; Biham, O.; Avnir, D. Scaling range and cutoffs in empirical fractals. *Phys. Rev. E* **1997**, *56*, 2817–2828. [CrossRef]
75. Mandelbrot, B.B. Is Nature fractal? *Science* **1998**, *279*, 783. [CrossRef]
76. Biham, O.; Malcai, O.; Lidax, D.A.; Avnir, D. Fractality in Nature—Response. *Science* **1998**, *279*, 1615–1616.
77. Whitehouse, D.J. Fractal or fiction. *Wear* **2001**, *249*, 345–353. [CrossRef]
78. Borodich, F.M. Parametric homogeneity and non-classical self-similarity. I. Mathematical background. *Acta Mech.* **1998**, *131*, 27–45. [CrossRef]
79. Borodich, F.M. Parametric homogeneity and non-classical self-similarity. II. Some applications. *Acta Mech.* **1998**, *131*, 47–67. [CrossRef]
80. Borodich, F.M.; Galanov, B.A. Self-similar problems of elastic contact for non-convex punches. *J. Mech. Phys. Solids* **2002**, *50*, 2441–2461. [CrossRef]
81. Borodich, F.M. Similarity properties of discrete contact between a fractal punch and an elastic medium. *C. R. L'Acad́emie Sci. Ser. 2* **1993**, *316*, 281–286.
82. Persson, B.N.J. Theory of rubber friction and contact mechanics. *J. Chem. Phys.* **2001**, *115*, 3840–3861. [CrossRef]
83. Persson, B.N.J. Contact mechanics for randomly rough surfaces. *Surf. Sci. Rep.* **2006**, *61*, 201–227. [CrossRef]
84. Falconer, K.J. *Fractal Geometry: Mathematical Foundations and Applications*; John Wiley: Chichester, UK, 1990.
85. Persson, B.N.J. Elastoplastic Contact between Randomly Rough Surfaces. *Phys. Rev. Lett.* **2001**, *87*, 116101. [CrossRef] [PubMed]
86. Borodich, F.M. Comment on "Elastoplastic Contact between Randomly Rough Surfaces". *Phys. Rev. Lett.* **2002**, *88*. [CrossRef] [PubMed]
87. Persson, B.N.J. Functional properties of rough surfaces from an analytical theory of mechanical contact. *MRS Bull.* **2023**, *47*, 1211–1219. [CrossRef]
88. Marshak, S. *Collection of Works in 8 Volumes, 4*; Khudozhestvennaya Literature: Moscow, Russia, 1969.
89. Chen, J.; Tang, J.; Shao, W.; Sun, Z.; Zhang, H.; Li, X.; Zhao, B. Numerical simulation method for three-dimensional rough surface of vibratory polishing parts. *Tribol. Int.* **2024**, *193*, 109417. [CrossRef]
90. Pawlus, P. Simulation of stratified surface topographies. *Wear* **2008**, *264*, 457–463. [CrossRef]
91. Tang, J.; Chen, J.; Yang, D.; Li, L.; Zhao, J.; Guo, M. A new method of layered superposition reconstruction modeling on grinding-shot peening surfaces. *Surf. Topogr. Metrol. Prop.* **2020**, *10*, 045010. [CrossRef]
92. Pawlus, P.; Reizer, R.; Wieczorowski, M. A review of methods of random surface topography modeling. *Tribol. Int.* **2020**, *152*, 106530. [CrossRef]
93. Borodich, F.M.; Pepelyshev, A.; Savencu, O. Micro and nano scale statistical properties of rough surfaces of significance in their friction. In Proceedings of the 6th Vienna International Conference Nano-Technology—Viennano'15, Wiener Neustadt, Austria, 24–25 November 2015; Bartz, W.J., Franek, F., Eds.; The Austrian Tribology Society, (OETG): Wiener Neustadt, Austria, 2015; pp. 13–14.
94. Golyandina, N.; Zhigljavsky, A. *Singular Spectrum Analysis for Time Series*; Springer: Berlin/Heidelberg, Germany, 2020.

95. Borodich Suarez, S.; Pepelyshev, A. Study of impact of COVID-19 on industrial production indices using singular spectrum analysis. *Stat. Its Interface* **2023**, *16*, 181–188. [CrossRef]
96. Borodich Suarez, S.; Heravi, S.; Pepelyshev, A. Forecasting industrial production indices with a new singular spectrum analysis forecasting algorithm. *Stat. Its Interface* **2023**, *16*, 31–42. [CrossRef]
97. Borodich, F.M.; Pepelyshev, A. Synthesis of Engineering Surfaces using Representative Elementary Patterns of Roughness. 2024, *submitted*.
98. Polonsky, I.A.; Keer, L.M. A numerical method for solving rough contact problems based on the multi-level multi-summation and conjugate gradient techniques. *Wear* **1999**, *231*, 206–219. [CrossRef]
99. Polonsky, I.A.; Keer, L.M. Fast methods for solving rough contact problems: A comparative study. *J. Tribol.* **2000**, *122*, 36–41. [CrossRef]
100. Zhou, Q.; Jin, X.; Wang, Z.; Wang, J.; Keer, L.M.; Wang, Q. Numerical EIM with 3D FFT for the contact with a smooth or rough surface involving complicated and distributed inhomogeneities. *Tribol. Int.* **2016**, *93*, 91–103. [CrossRef]

mathematics MDPI

Article

Asymptotic Analysis of an Elastic Layer under Light Fluid Loading

Sheeru Shamsi and Ludmila Prikazchikova *

School of Computer Science and Mathematics, Keele University, Keele ST5 5BG, UK; s.s.shamsi@keele.ac.uk
* Correspondence: l.prikazchikova@keele.ac.uk

Abstract: Asymptotic analysis for an elastic layer under light fluid loading was developed. The ratio of fluid and solid densities was chosen as the main small parameter determining a novel scaling. The leading- and next-order approximations were derived from the full dispersion relation corresponding to long-wave, low-frequency, antisymmetric motions. The asymptotic plate models, including the equations of motion and the impenetrability condition, motivated by the aforementioned shortened dispersion equations, were derived for a plane-strain setup. The key findings included, in particular, the necessity of taking into account transverse plate inertia at the leading order, which is not the case for heavy fluid loading. In addition, the transverse shear deformation, rotation inertia, and a number of other corrections appeared at the next order, contrary to the previous asymptotic developments for fluid-loaded plates not assuming a light fluid loading scenario.

Keywords: fluid–structure interaction; elastic layer; asymptotic analysis; light fluid loading; refined plate models

MSC: 74F10

1. Introduction

Numerous problems involving vibrating structures under fluid loading have been studied in the literature since the late nineteenth century; for example, see "Theory of Sound" [1] by Lord Rayleigh. The theoretical analysis on the subject most often makes use of 2D structural theories such as the classical Kirchhoff theory for thin elastic plates, e.g., see the 1988 Rayleigh medal lecture [2] and the books [3,4]. Ideally, approximate coupled models in fluid–structure interaction have to be mathematically justified starting from 3D dynamic theory in linear elasticity. At the same time, until recently, there have been limited attempts to address this matter, mainly using the traditional formulations based on 2D engineering structural theories using analytical methods [5–9], as well as [10–14] promoting experimental and numerical techniques. As an exception, we mention the paper by Johansson et al. [15], attempting to address this problem in the context of a fluid-loaded elastic plate but lacking an asymptotic consistency. We also mention a previous effort to adapt refined asymptotic setups for thin elastic shells not in contact with fluid to fluid–structure interaction problems; e.g., see [16]. A fresh work by Kaplunov et al. [17] developed a hierarchy of asymptotic models for a fluid-loaded elastic layer, emphasising the point that such a layer requires a special treatment, extending the well-established setup of Neumann boundary conditions for a layer with traction-free or mechanically loaded faces; e.g., see [18]. In this case, the effect of fluid loading supports the so-called fluid-borne bending wave; e.g., see [19], which assumed a novel asymptotic scaling. The methodology developed in [17] was next extended to low-frequency acoustic wave scattering by a circular cylindrical shell in [20].

It is worth noting that the aforementioned scaling does not cover the important scenario of light fluid loading, for which the ratio of the density of the fluid to the density of the solid is small. The general understanding of the related asymptotic limit from the prospects of dynamic elasticity seems to be of significant interest for modelling various fluid-loaded elastic structures. A special focus on the light fluid loading limit is given by Craster [21], establishing a perturbation scheme to obtain a robust approximate solution for a fluid-loaded elastic solid; see also [22,23], tackling both light and heavy fluid loading limits. A detailed asymptotic analysis of a Kirchhoff plate under light fluid loading was reported by Chapman and Sorokin [24] using the same small parameter as in the cited paper [21], involving the product of not only the ratio of densities of fluid and plate, but also the related wave speeds. The aforementioned small parameter was also adopted in [25], dealing with acoustic radiation due to harmonic vibrations of an elastic layer over a broad frequency range.

The proposed analysis is motivated by a lack of consistent fluid-structure models taking into account the effect of fluid loading. In particular, the dimension reduction for immersed thin-walled structures has not been yet asymptotically validated. To this end, the considerations in [17,20] have to be extended to the case of light fluid loading.

As an example, in this paper, we study a plane-strain time-harmonic problem for an elastic layer immersed into compressible fluid. The ratio of densities is chosen as the main small parameter. It seems to be more appropriate for the considered framework than that adopted in [21,24]. Below, for the sake of simplicity, we determine the long-wave scale through the above-mentioned small parameter. Its presence affects the asymptotic analysis of fluid-borne wave dictation taking into account the plate transverse inertia at the leading order. The transverse shear deformation, rotation inertia, and similar corrections, including those in the impenetrability condition, also appear at lower-order approximations in comparison with the treatment in paper [17].

This paper is organised as follows. The governing equations are given in Section 2. Section 3 is concerned with the derivation of the leading and first-order approximations to the full dispersion relation over the low-frequency range in the light fluid loading limit. This motivates the scaling for further asymptotic analysis of the problem. Section 4 aims at the formulation of the asymptotic models, supporting the above-mentioned approximations of the dispersion relation. The presentation in this section is structured similar to that in [17], which does not assume light fluid loading. The concluding remarks are summarised in Section 5.

2. Governing Equations

Consider small-amplitude free vibrations of an isotropic linearly elastic layer of thickness $2h$ immersed in a non-viscous compressible fluid. The Cartesian coordinate system is set up in such a way that axis x_1 goes through the midplane of the layer; see Figure 1. The axis x_2, perpendicular to the plane (x_1, x_3), is not shown in the figure. The following notation is used throughout the paper: E is Young's modulus, ν is the Poisson's ratio, ρ and ρ_0 are the solid and fluid densities, respectively, c_0 is the wave speed in fluid, and $c_2 = \sqrt{E/2\rho(1+\nu)}$ is the shear wave speed in solid.

We limit ourselves to a plane-strain problem in the coordinates (x_1, x_3). Thus, the equations of motion in linear elasticity may be written as

$$\begin{aligned} \frac{\partial \sigma_{11}}{\partial x_1} + \frac{\partial \sigma_{31}}{\partial x_3} - \rho \frac{\partial^2 v_1}{\partial t^2} &= 0, \\ \frac{\partial \sigma_{13}}{\partial x_1} + \frac{\partial \sigma_{33}}{\partial x_3} - \rho \frac{\partial^2 v_3}{\partial t^2} &= 0. \end{aligned} \tag{1}$$

Here and below, v_k $(k = 1,3)$ are displacements, σ_{mn} $(m,n = 1,2,3)$ are stresses, and t denotes time. The stresses and displacements given in the above equations satisfy the following relations, e.g., see [18]

$$
\begin{aligned}
\sigma_{11} &= \frac{E}{1-\nu^2}\frac{\partial v_1}{\partial x_1} + \frac{\nu}{1-\nu}\sigma_{33}, \\
\sigma_{22} &= \frac{E\nu}{1-\nu^2}\frac{\partial v_1}{\partial x_1} + \frac{\nu}{1-\nu}\sigma_{33}, \\
\frac{\partial v_3}{\partial x_3} &= \frac{1}{E}\left(\sigma_{33} - \nu(\sigma_{11}+\sigma_{22})\right), \\
\frac{\partial v_1}{\partial x_3} &= -\frac{\partial v_3}{\partial x_1} + \frac{2(1+\nu)}{E}\sigma_{31},
\end{aligned} \tag{2}
$$

adapted for the forthcoming asymptotic analysis.

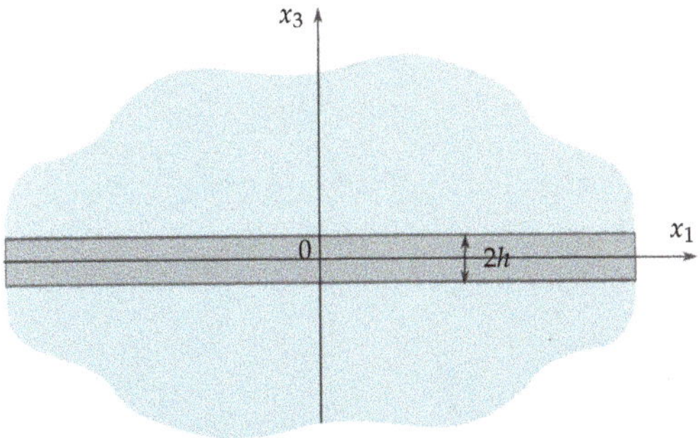

Figure 1. Elastic layer immersed in fluid.

In addition, the fluid velocity potential $\varphi(x_1, x_3, t)$, see for example [25–27], satisfies the wave equation

$$
\frac{\partial^2 \varphi}{\partial x_1^2} + \frac{\partial^2 \varphi}{\partial x_3^2} - \frac{1}{c_0^2}\frac{\partial^2 \varphi}{\partial t^2} = 0, \tag{3}
$$

with the interfacial conditions at $x_3 = \pm h$, given by

$$
\sigma_{31} = 0, \quad \sigma_{33} = \rho_0 \frac{\partial^2 \varphi}{\partial t^2}, \quad v_3 = \frac{\partial \varphi}{\partial x_3}. \tag{4}
$$

The main objective of this paper is to derive asymptotic models of the formulated problem over a long-wave low-frequency region under the conditions of light fluid loading. For this, $\rho_0 \ll \rho$, and $h \ll L$ and $h/c_2 \ll T$, with L and T denoting a typical wavelength and time scale, respectively. We also restrict ourselves to bending vibrations. Prior to proceeding with the asymptotic treatment of the equations of motion, we first study the associated shortened forms of the dispersion relation corresponding to (1)–(4).

3. Asymptotic Analysis of the Dispersion Relation

In this section, we analyse the antisymmetric dispersion relation, e.g., see [28]

$$
\Omega^4 r \cosh(A)\cosh(B) + H\left((2K^2-\Omega^2)^2 \frac{\sinh(A)}{A}\cosh(B) - 4K^2B^2\frac{\sinh(B)}{B}\cosh(A)\right) = 0, \tag{5}
$$

with

$$
A = \sqrt{K^2 - \Omega^2\kappa^2}, \quad B = \sqrt{K^2 - \Omega^2}, \quad H = \sqrt{K^2 - \Omega^2\delta^2}, \tag{6}
$$

and

$$
\Omega = \frac{\omega h}{c_2}, \qquad K = kh, \tag{7}
$$

where ω and k are angular frequency and wavenumber, respectively; $\delta = \frac{c_2}{c_0}$ and $\kappa = \sqrt{\frac{1-2\nu}{2-2\nu}}$.

Let us assume that $r = \rho_0/\rho \ll 1$ for a light fluid loading scenario and try the long-wave low-frequency scaling

$$\Omega = \Omega_* r^2, \qquad K = K_* r, \tag{8}$$

where Ω_* and K_* are assumed so far to be of order unity. In this case, the relation $\Omega \sim K^2$, characteristic of the bending wave on a free elastic plate, is satisfied; e.g., see [18].

Expanding Equation (5) into a Taylor series and taking into account the scaling (8), we obtain a two-term expansion in the small parameter, r,

$$\begin{aligned}&\left((K_*+1)\Omega_*^2 + \frac{2K_*^5}{3(\nu-1)}\right)\\&+\frac{(-15\delta^2(\nu-1)\Omega_*^4 + ((-10\delta^2+20(\nu-2))K_* + 30(\nu-1))K_*^3\Omega_*^2 + 4K_*^8)}{30K_*(\nu-1)}r^2 = 0.\end{aligned} \tag{9}$$

At the leading order, it yields

$$\Omega_*^2 = \frac{2K_*^5}{3(K_*+1)(1-\nu)}. \tag{10}$$

Now, by replacing the Ω_*^2 and Ω_*^4 terms in the coefficient at r^2 in (9) using (10), we obtain an $O(r^2)$ correction to the leading-order estimation. It takes the form of

$$\Omega_*^2 = \frac{2K_*^5}{3(K_*+1)(1-\nu)} + \frac{2((7\nu-17)K_*^2 - (5\delta^2 - 19\nu + 29)K_* + 12(\nu-1))K_*^7}{45(K_*+1)^3(1-\nu)^2}r^2. \tag{11}$$

Next, consider two limiting behaviours, for which $K_* \ll 1$ and $K_* \gg 1$, in order to elucidate the relation with previous results. For the first case, we obtain from (10) and (11)

$$\Omega_*^2 = \frac{2K_*^5}{3(1-\nu)}, \tag{12}$$

and

$$\Omega_*^2 = \frac{2K_*^5}{3(1-\nu)} - \frac{2K_*^6}{3(1-\nu)}, \tag{13}$$

respectively. In terms of parameters K and Ω, the latter becomes

$$\Omega^2 = \frac{2K^5}{3r(1-\nu)} - \frac{2K^6}{3r^2(1-\nu)}. \tag{14}$$

It is worth noting that it is identical to the two-term expansion in [28], not assuming light fluid loading.

For $K_* \gg 1$ we have from (10) and (11)

$$\Omega_*^2 = \frac{2K_*^4}{3(1-\nu)}, \tag{15}$$

and

$$\Omega_*^2 = \frac{2K_*^4}{3(1-\nu)} + \frac{2(7\nu-17)r^2K_*^6}{45(1-\nu)^2}. \tag{16}$$

We remark that for a plate without fluid loading, the two-term expansion of the Rayleigh–Lamb dispersion relation for the bending wave in the long-wave low-frequency region ($\Omega \sim K^2, K \ll 1$) is given by, see [18],

$$\Omega^2 = \frac{2K^4}{3(1-\nu)} + \frac{2(7\nu - 17)K^6}{45(1-\nu)^2}. \quad (17)$$

The last formula, rewritten in terms of K_* and Ω_*, coincides with (16).

The rest of this section reports on numerical results for a steel layer immersed in water, illustrated by Figures 2 and 3. The graphs presented in this paper have been produced using Python and Maple 2021 software. The problem parameters utilised are $c_0 = 1480$ m s^{-1}, $\nu = 0.2$, $c_2 = 3156$ ms^{-1}, $\rho = 7800$ kgm^{-3}, and $\rho_0 = 1000$ kgm^{-3}. Hence, the small parameter, characteristic of the light fluid loading, is $r = \rho_0/\rho \approx 0.128$. Indeed, this type of small parameter could be relevant to other materials, not just a steel/water combination. The latter is chosen in this paper to illustrate the behaviour corresponding to light fluid loading. For other materials with the same parameter r, the graphs will look similar.

The dispersion curves for leading-order (10) and first-order (11) approximations are shown in Figure 2, along with that for the full dispersion relation (5), rewritten in terms of (8). As can be observed, the first-order approximation works better in comparison with the leading order one due to the fact that it takes into account higher-order terms in the expansions and hence is in better agreement with the full dispersion relation.

Figure 3 demonstrates the limiting cases $K_* \ll 1$ and $K_* \gg 1$ of the leading-order approximation (10). It shows the comparison of the leading-order approximation (10) with the one-term expansions (12) and (15), oriented to the regions $K_* \ll 1$ and $K_* \gg 1$, respectively. The intersection of the blue and red curves in Figure 3 arises due to the approximation error being equal for both dispersion curves.

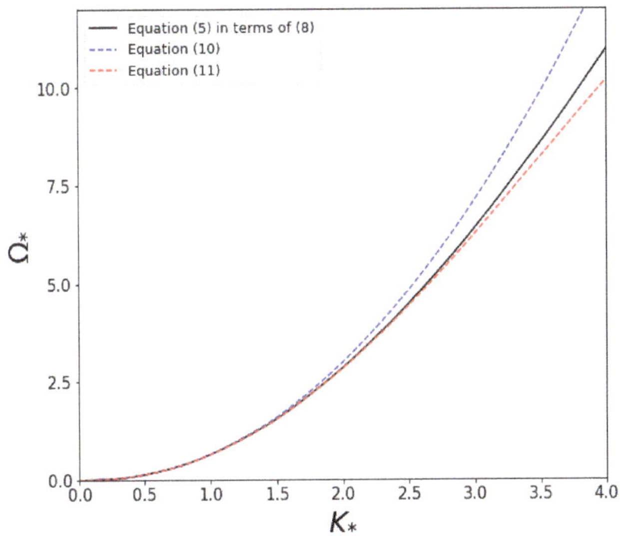

Figure 2. Comparison of the exact (solid black line) dispersion relation ((5) rewritten in terms of (8)) with the leading order-approximation (10) (dashed blue line) and the first-order approximation (11) (dashed red line).

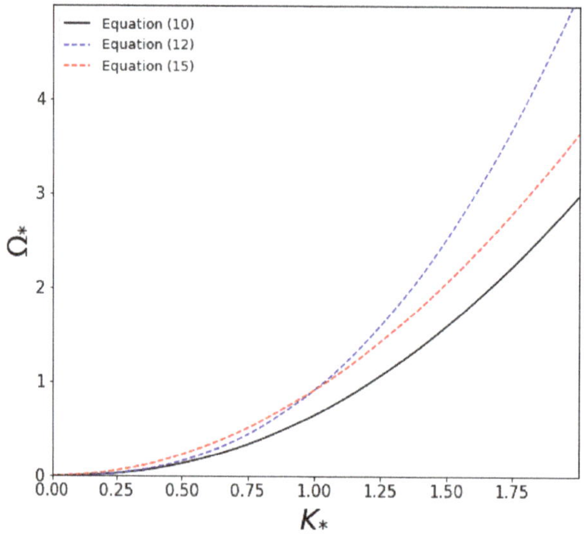

Figure 3. Comparison of the leading-order approximation (10) (solid black line) with one-term expansion (12) for $K_* \ll 1$ (dashed blue line) and one-term expansion (15) for $K_* \gg 1$ (dashed red line).

4. Asymptotic Models

4.1. Scaling

Below, for the first time, we proceed with the asymptotic dimension reduction for the dynamic equations in linear elasticity using the small ratio of densities r, typical for light fluid loading, as the main small parameter. The developed procedure amends the approach delivered in the recent paper [17].

First, scale the independent variables specified in the previous section by

$$t = T\tau, \qquad x_1 = L\xi, \qquad x_3 = \begin{cases} Lr\zeta, & \text{if } |x_3| < h, \\ L\gamma, & \text{otherwise,} \end{cases} \tag{18}$$

where for the sake of simplicity $L = h/r$, $(L \gg h)$ is a typical wavelength and $T = \dfrac{h}{r^2 c_2}$, $(T \gg h/c_2)$, which is motivated by the relation $\omega h/c_2 \sim (kh)^{5/2}$ between angular frequency ω and wave number k for a fluid-borne bending wave; see [28] as well as Formula (8) in Section 3.

Next, introduce the dimensionless stresses, displacements, and fluid potential setting

$$\begin{aligned} &\sigma_{11} = Er\sigma_{11}^*, \quad \sigma_{22} = Er\sigma_{22}^*, \quad \sigma_{31} = Er^2\sigma_{31}^*, \quad \sigma_{33} = Er^3\sigma_{33}^*, \\ &v_1 = hv_1^*, \quad v_3 = \frac{h}{r}v_3^*, \quad \varphi = \frac{h^2}{r^2}\varphi^*. \end{aligned} \tag{19}$$

The starred quantities above are assumed to be of order unity. Hence, Equations (1) and (2) can be written in the following dimensionless form

$$\begin{aligned} \frac{\partial \sigma_{31}^*}{\partial \zeta} &= -\frac{\partial \sigma_{11}^*}{\partial \xi} + \frac{1}{2(1+\nu)} r^2 \frac{\partial^2 v_1^*}{\partial \tau^2}, \\ \frac{\partial \sigma_{33}^*}{\partial \zeta} &= -\frac{\partial \sigma_{31}^*}{\partial \xi} + \frac{1}{2(1+\nu)} \frac{\partial^2 v_3^*}{\partial \tau^2}, \end{aligned} \tag{20}$$

and

$$\begin{aligned}
\sigma_{11}^{*} &= \frac{1}{1-\nu^{2}}\frac{\partial v_{1}^{*}}{\partial \xi} + \frac{\nu}{1-\nu} r^{2}\sigma_{33}^{*}, \\
\sigma_{22}^{*} &= \frac{\nu}{1-\nu^{2}}\frac{\partial v_{1}^{*}}{\partial \xi} + \frac{\nu}{1-\nu} r^{2}\sigma_{33}^{*}, \\
\frac{\partial v_{3}^{*}}{\partial \zeta} &= r^{4}\sigma_{33}^{*} - \nu r^{2}(\sigma_{11}^{*} + \sigma_{22}^{*}), \\
\frac{\partial v_{1}^{*}}{\partial \zeta} &= -\frac{\partial v_{3}^{*}}{\partial \xi} + 2(1+\nu) r^{2}\sigma_{31}^{*}.
\end{aligned} \tag{21}$$

In addition, the dimensionless form of relations (3) and (4) becomes

$$\frac{\partial^{2}\varphi^{*}}{\partial \xi^{2}} + \frac{\partial^{2}\varphi^{*}}{\partial \gamma^{2}} - r^{2}\delta^{2}\frac{\partial^{2}\varphi^{*}}{\partial \tau^{2}} = 0, \tag{22}$$

and

$$\begin{aligned}
&\sigma_{31}^{*}\Big|_{\zeta=\pm 1} = 0, \qquad \sigma_{33}^{*}\Big|_{\zeta=\pm 1} = \frac{1}{2(1+\nu)}\frac{\partial^{2}\varphi^{*}}{\partial \tau^{2}}\Big|_{\gamma=\pm r}, \\
&\frac{\partial \varphi^{*}}{\partial \gamma}\Big|_{\gamma=\pm r} = v_{3}^{*}\Big|_{\zeta=\pm 1}.
\end{aligned} \tag{23}$$

Next, expand the starred displacement and stress components as well as the fluid displacement potential in an asymptotic series as

$$\begin{aligned}
v_{1}^{*} &= v_{1}^{(0)} + r^{2}v_{1}^{(1)} + r^{4}v_{1}^{(2)} + \ldots \\
v_{3}^{*} &= v_{3}^{(0)} + r^{2}v_{3}^{(1)} + r^{4}v_{3}^{(2)} + \ldots \\
\sigma_{11}^{*} &= \sigma_{11}^{(0)} + r^{2}\sigma_{11}^{(1)} + r^{4}\sigma_{11}^{(2)} + \ldots \\
\sigma_{22}^{*} &= \sigma_{22}^{(0)} + r^{2}\sigma_{22}^{(1)} + r^{4}\sigma_{22}^{(2)} + \ldots \\
\sigma_{33}^{*} &= \sigma_{33}^{(0)} + r^{2}\sigma_{33}^{(1)} + r^{4}\sigma_{33}^{(2)} + \ldots \\
\sigma_{31}^{*} &= \sigma_{31}^{(0)} + r^{2}\sigma_{31}^{(1)} + r^{4}\sigma_{31}^{(2)} + \ldots \\
\varphi^{*} &= \varphi^{(0)} + r^{2}\varphi^{(1)} + r^{4}\varphi^{(2)} + \ldots
\end{aligned} \tag{24}$$

leading to shortened forms of the original plane-strain problem in hydro-elasticity.

In what follows, we restrict ourselves to antisymmetric motion about the midplane $x_3 = 0$. In this case, due to the symmetry of the problem, only the interfacial conditions along the upper face $x_3 = h$ are considered.

4.2. Leading-Order Approximation

Here, we focus on the leading-order approximation of the problem formulated in the previous subsection, retaining only the terms with the suffix (0) in the asymptotic series (24). Firstly, integrating $(21)_3$–$(21)_4$ along the thickness variable ζ, we obtain, respectively,

$$v_{3}^{(0)} = V_{3}^{(0)}(\xi,\tau) \quad \text{and} \quad v_{1}^{(0)} = -\zeta\frac{\partial V_{3}^{(0)}}{\partial \xi}. \tag{25}$$

Upon substituting the displacements (25) into $(21)_1$–$(21)_2$, we obtain

$$\sigma_{11}^{(0)} = -\zeta \frac{1}{1-\nu^2} \frac{\partial^2 V_3^{(0)}}{\partial \xi^2} \quad \text{and} \quad \sigma_{22}^{(0)} = -\zeta \frac{\nu}{1-\nu^2} \frac{\partial^2 V_3^{(0)}}{\partial \xi^2}. \tag{26}$$

Next, integrating $(20)_1$ along ζ, taking into account (25) and (26), leads to

$$\sigma_{31}^{(0)} = \zeta^2 \frac{1}{2(1-\nu^2)} \frac{\partial^3 V_3^{(0)}}{\partial \xi^3} + C^{(0)}(\xi, \tau), \tag{27}$$

where $C^{(0)}$ is an arbitrary function. Finally, inserting (25)–(27) into $(20)_2$ and integrating along ζ yields

$$\sigma_{33}^{(0)} = -\zeta^3 \frac{1}{6(1-\nu^2)} \frac{\partial^4 V_3^{(0)}}{\partial \xi^4} - \zeta \frac{\partial C^{(0)}}{\partial \xi} + \zeta \frac{1}{2(1+\nu)} \frac{\partial^2 V_3^{(0)}}{\partial \tau^2}. \tag{28}$$

Furthermore, the fluid potential at the leading order can be obtained by inserting $(24)_7$ into (22), arriving at the Laplace equation

$$\frac{\partial^2 \varphi^{(0)}}{\partial \xi^2} + \frac{\partial^2 \varphi^{(0)}}{\partial \gamma^2} = 0, \tag{29}$$

which governs incompressible fluid. On the other hand, inserting (24) into (23), leads to the interfacial conditions

$$\sigma_{31}^{(0)} \Big|_{\zeta=1} = 0, \quad \sigma_{33}^{(0)} \Big|_{\zeta=1} = \frac{1}{2(1+\nu)} \frac{\partial^2 \varphi^{(0)}}{\partial \tau^2} \Bigg|_{\gamma=r}, \quad \frac{\partial \varphi^{(0)}}{\partial \gamma} \Bigg|_{\gamma=r} = v_3^{(0)}. \tag{30}$$

By applying $(30)_1$ and $(30)_2$ to (27) and (28), respectively, we arrive at

$$C^{(0)} = -\frac{1}{2(1-\nu^2)} \frac{\partial^3 V_3^{(0)}}{\partial \xi^3}, \tag{31}$$

and

$$\frac{2}{3(1-\nu)} \frac{\partial^4 V_3^{(0)}}{\partial \xi^4} + \frac{\partial^2 V_3^{(0)}}{\partial \tau^2} - \frac{\partial^2 \varphi^{(0)}}{\partial \tau^2} \Bigg|_{\gamma=r} = 0. \tag{32}$$

The derived equation corresponds to a fluid-loaded Kirchhoff plate, which is similar to the one derived at the first (not leading) order in [17]. It is worth noting that the leading-order approximation in [17], dealing with a non-contrast case, does not contain transverse plate inertia, given by the second term in (32).

4.3. First-Order Approximation

In this subsection, the first-order approximation is derived, also retaining the terms with the suffix (1) in the asymptotic series (24). The procedure is essentially the same as in the previous subsection; hence, some of the intermediate computations are omitted. Now, we have

$$\frac{\partial^2 \varphi^{(1)}}{\partial \xi^2} + \frac{\partial^2 \varphi^{(1)}}{\partial \gamma^2} - \delta^2 \frac{\partial^2 \varphi^{(0)}}{\partial \tau^2} = 0, \tag{33}$$

and

$$
\begin{aligned}
v_3^{(1)} &= \zeta^2 \frac{\nu}{2(1-\nu)} \frac{\partial^2 V_3^{(0)}}{\partial \xi^2} + V_3^{(1)}(\xi, \tau), \\
v_1^{(1)} &= \frac{1}{6(1-\nu)}((2-\nu)\zeta^3 - 6\zeta) \frac{\partial^3 V_3^{(0)}}{\partial \xi^3} - \zeta \frac{\partial V_3^{(1)}}{\partial \xi}, \\
\sigma_{11}^{(1)} &= \frac{1}{6(1-\nu)(1-\nu^2)}(2(1-\nu)\zeta^3 + 3(\nu-2)\zeta) \frac{\partial^4 V_3^{(0)}}{\partial \xi^4} \\
&\quad - \zeta \frac{1}{1-\nu^2} \frac{\partial^2 V_3^{(1)}}{\partial \xi^2} + \zeta \frac{\nu}{2(1-\nu^2)} \frac{\partial^2 V_3^{(0)}}{\partial \tau^2}, \\
\sigma_{22}^{(1)} &= \frac{\nu}{6(1-\nu)(1-\nu^2)}((1-\nu)\zeta^3 - 3\zeta) \frac{\partial^4 V_3^{(0)}}{\partial \xi^4} \\
&\quad - \zeta \frac{\nu}{1-\nu^2} \frac{\partial^2 V_3^{(1)}}{\partial \xi^2} + \zeta \frac{\nu}{2(1-\nu^2)} \frac{\partial^2 V_3^{(0)}}{\partial \tau^2}, \\
\sigma_{31}^{(1)} &= -\frac{1}{12(1-\nu)(1-\nu^2)}((1-\nu)\zeta^4 + 3(\nu-2)\zeta^2) \frac{\partial^5 V_3^{(0)}}{\partial \xi^5} \\
&\quad + \zeta^2 \frac{1}{2(1-\nu^2)} \frac{\partial^3 V_3^{(1)}}{\partial \xi^3} - \zeta^2 \frac{1}{4(1-\nu^2)} \frac{\partial^3 V_3^{(0)}}{\partial \xi \partial \tau^2} + C^{(1)}(\xi, \tau), \\
\sigma_{33}^{(1)} &= \frac{1}{60(1-\nu)(1-\nu^2)}((1-\nu)\zeta^5 + 5(\nu-2)\zeta^3) \frac{\partial^6 V_3^{(0)}}{\partial \xi^6} \\
&\quad - \zeta^3 \frac{1}{6(1-\nu^2)} \frac{\partial^4 V_3^{(1)}}{\partial \xi^4} + \zeta^3 \frac{1}{12(1-\nu)} \frac{\partial^4 V_3^{(0)}}{\partial \xi^2 \partial \tau^2} \\
&\quad - \zeta \frac{\partial C^{(1)}}{\partial \xi} + \zeta \frac{1}{2(1+\nu)} \frac{\partial^2 V_3^{(1)}}{\partial \tau^2},
\end{aligned}
\tag{34}
$$

where $C^{(1)}$ is an arbitrary function; the interfacial conditions are given by

$$
\sigma_{31}^{(1)}\Big|_{\zeta=1} = 0, \quad \sigma_{33}^{(1)}\Big|_{\zeta=1} = \frac{1}{2(1+\nu)} \frac{\partial^2 \varphi^{(1)}}{\partial \tau^2}\Big|_{\gamma=r}, \quad \frac{\partial \varphi^{(1)}}{\partial \gamma}\Big|_{\gamma=r} = v_3^{(1)}. \tag{35}
$$

Furthermore, applying $(35)_1$ and $(35)_2$ to $(34)_5$ and $(34)_6$, respectively, we obtain

$$
C^{(1)} = \frac{2\nu-5}{12(1-\nu)(1-\nu^2)} \frac{\partial^5 V_3^{(0)}}{\partial \xi^5} - \frac{1}{2(1-\nu^2)} \frac{\partial^3 V_3^{(1)}}{\partial \xi^3} + \frac{1}{4(1-\nu^2)} \frac{\partial^3 V_3^{(0)}}{\partial \xi \partial \tau^2}, \tag{36}
$$

and

$$
\begin{aligned}
&\frac{2}{3(1-\nu)} \frac{\partial^4 V_3^{(1)}}{\partial \xi^4} + \frac{\partial^2}{\partial \tau^2}\left(V_3^{(1)} + \frac{7\nu-17}{15(1-\nu)} \frac{\partial^2 V_3^{(0)}}{\partial \xi^2} \right) \\
&+ \frac{\partial^2}{\partial \tau^2}\left(\frac{8-3\nu}{10(1-\nu)} \frac{\partial^2 \varphi^{(0)}}{\partial \xi^2} - \varphi^{(1)} \right)\Bigg|_{\gamma=r} = 0.
\end{aligned}
\tag{37}
$$

The comparison of (37) with the refined equation for an elastic plate in the absence of fluid but subject to prescribed mechanical loading, as articulated in [18,29], manifests that the two equations are in complete agreement. At this order, the terms involving transverse shear deformation, plate rotatory inertia, and fluid compressibility have to be kept, unlike in [17], where these terms did not appear until the third-order approximation.

4.4. Asymptotically Consistent Equations

This subsection is concerned with the formulation of asymptotic models, i.e., shortened equations of motion for a thin elastic plate immersed in fluid together with the impenetrability conditions along the interfaces, originating from the results entrenched in the previous section. At the leading order, we obtain from (29), $(30)_3$ and (32)

$$\frac{2}{3(1-\nu)}\frac{\partial^4 v_3^*}{\partial \xi^4}+\frac{\partial^2 v_3^*}{\partial \tau^2}-\left.\frac{\partial^2 \varphi^*}{\partial \tau^2}\right|_{\gamma=r}=0, \tag{38}$$

$$\frac{\partial^2 \varphi^*}{\partial \xi^2}+\frac{\partial^2 \varphi^*}{\partial \gamma^2}=0, \tag{39}$$

and

$$\left.\frac{\partial \varphi^*}{\partial \gamma}\right|_{\gamma=r}=v_3^*, \tag{40}$$

where $v_3^*(\xi,0,\tau) = V_3^{(0)}(\xi,\tau)$ and $\varphi^*(\xi,\gamma,\tau) = \varphi^{(0)}(\xi,\gamma,\tau)$. In original variables, Equations (38)–(40) become

$$\frac{Eh^3}{3(1-\nu^2)}\frac{\partial^4 v_3}{\partial x_1^4}+\rho h\frac{\partial^2 v_3}{\partial t^2}-\rho_0\left.\frac{\partial^2 \varphi}{\partial t^2}\right|_{x_3=h}=0, \tag{41}$$

$$\frac{\partial^2 \varphi}{\partial x_1^2}+\frac{\partial^2 \varphi}{\partial x_3^2}=0, \tag{42}$$

and

$$\left.\frac{\partial \varphi}{\partial x_3}\right|_{x_3=h}=v_3. \tag{43}$$

It is important to note that here and for the rest of this section, v_3 (the transverse displacement) is taken at the midplane $x_3 = 0$, i.e., $v_3 = v_3(x_1, 0, t)$. Hence, Equations (41)–(43) correspond to the traditional setup of a Kirchhoff plate submerged in incompressible fluid.

Next, consider the sum of Equations (32) and (37), multiplied by the small parameter r^2, to obtain

$$\begin{aligned}
&\frac{2}{3(1-\nu)}\frac{\partial^4}{\partial \xi^4}\left(V_3^{(0)}+r^2V_3^{(1)}\right)+\frac{\partial^2}{\partial \tau^2}\left(V_3^{(0)}+r^2V_3^{(1)}\right)\\
&-\frac{\partial^2}{\partial \tau^2}\left.\left(\varphi^{(0)}+r^2\varphi^{(1)}\right)\right|_{\gamma=r}+r^2\frac{7\nu-17}{15(1-\nu)}\frac{\partial^4}{\partial \xi^2\partial \tau^2}\left(V_3^{(0)}+r^2V_3^{(1)}\right)\\
&+r^2\frac{8-3\nu}{10(1-\nu)}\frac{\partial^4}{\partial \xi^2\partial \tau^2}\left.\left(\varphi^{(0)}+r^2\varphi^{(1)}\right)\right|_{\gamma=r}+O(r^4)=0.
\end{aligned} \tag{44}$$

In a similar manner, we have from Equations (29), (33), $(30)_3$ and $(35)_3$, respectively

$$\frac{\partial^2}{\partial \xi^2}\left(\varphi^{(0)}+r^2\varphi^{(1)}\right)+\frac{\partial^2}{\partial \gamma^2}\left(\varphi^{(0)}+r^2\varphi^{(1)}\right)-r^2\delta^2\frac{\partial^2}{\partial \tau^2}\left(\varphi^{(0)}+r^2\varphi^{(1)}\right)+O(r^4)=0, \tag{45}$$

and

$$\frac{\partial}{\partial \gamma}\left.\left(\varphi^{(0)}+r^2\varphi^{(1)}\right)\right|_{\gamma=r}=V_3^{(0)}+r^2V_3^{(1)}+r^2\frac{\nu}{2(1-\nu)}\frac{\partial^2}{\partial \xi^2}\left(V_3^{(0)}+r^2V_3^{(1)}\right)+O(r^4). \tag{46}$$

By neglecting terms of $O(r^4)$, Formulas (44)–(46) can be rewritten as

$$\begin{aligned}\frac{2}{3(1-\nu)}\frac{\partial^4 v_3^*}{\partial\xi^4}+\frac{\partial^2 v_3^*}{\partial\tau^2}-\frac{\partial^2\varphi^*}{\partial\tau^2}\bigg|_{\gamma=r}&+r^2\frac{7\nu-17}{15(1-\nu)}\frac{\partial^4 v_3^*}{\partial\xi^2\partial\tau^2}\\ &+r^2\frac{8-3\nu}{10(1-\nu)}\frac{\partial^4\varphi^*}{\partial\xi^2\partial\tau^2}\bigg|_{\gamma=r}=0,\end{aligned}\tag{47}$$

and

$$\frac{\partial^2\varphi^*}{\partial\xi^2}+\frac{\partial^2\varphi^*}{\partial\gamma^2}-r^2\delta^2\frac{\partial^2\varphi^*}{\partial\tau^2}=0,\tag{48}$$

with

$$\frac{\partial\varphi^*}{\partial\gamma}\bigg|_{\gamma=r}=v_3^*+r^2\frac{\nu}{2(1-\nu)}\frac{\partial^2 v_3^*}{\partial\xi^2},\tag{49}$$

where $v_3^* = V_3^{(0)} + r^2 V_3^{(1)}$ and $\varphi^* = \varphi^{(0)} + r^2\varphi^{(1)}$. In terms of original variables, Equations (47) and (49) take the form

$$\begin{aligned}\frac{Eh^3}{3(1-\nu^2)}\frac{\partial^4 v_3}{\partial x_1^4}&+\rho h\left(1+h^2\frac{7\nu-17}{15(1-\nu)}\frac{\partial^2}{\partial x_1^2}\right)\frac{\partial^2 v_3}{\partial t^2}\\ &-\rho_0\left(1-h^2\frac{8-3\nu}{10(1-\nu)}\frac{\partial^2}{\partial x_1^2}\right)\frac{\partial^2\varphi}{\partial t^2}\bigg|_{x_3=h}=0,\end{aligned}\tag{50}$$

with

$$\frac{\partial\varphi}{\partial x_3}\bigg|_{x_3=h}=\left(1+\frac{\nu h^2}{2(1-\nu)}\frac{\partial^2}{\partial x_1^2}\right)v_3,\tag{51}$$

together with Equation (3), governing the compressible fluid motion (Equation (48) in original variables).

Hence, Equations (3), (50) and (51) correspond to the first-order asymptotic model, which incorporates three corrections, including transverse shear deformation, plate rotation inertia, and fluid compressibility. We re-iterate that the aforementioned first-order model coincides with the third-order model derived in [17], which does not assume that r is a small parameter.

4.5. Comparison of Dispersion Relations

The focus of this subsection is to derive the dispersion relations corresponding to the approximate formulations established in the previous Section 4.4 and to establish the link between these and the leading- and first-order approximations obtained in Section 3.

Begin with the travelling wave solution of the leading-order problem (41)–(43), setting

$$\begin{aligned}v_3&=e^{i(kx_1-\omega t)},\\ \varphi&=-\frac{1}{k}e^{-k(x_3-h)+i(kx_1-\omega t)},\end{aligned}\tag{52}$$

Upon substituting (52) into the aforementioned formulae, we arrive at the dispersion relation

$$\Omega^2=\frac{2}{3(1-\nu)}\frac{K^5}{r+K},\tag{53}$$

where K and Ω are dimensionless wavenumber and frequency, respectively, defined by (7).

Next, we set

$$v_3(x_1, 0) = e^{i(kx_1 - \omega t)},$$
$$\varphi = -\left(k^2 - \frac{\omega^2}{c_0^2}\right)^{-\frac{1}{2}} \left(1 - \frac{\nu h^2 k^2}{2(1-\nu)}\right) e^{-\left(k^2 - \frac{\omega^2}{c_0^2}\right)^{\frac{1}{2}}(x_3 - h) + i(kx_1 - \omega t)}, \tag{54}$$

in Equations (3), (50) and (51), resulting in the dispersion relation for the first-order model, given by

$$\Omega^2\left(60(1-\nu)^2 H + 3r(3\nu - 8)(2(\nu - 1) + \nu K^2)K^2 + 4(1-\nu)(17 - 7\nu)K^2 H - 30r(1-\nu)(2(\nu - 1) + \nu K^2)\right) = 40(1-\nu)K^4 H, \tag{55}$$

where H is defined in (6).

It can be easily verified that substituting (8) into (53) leads to the leading-order approximation given by Equation (10). On the other hand, when substituting (8) into (55) and expanding for small r and retaining of two terms leads to the first-order approximation given by (11). This clearly demonstrates the link between the asymptotic models and the approximate equations given in Section 3. The comparison of (11) and (55) is displayed in Figure 4, showing excellent agreement.

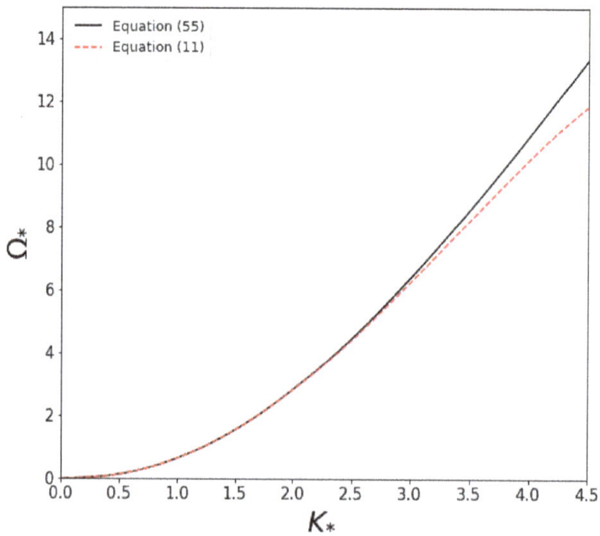

Figure 4. Comparison of the first-order dispersion relation (55) (solid black line) with the first-order approximation (11) (dashed red line) for the same parameters as in Section 3.

5. Concluding Remarks

In this paper, the two-step asymptotic analysis of light fluid loading has been developed. The first step is concerned with the derivation of the leading- and first-order approximations of the full dispersion relation. The second step is oriented towards the leading- and first-order asymptotic models, including the equations of motion and impenetrability condition. The presence of a small parameter, expressing light fluid loading, assumes the plate inertia to be incorporated into the leading-order model. At the same time, transverse shear deformation, plate rotation inertia, fluid compressibility, and other similar corrections appear at the next order.

The model problem considered in this paper can be readily extended to more elaborated setups, including one-side contact, fluid-loaded thin elastic shells, and radiation and scattering by submerged structures.

Author Contributions: S.S.: computation, investigation, methodology, problem statement, writing—original draft, writing—review and editing. L.P.: conceptualization, methodology, problem statement, supervision, writing—review and editing. All authors have read and agreed to the published version of the manuscript.

Funding: This research received no external funding.

Data Availability Statement: Data are contained within the article.

Acknowledgments: Fruitful discussions with J. Kaplunov are greatly acknowledged.

Conflicts of Interest: The authors declare no conflicts of interest.

References

1. Rayleigh, J.W.S.B. *The Theory of Sound*; Macmillan: New York, NY, USA, 1896; Volume 2.
2. Crighton, D. The 1988 Rayleigh medal lecture: Fluid loading—The interaction between sound and vibration. *J. Sound Vib.* **1989**, *133*, 1–27. [CrossRef]
3. Crighton, D.G.; Dowling, A.P.; Ffowcs-Williams, J.; Heckl, M.; Leppington, F.; Bartram, J.F. *Modern Methods in Analytical Acoustics Lecture Notes*; Springer: London, UK, 1992.
4. Junger, M.C.; Feit, D. *Sound, Structures, and Their Interaction*; MIT Press: Cambridge, MA, USA, 1986; Volume 225.
5. Norris, A.N.; Rebinsky, D.A. Acoustic coupling to membrane waves on elastic shells. *J. Acoust. Soc. Am.* **1994**, *95*, 1809–1829. [CrossRef]
6. Blonigen, F.J.; Marston, P.L. Leaky helical flexural wave scattering contributions from tilted cylindrical shells: Ray theory and wave-vector anisotropy. *J. Acoust. Soc. Am.* **2001**, *110*, 1764–1769. [CrossRef]
7. Caresta, M.; Kessissoglou, N.J. Structural and acoustic responses of a fluid-loaded cylindrical hull with structural discontinuities. *Appl. Acoust.* **2009**, *70*, 954–963. [CrossRef]
8. Titovich, A.S.; Norris, A.N. Acoustic scattering from an infinitely long cylindrical shell with an internal mass attached by multiple axisymmetrically distributed stiffeners. *J. Sound Vib.* **2015**, *338*, 134–153. [CrossRef]
9. Ruotolo, R. A comparison of some thin shell theories used for the dynamic analysis of stiffened cylinders. *J. Sound Vib.* **2001**, *243*, 847–860. [CrossRef]
10. Homm, A.; Ehrlich, J.; Peine, H.; Wiesner, H. Experimental and numerical investigation of a complex submerged structure. Part I: Modal analysis. *Acta Acust. United Acust.* **2003**, *89*, 61–70.
11. Forrest, J.A. Measured dynamics of a thin cylindrical shell subject to axial excitation. In Proceedings of the Acoustics 2005, Busselton, WA, Australia, 9–11 November 2005; pp. 9–11.
12. Slepyan, L.; Sorokin, S. Analysis of structural-acoustic coupling problems by a two-level boundary integral method: Part 1: A general formulation and test problems. *J. Sound Vib.* **1995**, *184*, 195–211. [CrossRef]
13. Cole, J.E., III. Vibrations of a framed cylindrical shell submerged in and filled with acoustic fluids: Spectral solution. *Comput. Struct.* **1997**, *65*, 385–393. [CrossRef]
14. Liu, C.H.; Chen, P.T. Numerical analysis of immersed finite cylindrical shells using a coupled BEM/FEM and spatial spectrum approach. *Appl. Acoust.* **2009**, *70*, 256–266. [CrossRef]
15. Johansson, M.; Folkow, P.D.; Hägglund, A.; Olsson, P. Approximate boundary conditions for a fluid-loaded elastic plate. *J. Acoust. Soc. Am.* **2005**, *118*, 3436–3446. [CrossRef]
16. Belov, A.; Kaplunov, J.; Nolde, E. A refined asymptotic model of fluid-structure interaction in scattering by elastic shells. *Flow Turbul. Combust.* **1998**, *61*, 255–267. [CrossRef]
17. Kaplunov, J.; Prikazchikova, L.; Shamsi, S. A hierarchy of asymptotic models for a fluid-loaded elastic layer. *Math. Mech. Solids* **2023**, *29*, 560–576. [CrossRef]
18. Kaplunov, J.D.; Kossovitch, L.Y.; Nolde, E. *Dynamics of Thin Walled Elastic Bodies*; Academic Press: Cambridge, MA, USA, 1998.
19. Talmant, M.; Überall, H.; Miller, R.D.; Werby, M.F.; Dickey, J. Lamb waves and fluid-borne waves on water-loaded, air-filled thin spherical shells. *J. Acoust. Soc. Am.* **1989**, *86*, 278–289. [CrossRef]
20. Yücel, H.; Ege, N.; Erbaş, B.; Kaplunov, J. A revisit to the plane problem for low-frequency acoustic scattering by an elastic cylindrical shell. *Math. Mech. Solids* **2024**. [CrossRef]
21. Craster, R. The light fluid loading limit for fluid/solid interactions. *Eur. J. Appl. Math.* **1997**, *8*, 485–505. [CrossRef]
22. Sorokin, S.V. Analysis of wave propagation in sandwich plates with and without heavy fluid loading. *J. Sound Vib.* **2004**, *271*, 1039–1062. [CrossRef]

23. Sorokin, S.V. Vibrations of and sound radiation from sandwich plates in heavy fluid loading conditions. *Compos. Struct.* **2000**, *48*, 219–230. [CrossRef]
24. Chapman, C.J.; Sorokin, S.V. The forced vibration of an elastic plate under significant fluid loading. *J. Sound Vib.* **2005**, *281*, 719–741. [CrossRef]
25. Kaplunov, J.; Markushevich, D. Plane vibrations and radiation of an elastic layer lying on a liquid half-space. *Wave Motion* **1993**, *17*, 199–211. [CrossRef]
26. Howe, M.S. *Acoustics of Fluid-Structure Interactions;* Cambridge University Press: Cambridge, UK, 1998.
27. Nakayama, Y. *Introduction to Fluid Mechanics;* Butterworth-Heinemann: Oxford, UK, 2018.
28. Kaplunov, J.; Prikazchikova, L.; Shamsi, S. Dispersion of the Bending Wave in a Fluid-loaded Elastic Layer. In *Advances in Solid and Fracture Mechanics: A Liber Amicorum to Celebrate the Birthday of Nikita Morozov;* Springer: Berlin/Heidelberg, Germany, 2022; pp. 127–134.
29. Goldenveizer, A.; Kaplunov, J.; Nolde, E. On Timoshenko-Reissner type theories of plates and shells. *Int. J. Solids Struct.* **1993**, *30*, 675–694. [CrossRef]

mathematics MDPI

Article

On Aspects of Continuous Approximation of Diatomic Lattice

Igor V. Andrianov [1,*], Lelya A. Khajiyeva [2], Askar K. Kudaibergenov [2] and Galina A. Starushenko [3]

[1] Chair and Institute of General Mechanics, RWTH Aachen University, Eilfschornsteinstrasse 18, D-52062 Aachen, Germany
[2] Department of Mathematical and Computer Modeling, Al-Farabi Kazakh National University, 71 Al-Farabi Ave., 050040 Almaty, Kazakhstan; khadle@mail.ru (L.A.K.); askarkud@gmail.com (A.K.K.)
[3] Department of Information Technology and Information Systems, Dnipro University of Technology, 19 Dmytra Yavornytskoho Ave., 49005 Dnipro, Ukraine; gs.gala.star@gmail.com
* Correspondence: igor.andrianov@gmail.com

Abstract: This paper is devoted to the continualization of a diatomic lattice, taking into account natural intervals of wavenumber changes. Continualization refers to the replacement of the original pseudo-differential equations by a system of PDEs that provides a good approximation of the dispersion relations. In this regard, the Padé approximants based on the conditions for matching the values of the dispersion relations of the discrete and continuous models at several characteristic points are utilized. As a result, a sixth-order unconditionally stable system with modified inertia is obtained. Appropriate boundary conditions are formulated. The obtained continuous approximation accurately describes the amplitude ratios of neighboring masses. It is also shown that the resulting continuous system provides a good approximation for the natural frequencies.

Keywords: diatomic lattice; continualization; Padé approximants; model with modified inertia; normal dispersion; anomalous dispersion

MSC: 39A06; 39A23; 41A21

1. Introduction

A diatomic lattice was proposed as a useful model in the theory of heat capacity by Born and von Kármán [1,2]. This model is currently used to describe various metamaterials [3–6]. Therefore, the dynamics of the diatomic lattice has become a subject of extensive investigations, in which both discrete model and its continuous approximations have been considered [3,5,7–12].

It is worth mentioning the relationship between the discrete and continuous models. This issue was first discussed by the ancient Greeks. After Newton's Calculus, the continuous models had been a priority for a long time, and continuous mathematics accumulated the huge number of tools for their analysis. The situation began to change with the advent of modern computers. Kolmogorov wrote about the relations between discrete and continuous mathematics [13]: "It is very likely that with the development of modern computational techniques it will be understood that in many cases it makes sense to study real phenomena without using the intermediate step of their stylization in the form of infinite and continuous mathematics and proceeding directly to discrete models. Pure mathematics was successfully developed mainly as a science of the infinite. Obviously, this state of affairs is deeply rooted in our consciousness, which works with great ease with an intuitively clear understanding of unbounded sequences, limiting processes, continuous and even smooth manifolds, etc.".

Continuous models can be treated as an asymptotic approximation to discrete models. Their strength is that, as Kolmogorov noted, they are most accessible to our understanding (or perhaps our understanding is distorted by traditional mathematical education, which tends to bias thinking in the direction of continuous mathematics). The advantage of continuous models is the possibility of using the powerful arsenal of continuous mathematics for their analysis. At the same time, when using the continuous models to describe a discrete structure, one must consider the natural limitations of these approximations to avoid various "paradoxes" and artifacts [14–17].

Interestingly, continualization can lead to important insights, even in cases where it does not allow a complete understanding of a discrete model. For example, "When Kruskal derived the Korteweg–de Vries (K-dV) equation from the FPU lattice, he bypassed the blockade imposed by the discrete to reach the enlightenment of the continuum. In this moment of the journey of the FPU problem, the sweet fruit of solitonic integrability, one of the most beautiful scientific discoveries of the second half of the last century, almost made us forget that bypassing the blockade did not eliminate it. As later studies have clearly shown, crossing the weak excitation threshold causes discretion to re-emerge in full force, or in all its glory if you will, displacing K-dV and its siblings. The maturity of fifty years of experience now makes us realize that Kruskal's seminal leap is not a license to ignore the discrete system, but rather a bridge to a synergetic methodology wherein we mediate between this true system, for which our analytical means are quite poor, and an idealized continuum system about which we can say many things." [18].

The connection between discrete and continuous approximations is established by the methods of discretization and continualization (homogenization) [3,4,8,14,15,19–22]. Discrete and continuous models are fundamentally different from each other. Thus, when we replace one model by another, we can usually only achieve a good approximation of some properties of the original object. Continualization often focuses on the comparison of dispersion relations [3,14,15,22]. In addition, one can assume the correspondence of the asymptotic behavior of wave processes in discrete and continuous media [19], the equality of the group impedance matrices of the model and the modeled structure [4]. Homogenization of the discrete structure can also mean that a periodic structure is replaced by a periodic structure of the same dimension and geometry whose periodicity cell is the simplest discrete mechanical oscillatory system with the smallest possible number of discrete parameters [3,4].

In the paper [20], the possibilities of deterministic discretization and chaotic continualization were shown using the example of the logistic equation. This confirms the ambiguity of the discretization and continualization processes.

Continualization is commonly understood as the approximation of non-local (pseudo-differential or integral) operators by local (differential) ones. The simplest possibility is a polynomial approximation [16] (the term "standard continuation" is also used for this approximation [22]). It can be based on the Taylor series expansion of non-local operators or the analysis of dispersion relations [16,17]. A continuous approximation based on the first roots of the dispersion equation is also possible [16,17]. These procedures are only suitable for describing the lower part of the spectrum or long waves and can lead to ill-posed problems. In this regard, methods of regularization and extension for the range of applicability of continuum models have become widespread. Padé approximants (non-standard continualization [22]) [5,8,14,15,18–21] are often used for this purpose. They allow the construction of improved long-wave approximations that approximate well the asymptotic behavior of discrete wave processes or describe frequencies in the first Brillouin zone. However, it must be taken into account that the use of such approximations has significant limitations. As mentioned in [16], p. 65, "other approximation models that are not associated with the polynomial approximation are also possible. An approximate description of the dispersion curve over a wide range of wavelengths using an appropriate function of the wavenumber k (for example when interpolating experimental data) can be of interest in a number of cases. This model is good for describing non-decaying waves, but in general it is incorrect to continue it in the complex region and use it for boundary problems".

Our paper is devoted to the linear diatomic lattice. This classical problem still arouses interest despite its long history [23–27]. In particular, [23–25] analyzed the question of the range of wavenumbers that allow a correct description of the wave processes in the diatomic lattice. The peculiarities of diatomic lattice modeling mentioned in [24,25] are considered in our work.

There is extensive literature dealing with diatomic lattices. We only refer to studies related to continualization using Padé approximants [8,21] and the construction of higher-order continuous models [6].

Our paper is concerned with the continuization of diatomic lattices based on the multipoint Padé approximants (Padé approximants of the second kind [28,29]). The proximity of the dispersion curves of discrete and continuous models is used as a criterion for approximation. It is shown that each natural frequency is also well approximated in this case. An important feature of the continuous approximation constructed in this work is that it is considered in the most natural periodic interval of wavenumbers.

2. Discrete System

Consider a finite lattice consisting of $2N + 2$ alternating masses m_1 and m_2 connected by springs of stiffness c (see Figure 1). We construct the final chain so that the number of particles with masses m_1 and m_2 is equal. Since the total number of particles is assumed to be large, this restriction is not important.

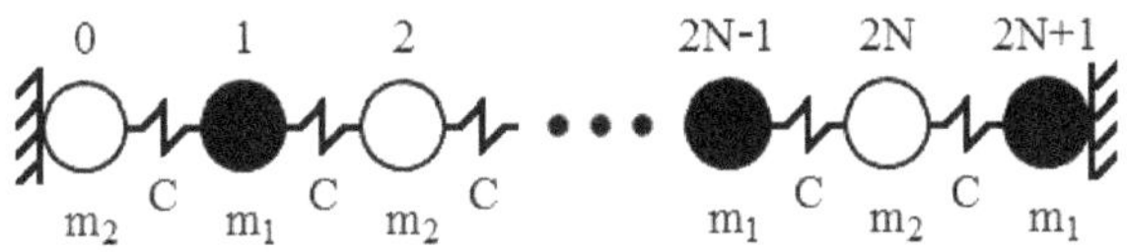

Figure 1. Diatomic lattice.

The oscillation of this system is described by a system of mixed differential-difference equations:

$$\begin{aligned} m_1\ddot{u}_n + c(2u_n - u_{n+1} - u_{n-1}) = 0,\ n = 1,3,5,\ldots,2N+1, \\ m_2\ddot{u}_n + c(2u_n - u_{n+1} - u_{n-1}) = 0,\ n = 0,2,4,\ldots,2N, \end{aligned} \tag{1}$$

where $(\dot{u}) \equiv \frac{du}{dt}$.

The lattice is fixed at the ends

$$u_0 = u_{2N+1} = 0. \tag{2}$$

Substituting the solution in the form

$$\begin{aligned} u_n = A\sin\beta e^{i\omega t},\ n = 1,3,5,\ldots,2N+1, \\ u_n = B\sin\beta e^{i\omega t},\ n = 0,2,4,\ldots,2N \end{aligned} \tag{3}$$

into Equation (1) and eliminating the amplitudes A and B, we find the relations:

$$\begin{aligned} \left(-\omega^2 + \omega_1^2\right)A - \tfrac{1}{2}\omega_1^2 \cos\tfrac{n\pi}{2N+1} B = 0, \\ \left(-\omega^2 + \omega_2^2\right)B - \tfrac{1}{2}\omega_2^2 \cos\tfrac{n\pi}{2N+1} A = 0, \end{aligned} \tag{4}$$

where $\beta = \frac{n\pi}{2N+1}$, $\omega_i = \sqrt{2c/m_i}$, $i = 1,2$.

It is worth noting that, for the infinite lattice, the variable β should be considered as continuously changing in the interval $0 \leq \beta \leq \pi$.

The dispersion equation can be written as follows

$$\begin{aligned}&\omega^4-(\omega_1^2+\omega_2^2)\omega^2+\omega_1^2\omega_2^2\sin^2[\pi n/(2N+1)]=0,\\&n=0,1,2,\ldots,2N-1,2N,2N+1.\end{aligned} \tag{5}$$

The question of interest is as follows: in which interval of wavenumber $\beta=\frac{n\pi}{2N+1}$ changes should we consider solutions for the dispersion Equation (5)? As indicated in the articles [23–25], we can limit ourselves to the interval $0\leq\beta\leq\pi/2$ when determining the oscillation frequencies. For a complete study of the system, the interval $0\leq\beta\leq\pi$ should be taken into account.

Consider various limiting cases. In the following, we assume $2N\gg1$, which is justified since we consider the possibilities of a continuous approximation to the discrete system. For $1/N\ll1$, Equation (5) can be approximately replaced by the following:

$$\omega^4-\left(\omega_1^2+\omega_2^2\right)\omega^2+\frac{\pi^2n^2}{(2N+1)^2}\omega_1^2\omega_2^2=0. \tag{6}$$

For $n = N$, $n = [N/2]$ (here symbol [] means the integer part of the number) and $n = 2N + 1$, we obtain from (5)

$$\omega^2\approx\omega_1^2;\ \omega^2\approx\omega_2^2, \tag{7}$$

$$\omega^2\approx0.5\left(\omega_1^2+\omega_2^2\pm\sqrt{\omega_1^4+\omega_2^4}\right), \tag{8}$$

$$\omega^2=0 \text{ and } \omega^2=\omega_1^2+\omega_2^2, \tag{9}$$

respectively.

3. Continuous Approximations

We suppose that the masses are uniformly distributed along the interval of length L of the X-axis (Figure 1). Then, the distance between the masses is $H=L/(2N+1)$. Assuming that the number of masses is large, namely $2N\gg1$, we can introduce a natural small parameter $h=H/L\ll1$. Let us introduce dimensionless spatial coordinate $x=X/L$. We also introduce functions of spatial and time arguments $U_j(x,\,t),\ j=1,2$ as follows

$$\begin{aligned}&U_1(hk,t)=u_k(t),k=1,3,5,\ldots,\\&U_2(hk,t)=u_k(t),k=0,2,4,\ldots\end{aligned} \tag{10}$$

"The basic idea is to establish a one-to-one correspondence between the functions of discrete arguments and a class of analytical functions, as well as between the operations on them. Let the function $u(nh)$ be given by its values, which are generally complex, at the points nh (nodes) of the X-axis. Let us consider the problem of interpolation of $u(nh)$ by a smooth function $u(x)$. Obviously, $u(x)$ is defined within any $\Psi(x)$, which vanishes in all nodes, i.e., $u(x)+\Psi(x)$ will also be an interpolating function. It is obvious to try to select the smoothest interpolating functions from the entirety of interpolating functions filtering out rapidly oscillating components" [16]. (See also the explanation of this question in [30], p. 59). In [16] the concept of the quasicontinuum is introduced and the Whittaker–Shannon–Kotel'nikov function (sinc-function) of the form

$$sinc(x)=\frac{\sin(\pi x/h)}{\pi x} \tag{11}$$

is utilized.

The functions $U_j(x,t)$, $j=1,2$ are assumed to be infinitely differentiable with respect to the spatial coordinate. Hence, they can be expanded into the Taylor series

$$U_j(x_0 \pm h, t) = U_j(x_0, t) + \sum_{i=1}^{\infty} (-1)^i h^i \left. \frac{d^i U_j(x,t)}{dx^i} \right|_{x=x_0}. \tag{12}$$

Rewrite system (1) in the form

$$\begin{array}{l} m_1 \ddot{U}_1(kh,t) + 2c\,[U_1(kh,t) - U_2(kh,t)] \\ \quad -c\,[U_2((k+1)h,t) - 2U_2(kh,t) + U_2((k-1)h,t)] = 0,\ k = 1,3,5,\ldots \\ m_2 \ddot{U}_2(kh,t) + 2c\,[U_2(kh,t) - U_1(kh,t)] \\ \quad -c\,[U_1((k+1)h,t) - 2U_1(kh,t) + U_1((k-1)h,t)] = 0,\ k = 0,2,4,\ldots \end{array} \tag{13}$$

Using the pseudo-differential operators and sinc-function (11), we can represent system (13) as follows

$$\begin{array}{l} \left\{ m_1 \ddot{U}_1(x,t) + 2c\,[U_1(x,t) - U_2(x,t)] \right\} \Phi(x) + 4c \sin^2\left(-\frac{ih}{2}\frac{\partial}{\partial x}\right) U_2(x,t) = 0, \\ \left\{ m_2 \ddot{U}_2(x,t) + 2c\,[U_2(x,t) - U_1(x,t)] \right\} \Phi(x) + 4c \sin^2\left(-\frac{ih}{2}\frac{\partial}{\partial x}\right) U_1(x,t) = 0. \end{array} \tag{14}$$

Here, $4\sin^2\left(-\frac{ih}{2}\frac{\partial}{\partial x}\right)$ is the pseudo-differential operator [25], $\Phi(x) = \sum_{n=0}^{2N+1} \sin c(x - nh)$. For continualization, we use the expansion

$$\begin{array}{l} \sin^2\left(-\frac{ih}{2}\frac{\partial}{\partial x}\right) = -\frac{1}{2}\sum_{k=1} \frac{h^{2k}}{(2k)!}\frac{\partial^{2k}}{\partial x^{2k}} \\ = -\frac{h^2}{4}\frac{\partial^2}{\partial x^2}\left(1 + \frac{h^2}{12}\frac{\partial^2}{\partial x^2} + \frac{h^4}{360}\frac{\partial^4}{\partial x^4} + \frac{h^6}{20160}\frac{\partial^6}{\partial x^6} + \ldots\right) \end{array} \tag{15}$$

and the equality

$$\Phi(x) \to 1 \ \text{ at } h \to 0. \tag{16}$$

In the zero approximation, where only terms of h^0 order are considered, we obtain a system of two coupled oscillators

$$\begin{array}{l} \ddot{U}_1 + \omega_1^2(U_1 - U_2) = 0, \\ \ddot{U}_2 + \omega_2^2(U_2 - U_1) = 0. \end{array} \tag{17}$$

Keeping the terms up to h^4 in expansion (12), we obtain

$$\begin{array}{l} \ddot{U}_1 + \omega_1^2(U_1 - U_2) - \frac{1}{2}\omega_1^2 h^2 \left[U_{2xx} + (1/12)h^2 U_{2xxxx}\right] = 0, \\ \ddot{U}_2 + \omega_2^2 2c(U_2 - U_1) - \frac{1}{2}\omega_2^2 h^2 \left[U_{1xx} + (1/12)h^2 U_{1xxxx}\right] = 0. \end{array} \tag{18}$$

The comparison of the calculation results using Equation (18) with the solution of the dispersion Equation (5) demonstrates the low accuracy of this continuum approximation. In addition, as shown in [27,31], for Equation (18), the effect of the short wavelength instability take place. This artifact can lead to instability or the divergence of numerical algorithms when used to solve Equation (18).

Increasing the accuracy of continuous approximation and the short wavelength instability suppression can be achieved using Padé approximants [27,31]

$$1 + \frac{h^2}{12}\frac{\partial^2}{\partial x^2} \sim 1/\left(1 - \frac{h^2}{12}\frac{\partial^2}{\partial x^2}\right). \tag{19}$$

Hence, we obtain a "model with modified inertia" (the term was proposed in [32])

$$\begin{aligned}&\left(1-\frac{h^2}{12}\frac{\partial^2}{\partial x^2}\right)\left[\ddot{U}_1+\omega_1^2(U_1-U_2)\right]-\frac{1}{2}\omega_1^2h^2U_{2xx}=0,\\&\left(1-\frac{h^2}{12}\frac{\partial^2}{\partial x^2}\right)\left[\ddot{U}_2+\omega_2^2(U_2-U_1)\right]-\frac{1}{2}\omega_2^2h^2U_{1xx}=0.\end{aligned} \tag{20}$$

The boundary conditions for system (20) are formulated as follows

$$U_j=0 \text{ at } x=0,1. \tag{21}$$

Boundary value problem (20), (21) describes well-posed continuous approximations of original discrete systems.

The natural vibrations of the continuous system can be described as

$$U_j=C_j\sin(n\pi x)\exp(i\omega t). \tag{22}$$

System (20) approximates the oscillation frequencies of a discrete system quite well [8]. It is sufficient to consider the system on the interval $0\le\beta\le\pi/2$. However, as mentioned in [23–25], the original system should be considered on the interval $0\le\beta\le\pi$. In this case, system (20) does not provide a good approximation to the dispersion curve of the discrete system.

Let us construct a higher-order approximation system using the conditions for matching the solutions to the dispersion equations of the discrete and continuous systems at $n = N$, $n = [N/2]$ and $n = 2N + 1$ (see (7)–(9)). This rational approximation is sometimes called the Padé approximation of the second kind [28,29]. As a result, we obtain a continuous system with modified inertia

$$\begin{aligned}&\left(1+d_1h^2\frac{\partial^2}{\partial x^2}+d_2h^4\frac{\partial^4}{\partial x^4}+d_3h^6\frac{\partial^6}{\partial x^6}\right)\left[\frac{\partial^2}{\partial t^2}U_1+\omega_1^2(U_1-U_2)\right]-\frac{1}{2}\omega_1^2h^2\frac{\partial^2}{\partial x^2}U_2=0,\\&\left(1+d_1h^2\frac{\partial^2}{\partial x^2}+d_2h^4\frac{\partial^4}{\partial x^4}+d_3h^6\frac{\partial^6}{\partial x^6}\right)\left[\frac{\partial^2}{\partial t^2}U_2+\omega_2^2(U_2-U_1)\right]-\frac{1}{2}\omega_2^2h^2\frac{\partial^2}{\partial x^2}U_1=0,\end{aligned} \tag{23}$$

where

$$\begin{aligned}&d_1=\frac{73}{140}+\frac{61}{9\pi^2}-\frac{32\sqrt{2}}{35};\ d_2=\frac{1}{63\pi^4}\left[812+\left(327-288\sqrt{2}\right)\pi^2\right];\\&d_3=\frac{16}{315\pi^6}\left[140+\left(87-72\sqrt{2}\right)\pi^2\right].\end{aligned} \tag{24}$$

Unfortunately, system (22) requires additional boundary conditions. This is a typical difficulty that arises when higher-order models, derived originally for infinite media, are applied to bounded domains [5,14,15]. To overcome this fundamental difficulty, various approaches are used.

In [33], this problem was studied for 1D linear lattice, when the exact solution can be easily obtained. It is proposed to introduce boundary conditions so as to achieve the most suitable frequencies and mode approximations. The limitations of this method are clear.

An often used technique is the formation of some edge zone. Thus, in [34,35], for second-grade elasticity, it is proposed to introduce the so-called ortho-fiber—a thin, notional fiber of a material, starting at the surface and extending inward along the direction of normal. The length of the ortho-fiber is small compared to the macroscopic characteristic length of the problem and tends to zero.

For continuous systems (beams, plates), the correct formulation of boundary conditions requires the consideration of the methods of fastening the plates used in practice (for example, embedding in a groove). For the problem of bending a plate, when deriving and justifying the conditions of simply support, it is assumed that there is a narrow zone of pinching of the plate edge, with the length of it tending towards zero as some power the thickness of the plate [36].

Similar problems occur in grid methods (see overview in [5]), where they can be solved by introducing some fictitious points outside the considered domain.

All these techniques are based on the simple consideration that, in reality, the conditions of ends fixation have a fairly complex structure, which should be taken into account when studying continuous or discrete systems.

Using this ideology, we introduce fictitious masses at points $k = -1, -2, -3, \ldots$; $k = 2N + 2, 2N + 3, 2N + 4, \ldots$ (Figure 2).

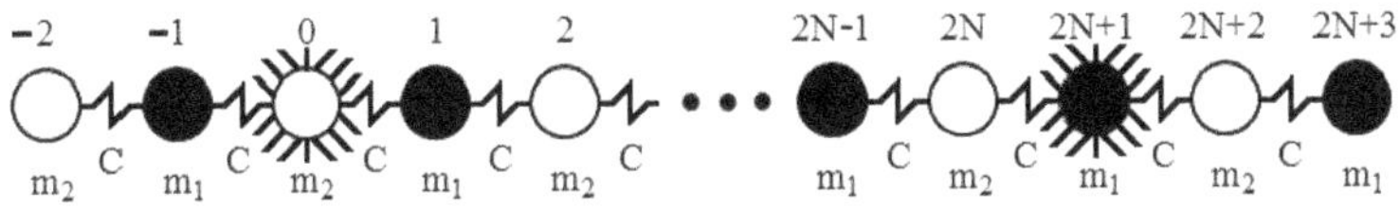

Figure 2. Lattice with fictitious masses, the introduction of which was caused by the need for additional boundary conditions.

Particularly in the case of a periodic spatial extension ("simple support"), one obtains (see Figure 2):

$$u_i = -u_{-i},\ i = 1, 2, 3, \ldots;\ u_{2N-j} = -u_{2N+2+j},\ j = 0, 1, 2, 3, \ldots \tag{25}$$

In the continualization problems, this technique was applied in [14,15]. Using it for the considered case, we arrive at the following boundary conditions

$$U_1 = U_{1xx} = U_{1xxxx} = 0,\ U_2 = U_{2xx} = U_{2xxxx} = 0 \text{ at } x = 0, 1. \tag{26}$$

From system (23), we obtain the dispersion equation for the continuous system

$$\begin{gathered} \omega^4 - (\omega_1^2 + \omega_2^2)\omega^2 + \omega_1^2\omega_2^2 \frac{\left(\frac{n\pi}{2N+1}\right)^2}{1 - d_1\left(\frac{n\pi}{2N+1}\right)^2 + d_2\left(\frac{n\pi}{2N+1}\right)^4 - d_3\left(\frac{n\pi}{2N+1}\right)^6} \times \\ \times \left[1 - \frac{\left(\frac{n\pi}{2N+1}\right)^2}{4\left[1 - d_1\left(\frac{n\pi}{2N+1}\right)^2 + d_2\left(\frac{n\pi}{2N+1}\right)^4 - d_3\left(\frac{n\pi}{2N+1}\right)^6\right]}\right] = 0. \end{gathered} \tag{27}$$

The analysis of Equation (27) shows that the system (23), (24) is unconditionally stable.

Note that the boundary value problem of the continuous approximation (23), (26) remains the same for the case of an infinite lattice.

4. Numerical Results

The results of comparing the dispersion curves for the discrete and continuous systems at $\omega_2^2/\omega_1^2 = 1.5$ are shown in Figure 3.

The solid curves show branches with normal dispersion, and the dotted curves show those with anomalous dispersion. The calculations were carried out at the points $n = 0, 1, 2, \ldots, 2N, 2N + 1$. The value $2\beta/\pi$ is plotted along the abscissa.

As can be seen from Figure 3, the continuous system (23), (24) provides a good approximation to the dispersion curve of the discrete system over the entire range of wavenumber changes. Recall that we assume $2N \gg 1$, and the distance between the masses is small. Hence, the dispersion curves in Figure 3 are shown as continuous lines. Note that the curves corresponding to the discrete and continuous models practically merge on the scale of the graph.

The variables U_1, U_2 have no obvious physical meaning (variables describing low-frequency oscillations of the centers of mass of the cells and high-frequency oscillations associated with internal degrees of freedom have physical meaning [9]).

The ratio of the amplitudes of mass vibrations C_1 and C_2 is given by the following relations:

$$C_1/C_2 = \frac{(-A_2 + A_1)\omega_1^2}{A_1(-\omega^2 + \omega_1^2)} \text{ at } \omega^2 \le \omega_1^2 \tag{28}$$

and

$$C_2/C_1 = \frac{(-A_2 + A_1)\omega_2^2}{A_1(-\omega^2 + \omega_2^2)} \text{ at } \omega_2^2 \le \omega^2, \tag{29}$$

where $A_1 = 1 - d_1\left(\frac{n\pi}{2N+1}\right)^2 + d_2\left(\frac{n\pi}{2N+1}\right)^4 - d_3\left(\frac{n\pi}{2N+1}\right)^6$; $A_2 = \frac{1}{2}\left(\frac{n\pi}{2N+1}\right)^2$.

The calculation results at $\omega_2^2/\omega_1^2 = 1.5$ are presented in Figure 4. The obtained curves qualitatively coincide with the corresponding curves for the discrete system [24] and make it possible to estimate the ratios of the amplitudes of the neighboring masses.

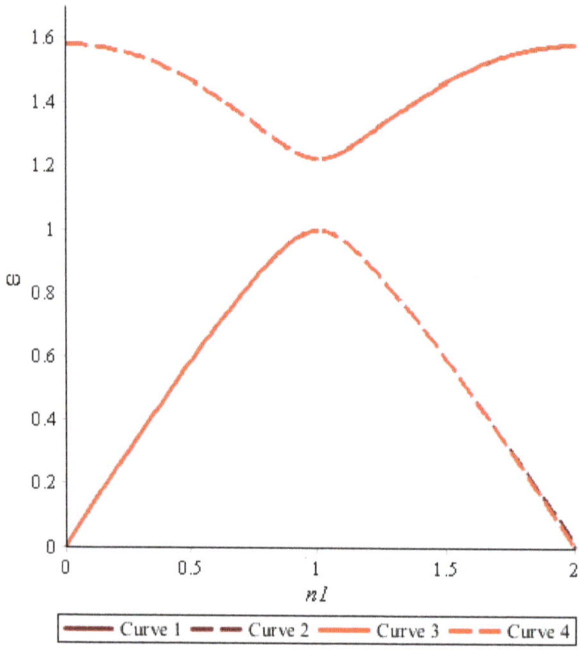

Figure 3. Comparison of the dispersion curves for the discrete (1) (curves 1 and 2) and continuous (23), (24) (curves 3 and 4) systems at $\omega_2^2/\omega_1^2 = 1.5$.

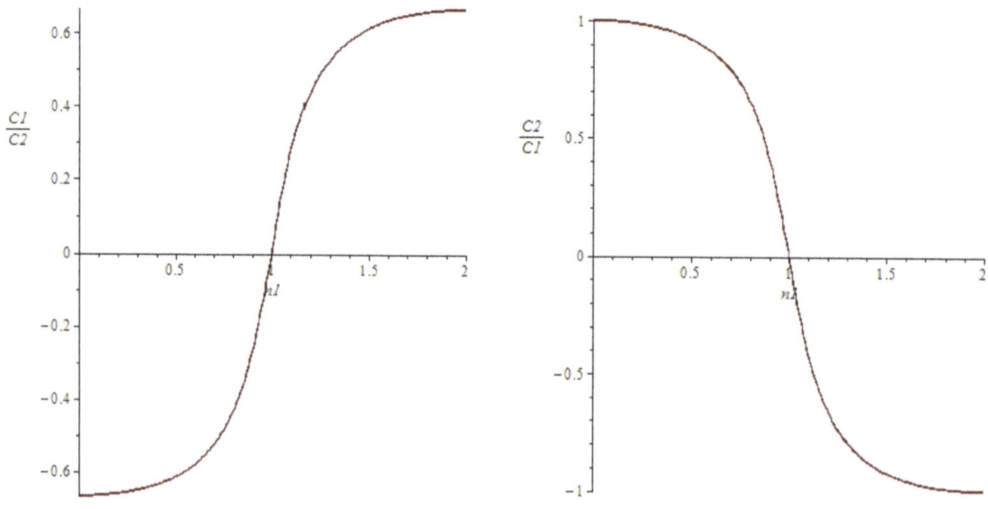

Figure 4. Dependence for the ratio of the oscillation amplitudes of the neighboring masses in the ranges: (**a**) $\omega^2 \le \omega_1^2$ and (**b**) $\omega_2^2 \le \omega^2$ at $\omega_2^2/\omega_1^2 = 1.5$. The value $2\beta/\pi$ is plotted along the abscissa.

In the limiting case $m_1 = m_2$, the dispersion curve for the Lagrange lattice is obtained (solid curve in Figure 5). The dotted curve corresponds to the displacements with zero amplitude [24,25].

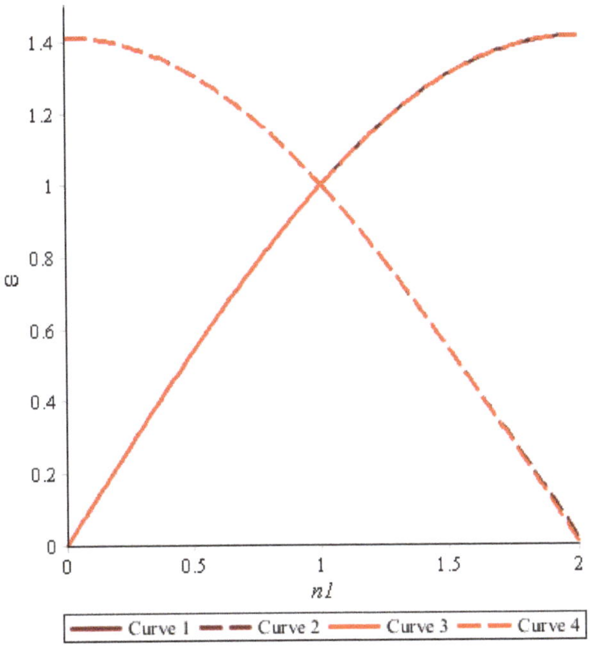

Figure 5. Limit transition of the Born–Kármán lattice to the Lagrange lattice at $m_1 = m_2$. The line of curve 1 almost completely coincides with the line for curve 3, so it seems to be invisible.

Dispersion curves for the discrete (1) (curves 1 and 2) and continuous (23), (24) (curves 3 and 4) systems are presented at $\omega_2^2/\omega_1^2 = 1$. Note that the curves corresponding to the discrete and continuous models practically merge on the scale of the graph.

We also compare the eigenfrequency dependence with respect to the mode number for the finite diatomic lattice, both with the discrete and the continuous approximation proposed in the paper. Such a comparison is uninformative because the frequency values are of different orders. Comparing the values of frequencies ω_d and ω_c in some averaged sense looks more reasonable. For that, we use the following expression for the average approximation error [37]:

$$A = \frac{100\%}{\overline{\omega}_d}\sqrt{\frac{\sum\limits_{i=0}^{2N+1}(\omega_{id} - \omega_{ic})^2}{2N+2}}, \tag{30}$$

where $\overline{\omega}_d = \frac{\sum\limits_{i=0}^{2N+1}\omega_{id}}{2N+2}$.

The calculations were carried out at $\omega_2^2/\omega_1^2 = 1.5$. The maximum error was obtained for the acoustic branch (Figure 3). The calculation results for various values of N are given in Table 1.

Table 1. The average error of the continuous approximations for frequencies.

N	5	10	15	20	25	30	35	40	45	50	∞
A, %	0.4465	0.5837	0.6272	0.6471	0.6579	0.6645	0.6688	0.6718	0.6740	0.6755	0.6824

When calculating the limit, summation is replaced by integration for a sufficiently large N as follows

$$\begin{gathered}\overline{\omega}_d \approx \frac{1}{2}\int\limits_0^2 \omega_d(n_1)\,dn_1,\\ \frac{\sum\limits_{i=0}^{2N+1}(\omega_{id}-\omega_{ic})^2}{2N+2} \approx \frac{1}{2}\int\limits_0^2 (\omega_{id}(n_1)-\omega_{ic}(n_1))^2\,dn_1,\ n_1=\frac{n\pi}{N+0.5}.\end{gathered} \tag{31}$$

Hence, we obtain the following estimate:

$$A \approx \frac{\sqrt{2\int\limits_0^2(\omega_{id}(n_1)-\omega_{ic}(n_1))^2\,dn_1}}{\int\limits_0^2 \omega_d(n_1)\,dn_1}\cdot 100\% \tag{32}$$

It is obvious that the approximation of frequencies using the continuous approximation is quite satisfactory.

5. Conclusions

The continualization of a diatomic lattice was considered in the natural interval of wavenumber changes. It provided a consistent limiting transition to the Lagrange lattice.

Note that the original system of $2N + 2$ difference-differential Equation (1), each of which contains the second-order derivative with respect to time, can be reduced to a system of $N + 1$ difference-differential equations with the fourth-order derivative with respect to time. After conducting its continualization, we obtain one continuous equation with the fourth-order derivative with respect to time. This result agrees with the fourth-order equation with respect to the time obtained from the continuous approximation (23).

The continualization consisted in the replacement of the original pseudo-differential equations by a system of PDEs that provided a good approximation for the dispersion relations. The use of the Padé approximants guaranteed the matching of the dispersion relation values of the discrete and continuous models at several characteristic points. It was not possible to limit ourselves to the second-order system of PDEs, which led to the need for the correct formulation of boundary conditions. An adequate result was obtained using a sixth-order system of PDEs with respect to spatial variables with modified inertia, which is unconditionally stable.

The application of the continuous approximations allowed correctly describing not only the frequency spectrum in the first Brillouin zone, but also the amplitude ratios of neighboring masses.

The technique proposed in this paper can be used to construct continuous models for flexural waves in the diatomic lattices [12]. The obtained continuous model can be also applied to the nonlinear problems for the diatomic lattice described in [9,11,26].

Author Contributions: Conceptualization, I.V.A.; methodology, I.V.A. and G.A.S.; investigation, I.V.A. and G.A.S.; writing and original draft preparation, I.V.A., L.A.K. and A.K.K.; software, G.A.S.; writing, review, and editing, I.V.A., L.A.K., A.K.K. and G.A.S.; formal analysis, I.V.A., L.A.K., A.K.K. and G.A.S. All authors have read and agreed to the published version of the manuscript.

Funding: This work was supported partly by a grant from the Simons Foundation (Award ID: 1160642, Project Title: Simons Foundation Support to Researchers in Ukraine, Program: Presidential Discretionary-Ukraine Support Grants).

Data Availability Statement: Data are contained within the paper.

Acknowledgments: We are grateful to F. Gómez-Silva, R. Zaera, I. Zeidis for valuable discussions.

Conflicts of Interest: The authors declare that they have no known competing financial interests or personal relationships that could have appeared to influence the work reported in this paper.

References

1. Born, M.; Huang, K. *Dynamical Theory of Crystal Lattices*; Oxford Press: Oxford, UK, 1988.
2. Born, M.; von Kármán, T. Über Schwingungen im Raumgittern. *Physik. Z* **1912**, *13*, 297–309.
3. Bobrovnitskii, Y.I. Effective parameters and energy of acoustic metamaterials and media. *Acoust. Phys.* **2014**, *60*, 134–141. [CrossRef]
4. Bobrovnitskii, Y.I. Impedance theory of wave propagation on infinite periodic structures. *J. Sound Vib.* **2022**, *525*, 116801. [CrossRef]
5. Li, Y.; Challamel, N.; Elishakoff, I. Effective mass and effective stiffness of finite and infinite metamaterial lattices. *Arch. Appl. Mech.* **2023**, *93*, 301–321. [CrossRef]
6. Zhou, Y.; Wie, P.; Li, Y.; Tang, Q. Continuum model of acoustic metamaterials with diatomic crystal lattice. *Mech. Adv. Mat. Struct.* **2017**, *24*, 1059–1073. [CrossRef]
7. Askar, A. *Lattice Dynamical Foundations of Continuum Theories: Elasticity, Piezoelectricity, Viscoelasticity, Plasticity*; World Scientific: Singapore, 1985.
8. Ghavanloo, E.; Fazelzadeh, S.A.; Rafii-Tabar, H. Formulation of an efficient continuum mechanics-based model to study wave propagation in one-dimensional diatomic lattices. *Mech. Res. Commun.* **2020**, *103*, 103467. [CrossRef]
9. Leamy, M.; Fronk, M.D. Higher-order dispersion, stability, and waveform invariance in nonlinear monoatomic and diatomic systems. *J. Vib. Acoust.* **2017**, *139*, 051003-1–051003-13.
10. Porubov, A.V.; Antonov, I.D.; Fradkov, A.L. Boundary control of nonlinear strain waves in di-atomic crystal layer. *Wave Motion* **2019**, *91*, 102400. [CrossRef]
11. Wang, Y.Z.; Wang, Y.S. Active control of elastic wave propagation in nonlinear phononic crystals consisting of diatomic lattice chain. *Wave Motion* **2018**, *78*, 1–8. [CrossRef]
12. Wu, H.; Huang, S.; Luo, J.; He, L.; Kuang, Y. Lattice dynamics models to predict transmission properties of flexural waves in one-dimensional atom chains with defects. *Acta Mech. Solida Sin.* **2023**, *36*, 105–115. [CrossRef]
13. Kolmogorov, A.N. Combinatorial foundations of information theory and the calculus of probabilities. *Russ. Math. Surv.* **1983**, *38*, 29–40. [CrossRef]
14. Andrianov, I.V.; Awrejcewicz, J.; Danishevskyy, V.V. *Linear and Nonlinear Waves in Microstructured Solids: Homogenization and Asymptotic Approaches*; Taylor & Francis: Boca Raton, FL, USA, 2021.
15. Andrianov, I.V.; Awrejcewicz, J.; Weichert, D. Improved continuous models for discrete media. *Math. Probl. Eng.* **2010**, *2010*, 986242. [CrossRef]
16. Kunin, I.A. *Elastic Media with Microstructure*; 1. One-Dimensional Models; Springer: Berlin, Germany, 1982.
17. Kunin, I.A. *Elastic Media with Microstructure*; 2. Three-Dimensional Models; Springer: Berlin, Germany, 1983.
18. Rosenau, P.; Schochet, S. Compact and almost compact breathers: A bridge between an anharmonic lattice and its continuum limit. *Chaos* **2005**, *15*, 015111. [CrossRef] [PubMed]
19. Andrianov, I.; Koblik, S.; Starushenko, G. Transition from discrete to continuous media: The impact of symmetry changes on asymptotic behavior of waves. *Symmetry* **2021**, *13*, 1008. [CrossRef]
20. Andrianov, I.; Starushenko, G.; Kvitka, S.; Khajiyeva, L. The Verhulst-like equations: Integrable OΔE and ODE with chaotic behavior. *Symmetry* **2019**, *11*, 1446. [CrossRef]
21. Collins, M.A. Solitons in the diatomic chain. *Phys. Rev.* **1985**, *31*, 1754–1762. [CrossRef] [PubMed]
22. Gómez-Silva, F.; Zaera, R. Low-order non-classical continuum models for the improved prediction of an anisotropic membrane lattice's dynamics. *Thin-Walled Struct.* **2022**, *179*, 109632. [CrossRef]
23. Bobrovnitskii, Y.I. Features of normal wave dispersion in periodic structures. *Acoust. Phys.* **2011**, *57*, 442–446. [CrossRef]
24. Landa, P.S. *Regular and Chaotic Oscillations*; Springer: Berlin/Heidelberg, Germany, 2001.
25. Landa, P.S.; Marchenko, V.F. On the linear theory of waves in media with periodic structures. *Sov. Phys. Usp.* **1991**, *34*, 830–834. [CrossRef]
26. Porubov, A.V.; Andrianov, I.V. Nonlinear waves in diatomic crystals. *Wave Motion* **2013**, *50*, 1153–1160. [CrossRef]
27. Wattis, J.A.D. Solitary waves in a diatomic lattice: Analytic approximations for a wide range of speeds by quasi-continuum methods. *Phys. Lett. A* **2001**, *284*, 16–22. [CrossRef]
28. Badikov, S.A.; Vinogradov, V.A.; Gai, E.V.; Rabotnov, N.S. Analytic approximation of neutron physics data. *Sov. At. Energy* **1984**, *56*, 19–26. [CrossRef]

29. Vinogradov, V.N.; Gay, E.V.; Rabotnov, N.C. *Analytical Approximation of Data in Nuclear and Neutron Physics*; Energoatomizdat: Moscow, Russia, 1987. (In Russian)
30. Maslov, V.P. *Operational Methods*; Mir: Moscow, Russia, 1976.
31. Rosenau, P. Hamiltonian dynamics of dense chains and lattices: Or how to correct the continuum. *Phys. Lett. A* **2003**, *311*, 39–52. [CrossRef]
32. Kaplunov, J.; Kossovich, L.Y.; Nolde, E. *Dynamics of Thin Walled Elastic Bodies*; Academic Press: San Diego, CA, USA, 1998.
33. Kaplunov, J.; Pichugin, A. On rational boundary conditions for higher-order long-wave models. In Proceedings of the IUTAM Symposium on Scaling in Solid Mechanics, Cardiff, UK, 25–29 June 2007; Borodich, F., Ed.; IUTAM Book Series 10. Springer: Dordrecht, The Netherlands, 2009; pp. 81–90.
34. Froiio, F.; Zervos, A.; Vardoulakis, I. On natural boundary conditions in linear 2nd-grade elasticity. In *Mechanics of Generalized Continua: One Hundred Years after the Cosserats*; Springer: New York, NY, USA, 2010; pp. 211–221.
35. Froiio, F.; Zervos, A. Second-grade elasticity revisited. *Math. Mech. Solids* **2019**, *24*, 748–777. [CrossRef]
36. Nazarov, S.A. Boundary layers and the hinge-support conditions for thin plates. *J. Math. Sci.* **2002**, *108*, 806–850. [CrossRef]
37. Bolshakova, L.V.; Grachev, A.V. The use of average error of approximation as a criterion of adequacy of regression model. In *Regional Informatics and Information Security. Sbornik Trudov (Collection of Works)*; St. Petersburg Society of Informatics, Computer Science, Communications and Control Systems: St. Petersburg, Russia, 2016; Issue 2, pp. 162–164. (In Russian)

mathematics

MDPI

Article

Norm-Resolvent Convergence for Neumann Laplacians on Manifold Thinning to Graphs

Kirill D. Cherednichenko [1], **Yulia Yu. Ershova** [2] **and Alexander V. Kiselev** [1,*]

[1] Department of Mathematical Sciences, University of Bath, Claverton Down, Bath BA2 7AY, UK; k.cherednichenko@bath.ac.uk
[2] Department of Mathematics, Texas A&M University, College Station, TX 77843-3368, USA; yuershova@tamu.edu
* Correspondence: alexander.v.kiselev@gmail.com or ak2084@bath.ac.uk

Abstract: Norm-resolvent convergence with an order-sharp error estimate is established for Neumann Laplacians on thin domains in $\mathbb{R}^d$, $d \geq 2$, converging to metric graphs in the limit of vanishing thickness parameter in the "resonant" case. The vertex matching conditions of the limiting quantum graph are revealed as being closely related to those of the δ' type.

Keywords: PDE; quantum graphs; generalised resolvent; thin structures; norm-resolvent asymptotics

MSC: 35C20; 35J25; 47A10; 47A20; 81Q35; 74K99

1. Introduction

In the book [1] (see also references therein, in particular [2–4]), Neumann Laplacians A_ε on thin manifolds, converging to metric graphs as $\varepsilon \to 0$, were studied; see, e.g., Figure 1. Here, ε represents the "thickness" of the manifolds in those parts where they converge to the graph edges. The named works attacked the question of spectral (and, in the case of [1], norm-resolvent) convergence of such partial differential operators to a graph Laplacian with certain matching conditions at the graph vertices. This latter ordinary differential operator is introduced as follows. Denoting by E the set of edges e of the limiting graph G, each $e \in E$ can be identified with the interval $[0, l_e]$, where l_e is the length of e. We use notation $L^2(e)$ for the associated Hilbert space $L^2(0, l_e)$. Similarly, we denote $L^2(G) := \oplus_e L^2(e)$. The graph Laplacian A_G is generated by the differential expression $-u''$, $u \in H^2(e)$ on each edge $e \in E$ separately (see [5] for details), subject to the vertex conditions discussed below.

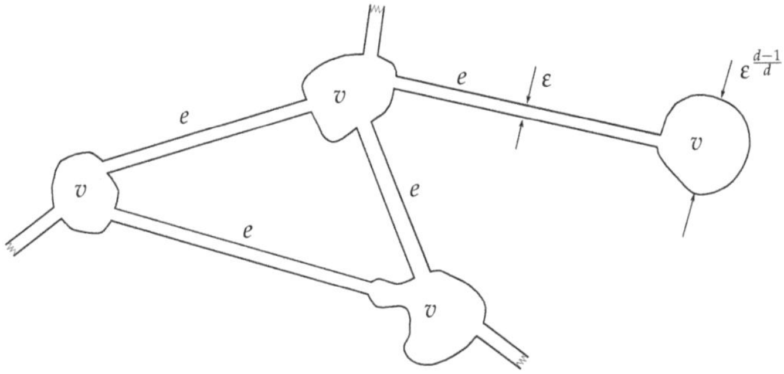

Figure 1. An example of a thin network in the case of a resonant scaling of the edge thickness and vertex diameter. The (non-dimensional) parameter ε is the ratio of the actual size to the wavelength. As $\varepsilon \to 0$, the structure converges to a graph; the related notation is introduced at the start of Section 2.

It was proved in [2–4] that, within any compact $K \Subset \mathbb{C}$, the spectra of A_ε converge in the Hausdorff sense to the spectrum of a graph Laplacian A_G. In the book [1], the claimed convergence was enhanced to the norm-resolvent type, with an explicit control of the error as $O(\varepsilon^\gamma)$, where $\gamma > 0$ depends on whether the ambient space is two-dimensional. The matching conditions at the vertices of the limiting graph turn out to be any one of the following:

(i) Kirchhoff (i.e., standard) if the vertex volumes are decaying, as $\varepsilon \to 0$, faster than the edge volumes;
(ii) "Resonant", which can be equivalently described in terms of δ-type matching conditions with coupling constants proportional to the spectral parameter z, see [2,3], if the vertex and edge volumes are of the same order;
(iii) "Dirichlet-decoupled" (i.e., the graph Laplacian becomes completely decoupled) if the vertex volumes vanish slower than the edge ones.

We also refer the reader to the frequently overlooked papers [6,7], where an alternative approach to the asymptotic analysis of thin networks, aimed at capturing the resonant and scattering features, is developed; see also the references therein. In the present paper we consider the case of scalar PDEs, although our approach can be immediately generalised to the case of PDE systems, in particular, in the context of thin elastic structures with applications to, e.g., pentamodes (see [8,9]) by utilising the results of [10].

In the present paper, we are primarily interested in the most non-trivial resonant case (ii). We provide a straightforward, alternative to [1], proof of the fact that the Neumann Laplacians A_ε in this case converge in a norm-resolvent sense to a linear operator acting in the Hilbert space $L^2(G) \oplus \mathbb{C}^N$, where N is the number of vertices. The mentioned limiting operator is in fact the one first pointed out in [4] as a self-adjoint operator whose spectrum coincides with the Hausdorff limit of spectra for the family A_ε.

On the technical side, our approach can be seen as a modification of the one developed by us in [10,11]; see also [12–14]. We specifically point out that the framework originally developed for high-contrast homogenisation admits a natural generalisation to setups where the "contrast" is achieved by purely geometric means, including (but not limited to) thin networks, as in the present paper. This seems to widen significantly the range of dimension-reduction-type models that are amenable to this kind of analysis and thus to establish a transparent connection between previously unrelated physical contexts, providing for a possibility to develop new types of media in materials science.

We obtain a better error bound than [1] (Section 6.7) (and skip a visit to the "plumber's shop" of [1,15]). Our estimate in the planar case is $O(\varepsilon|\log\varepsilon|)$ and in the case of $\mathbb{R}^d$, $d \geq 3$, it is $O(\varepsilon)$. Unlike that of [1], our method does not allow us to study the full set of asymptotic regimes in (iii). This is due to the fact that the argument of our paper [11] is based on the Dirichlet-to-Neumann (DN) machinery. There is a possibility to modify the approach by invoking Neumann-to-Dirichlet maps instead, which would have two advantages: one could consider all rates of vertex volume decay in (iii), and certain geometric smoothness requirements could be somewhat relaxed. Nevertheless, in this paper we stick with the DN version of the approach in order to align the exposition with that of [11]. The alternative strategy will be followed up elsewhere, both in the present context and in the setting of [11].

The above results of course imply the Hausdorff spectral convergence, at the same time yielding a sharp estimate on its rate. Moreover, in contrast to [1], our approach allows one to consider "high-frequency" regimes, i.e., setups where the spectral parameter (which in the wave propagation context may represent the square of the frequency) is no longer constrained to a compact set but is still constrained by some negative power of the small parameter ε. (In non-dimensional terms this corresponds to the wavelength being of the order of some positive power of ε). This leads to a sequence of "effective", dimensionally reduced, models of the thin structure, which are sequentially applicable for a set of (asymptotic) frequency intervals. The complexity of the dimension reduction process for these models increases along the sequence. While an initial result regarding the high-frequency situation is presented below (which suffices to reveal a metamaterial in a periodic thin network of [16]), we postpone the full analysis to a future publication.

Alongside the high-frequency analysis, yet another sequence of models will be revealed by a version of the same argument. This corresponds to the transition from the resonant regime to the Dirichet-decoupled one and will allow us to reconcile the asymptotic analysis of [1] with that of [7] by introducing "transitional" models of increasing complexity. In these transitional regimes, from the point of view of [1], one gets arbitrarily close to the Dirichlet-decoupled situation, see (iii) above, whereas from the point of view of [7] (where the vertex volumes do not decay at all), one faces a highly non-trivial picture of resonant scattering.

Aiming at better clarity, in the present paper we restrict ourselves to the case where (without much loss of generality) the edge subdomains are assumed straight and uniformly thin, whereas the vertex subdomains are smooth with, possibly, the exception of the points where they meet the edges (see Section 2 for further details). These assumptions appear very natural in view of the possible examples shown in Figures 1 and 2.

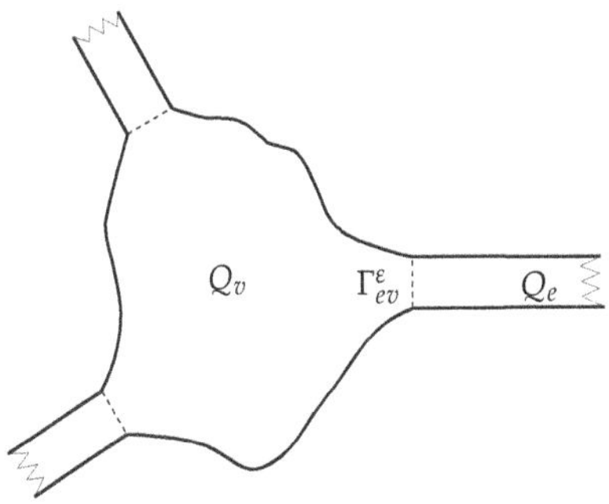

Figure 2. An example of a vertex subdomain Q_v and an edge subdomain Q_e separated by a contact plate $\Gamma^{\varepsilon}_{ev}$ of linear size ε. The volumes of Q_v and Q_e are assumed to be of the same order ε^{d-1}, which is also the area of the contact plate. As in Figure 1, the (non-dimensional) parameter ε is the ratio of the actual size to the wavelength.

The ultimate section of the paper is dedicated to the spectral analysis of the effective graph Laplacian. We show that effective matching conditions at graph vertices, having been commonly treated as impedance type (namely, linear in the spectral parameter), are in fact of δ' type after an application of a unitary gauge. This latter, however, introduces constant magnetic potentials on the edges of the limit graph. It is notable that the presence of such a magnetic field leads to a phase transition to a medium exhibiting double-negative metamaterial properties; see [16]. This observation will permit us to construct thin periodic networks with negative group velocity without the need for an external magnetic field, by relying instead on the resonant geometric properties of the network.

2. Problem Setup and Preliminaries

In what follows, we consider a prototypical setup only, which already presents all the challenges appearing in the general case. We also refer the reader to [1], where the most general setup is meticulously introduced.

For the limiting graph, the following notation will be used: the metric graph G will be identified with the set of edges E, so each individual edge is denoted by $e \in E$ and is associated with an interval $[0, l_e]$. We denote by V the set of graph vertices and treat each $v \in V$ as the set of edge endpoints meeting at v. The graph G is assumed to be oriented throughout.

Proceeding to the setup for the Neumann Laplacian on a thin-graph-convergent structure, let a connected (open) domain Q be the union of the "vertex" part Q_V, the "edge" part Q_E, and the interface boundary Γ^{ε}_v between the two, where Q_E will be assumed to be a finite collection of ε-thin cylinders (for $d=2$, rectangular boxes), $Q_E = \cup_e Q_e$ of lengths l_e. For each e, the domain Q_e is assumed to be, up to a linear spatial transform (i.e., shift and rotation), defined by (for $d=2$)

$$Q_e = \{x \in \mathbb{R}^2 : x_1 \in (0, l_e), x_2 \in (0, \varepsilon)\} = (0, l_e) \times (0, \varepsilon).$$

In dimensions $d \geq 3$, the set Q_e is defined as the direct product of the interval $(0, l_e)$ and a smooth cross-section Q_e^c of volume ε^{d-1}.

It is further assumed that $Q_V = \cup_v Q_v$, where each of the disjoint domains Q_v is connected and has piecewise smooth boundary ∂Q_v, which can be decomposed as $\partial Q_v = \overline{\Gamma_v^\varepsilon} \cup \overline{\widetilde{\Gamma}_v^\varepsilon}$, where Γ_v^ε is the interface between the vertex and edge domains and $\widetilde{\Gamma}_v^\varepsilon$ is the remaining part of ∂Q_v. Henceforth, we often drop the dependence on ε in the notation for those geometric elements where it can be removed by a rescaling or is otherwise irrelevant in the analysis to follow.

The contact part Γ_v^ε is further decomposed into a union of flat plates (for $d=2$, straight segments): $\Gamma_v^\varepsilon = \cup \Gamma_{ev}^\varepsilon$. Here, the union is taken over all edge domains Q_e connected to Q_v so that $\overline{Q}_e \cap \overline{Q}_v = \overline{\Gamma_{ev}^\varepsilon}$. In what follows, we will refer to the segments Γ_{ev}^ε as *contact plates*. Since operators of Zaremba (or mixed) boundary value problem [17] will be used below, we further require that the contact plates Γ_{ev}^ε meet $\widetilde{\Gamma}_v^\varepsilon$ at angles strictly less than π; see [18,19] for further details. We will further assume that the curves $\widetilde{\Gamma}_v^\varepsilon$ are smooth.

Furthermore, we assume that for all vertices v the domains Q_v have volumes of order ε^{d-1} and are obtained by suitable transformations of an ε-independent domain Q_v^0 containing the origin. More precisely, we assume that for all v and $\varepsilon > 0$ one has

$$Q_v = \varepsilon^{\frac{d-1}{d}} \mathcal{T}_v^\varepsilon(Q_v^0) + b_v^\varepsilon, \tag{1}$$

where $b_v^\varepsilon \in \mathbb{R}^d$, and the domain $\mathcal{T}_v^\varepsilon(Q_v^0)$ is obtained from Q_v^0 by a suitable homeomorphism $\mathcal{T}_v^\varepsilon$ that (a) preserves volume and (b) maintains the angles between the contact plates and the adjacent parts of the boundary to be uniformly less than π, cf. [17,18].

For example, if Q_v is star-shaped with respect to the origin (recall that a domain Q_v is said to be star-shaped with respect to $x_v \in Q_v$ if for all $x \in Q_v$ one has $\{x_v + t(x - x_v), t \in [0,1]\} \subset Q_v$), the deformation $\mathcal{T}_v^\varepsilon$ can be constructed by "cut-and-glue surgery" applied to Q_v^0 followed by a suitable "radial" scaling, as follows. The domain Q_v^0 is first split into several conical (sectorial for $d=2$) subdomains by making cuts alongside the cone "generatrices" ("radii" for $d=2$) from the origin to the boundary points (endpoints for $d=2$) of all subsets of ∂Q_v^0 pertaining to the contact plates of Q_v. Each cone is then transformed by a suitable change of variable so that (A) the images of the cone bases pertaining to the contact plates of Q_v have linear size $\varepsilon^{1/d}$ and (B) the image of the complementary part of Q_v^0 is reattached to the lateral boundaries of the cones in A. Note that this procedure in general changes the domain volume by a small quantity. In order to restore the volume, a suitable "radial" scaling (centred at the origin) is applied, ensuring that it does not affect any of the mentioned conical subsets.

We remark that the representation (1) guarantees that we are in the "resonant" (or "borderline") case of [2,3], i.e., that the volumes of the contact plates are proportional to the volumes of vertex domains.

The requirement that the homeomorphism $\mathcal{T}_v^\varepsilon$ in (1) be volume preserving is not essential and can be removed, as long as the ratios of the volumes of the vertex domains to the volumes of the contact plates are bounded below. In this case, the corresponding final convergence estimates, similar to those we derive below (see Theorem 3), will in general depend on the said ratios, as is evident from the formula (14).

On the domain Q, we consider a family of self-adjoint operators A_ε defined by their sesquilinear forms and corresponding to the differential expression $-\Delta$ subject to Neumann boundary conditions. Other types of boundary conditions can be considered as well, including those of Robin and Dirichlet [6,20], in which case the lower edge of the spectrum is ε-dependent, which in the context of homogenisation corresponds to the so-called high-frequency regime or, using the terminology of M. S. Birman, to "the neighbourhood of the edge of a gap" [21,22].

Consider $\sigma > 0$ and an arbitrary non-negative function $R = R(\varepsilon)$, $\varepsilon \in (0,1)$ such that $R(\varepsilon)\varepsilon^{1/2}|\log\varepsilon|^{\gamma} \to 0$ as $\varepsilon \to 0$, where $\gamma = 1$ for $d = 2$ and $\gamma = 0$ otherwise. In what follows, we will deal with the family of resolvents $(A_\varepsilon - z)^{-1}$. We shall always assume that $z \in \mathbb{C}$ is separated from the spectrum of the original operator family. In particular cases, we will assume that either z is constrained to the ε-growing compact set

$$K_\sigma^\varepsilon := B_{R(\varepsilon)}(0) \setminus \{z \in \mathbb{C} |\ \mathrm{dist}(z, \mathbb{R}) \geq \sigma\},$$

where $B_{R(\varepsilon)}(0)$ denotes the ball of radius $R(\varepsilon)$ centered at the origin, or

$$z \in K_\sigma := \{z \in \mathbb{C} |\ z \in K \text{ a compact set in } \mathbb{C},\ \mathrm{dist}(z, \mathbb{R}) \geq \sigma\}.$$

In particular, for all ε small enough, one has $K_\sigma \subset K_\sigma^\varepsilon$. After we have established the operator-norm asymptotics of $(A_\varepsilon - z)^{-1}$ for $z \in K_\sigma$, the result is extended by analyticity to a compact set K_σ^{ext} whose distance to the spectrum of the leading order of the asymptotics is bounded below by σ, and the same statement holds in the case of K_σ^ε.

Proceeding similarly to, e.g., [23] in the related area of critical-contrast homogenisation and facilitated by the abstract framework of [24], we consider A_ε as operators of transmission problems (see [25] and references therein) relative to the *internal boundary* $\Gamma^\varepsilon := \cup_{e,v}\Gamma_{ev}^\varepsilon$. The transmission problem is formulated as, given a function $f \in L^2(Q)$, finding the weak solution $u \in L^2(Q)$ of the boundary value problem

$$\begin{cases} -\Delta u(x) - zu(x) = f(x), & x \in Q_V \text{ or } x \in Q_E, \\ u_v(x) = u_e(x), \quad \dfrac{\partial u_v}{\partial n} - \dfrac{\partial u_e}{\partial n} = 0 & \text{on } \Gamma_{ev}^\varepsilon, \\ \dfrac{\partial u}{\partial n} = 0 & \text{on } \partial Q. \end{cases} \tag{2}$$

Here, $u_v := u|_{Q_v}$, $u_e := u|_{Q_e}$ for all admissible e and v (i.e. when $\Gamma_{ev}^\varepsilon \neq \varnothing$), and n represents the exterior normal on ∂Q and the "edge-inward" normal (i.e., directed from Q_v to Q_e) on any of contact plates Γ_{ev}^ε. By a classical argument, the solution of the above problem is shown to be equal to $(A_\varepsilon - z)^{-1}f$. Note that the boundary value problem for the Neumann Laplacian on Q is given by the first and third lines in (2). While the second line is, strictly speaking, redundant, it proves important in order to view the problem as a transmission one relative to Γ^ε.

It remains to be seen that the linear operator of the transmission problem (2) defined via the technique in [24], which we briefly recall below, is the same operator A_ε; the proof of this fact follows easily by combining [24] and the main estimate of [25].

Following the approach of [24] (cf. [26,27] and references therein for alternative approaches), which is based on the ideas of the classical Birman–Kreĭn–Višik theory (see [28–30]), the linear operator of the transmission boundary value problem is introduced as follows. Let $\mathcal{H} := L^2(\Gamma^\varepsilon) = \oplus_{e,v} L^2(\Gamma_{ev}^\varepsilon)$, and consider the harmonic lift operators Π_V and Π_E defined on $\phi \in \mathcal{H}$ via

$$\Pi_V\phi := u_\phi, \quad \text{where} \quad \begin{cases} \Delta u_\phi = 0, \ \ u_\phi \in L^2(Q_V), \\ u_\phi|_{\Gamma^\varepsilon} = \phi, \end{cases}$$

$$\Pi_E\phi := u_\phi, \quad \text{where} \quad \begin{cases} \Delta u_\phi = 0, \ \ u_\phi \in L^2(Q_E), \\ u_\phi|_{\Gamma^\varepsilon} = \phi, \end{cases}$$

subject to Neumann boundary conditions on ∂Q. These operators are first defined on $\phi \in C_0^2(\Gamma^\varepsilon)$, in which case the corresponding solutions u_ϕ can be seen as classical [17]. The results of [18] allow one to extend both harmonic lifts to bounded (in fact, compact) operators on $\mathcal{H}$, in which case u_ϕ are to be treated as distributional solutions of the respective boundary value problems. The solution operator $\Pi : \mathcal{H} \mapsto L^2(Q) = L^2(Q_V) \oplus L^2(Q_E)$ is defined as follows:

$$\Pi\phi := \Pi_V\phi \oplus \Pi_E\phi.$$

Consider the self-adjoint operator family A_0 (choosing not to reflect the ε-dependence in the notation) to be the Dirichlet decoupling of the operator family A_ε, i.e., the operator of the boundary value problem on both Q_V and Q_E, where the Dirichlet boundary conditions are imposed on Γ^ε together with Neumann boundary conditions on ∂Q. The operator A_0 is generated by the same differential expression as A_ε. Clearly, one has $A_0 = A_0^V \oplus A_0^E$ relative to the orthogonal decomposition $L^2(Q) = L^2(Q_V) \oplus L^2(Q_E)$; all three operators A_0, A_0^V and A_0^E are self-adjoint and positive-definite. Moreover, by [18,31] there exists a bounded inverse A_0^{-1}. Note that $\operatorname{dom} A_0 \cap \operatorname{ran} \Pi = \varnothing$; see [24].

Furthermore, denoting by $\widetilde{\Gamma}_0^V$ and $\widetilde{\Gamma}_0^E$ the left inverse of Π_V and Π_E, respectively, one introduces the trace operator Γ_0^V (respectively, Γ_0^E) as the null extension of $\widetilde{\Gamma}_0^V$ (respectively, $\widetilde{\Gamma}_0^E$) to the domain $\operatorname{dom} A_0^V \dotplus \operatorname{ran} \Pi_V$ (respectively, $\operatorname{dom} A_0^E \dotplus \operatorname{ran} \Pi_E$). In the same way, we introduce the operator $\widetilde{\Gamma}_0$ and its null extension Γ_0 to the domain $\operatorname{dom} A_0 \dotplus \operatorname{ran} \Pi$.

The solution operators S_z^V, S_z^E of the boundary value problems

$$\begin{cases} -\Delta u_\phi - z u_\phi = 0, & u_\phi \in \operatorname{dom} A_0^V \dotplus \operatorname{ran} \Pi_V, \\ \Gamma_0^V u_\phi = \phi, \end{cases}$$

$$\begin{cases} -\Delta u_\phi - z u_\phi = 0, & u_\phi \in \operatorname{dom} A_0^E \dotplus \operatorname{ran} \Pi_E, \\ \Gamma_0^E u_\phi = \phi \end{cases}$$

are defined as linear mappings from ϕ to u_ϕ, respectively. These operators are bounded from $L_2(\Gamma^\varepsilon)$ to $L_2(Q_V)$ and $L_2(Q_E)$, respectively, and admit the following representations:

$$S_z^E = (1 - z(A_0^E)^{-1})^{-1}\Pi_E, \quad S_z^V = (1 - z(A_0^V)^{-1})^{-1}\Pi_V.$$

The solution operator S_z from $L^2(\Gamma^\varepsilon)$ to $L^2(Q_V) \oplus L^2(Q_E)$ is now defined as $S_z = S_z^V \oplus S_z^E$; it admits the representation $S_z = (1 - z(A_0)^{-1})^{-1}\Pi$ and is bounded.

Having introduced orthogonal projections P_V and P_E from $L^2(Q)$ onto $L^2(Q_V)$ and $L^2(Q_E)$, respectively, one has the obvious identities

$$S_z^V = P_V S_z, \quad S_z^E = P_E S_z, \qquad \Pi_V = P_V \Pi, \quad \Pi_E = P_E \Pi.$$

Fix self-adjoint (and, in general, unbounded) operators Λ^E, Λ^V defined on domains $\operatorname{dom} \Lambda^E, \operatorname{dom} \Lambda^V \subset L^2(\Gamma^\varepsilon)$ (in what follows these operators will be chosen as DN maps of Zaremba problems on Q_E and Q_V, respectively, and well-defined on $H^1(\Gamma^\varepsilon)$, where $H^1(\Gamma^\varepsilon)$ is the standard Sobolev space pertaining to the internal boundary Γ^ε). Still following [24], we define the "second boundary operators" Γ_1^E and Γ_1^V to be linear operators on the domains

$$\operatorname{dom} \Gamma_1^E := \operatorname{dom} A_0^E \dotplus \Pi_E \operatorname{dom} \Lambda^E, \quad \operatorname{dom} \Gamma_1^V := \operatorname{dom} A_0^V \dotplus \Pi_V \operatorname{dom} \Lambda^V.$$

The action of $\Gamma_1^{E(V)}$ is set by:

$$\Gamma_1^E : (A_0^E)^{-1} f \dotplus \Pi_E \phi \mapsto \Pi_E^* f + \Lambda^E \phi, \quad \Gamma_1^V : (A_0^V)^{-1} f \dotplus \Pi_V \phi \mapsto \Pi_V^* f + \Lambda^V \phi$$

for all $f \in L^2(Q_E), \phi \in \operatorname{dom} \Lambda^E$ and $f \in L^2(Q_V), \phi \in \operatorname{dom} \Lambda^V$, respectively.

Alongside $\Gamma_1^{E(V)}$, introduce a self-adjoint operator Λ on $\operatorname{dom} \Lambda \subset \mathcal{H}$ and then the following "boundary" operator Γ_1 :

$$\operatorname{dom} \Gamma_1 := \operatorname{dom} A_0^{-1} \dotplus \Pi \operatorname{dom} \Lambda,$$

$$\Gamma_1 : \; A_0^{-1} f \dot{+} \Pi \phi \mapsto \Pi^* f + \Lambda \phi \qquad \forall f \in L^2(Q), \, \phi \in \operatorname{dom} \Lambda.$$

We remark that the operators $\Gamma_1, \Gamma_1^{E(V)}$ thus defined are assumed to be neither closed nor indeed closable.

In our setup, we make the following concrete choice of the operators $\Lambda^{E(V)}$: in what follows, they are the DN maps pertaining to the components Q_E and Q_V, respectively. More precisely, for the problem

$$\Delta u_\phi = 0, \quad u_\phi \in L^2(Q_E),$$
$$u_\phi|_{\Gamma^\varepsilon} = \phi, \quad \partial_n u_\phi|_{\partial Q} = 0,$$

the operator Λ^E maps the boundary values ϕ of u_ϕ to the negative traces of its normal derivative

$$\partial_n u_\phi|_{\Gamma^\varepsilon} := \nabla u \cdot n|_{\Gamma^\varepsilon},$$

where $n = -n_E$ is as above the "edge-inward" normal. This operator is well defined by its sesquilinear form as a self-adjoint operator on $L^2(\Gamma^\varepsilon)$ (see, e.g., [32,33]), and $\operatorname{dom} \Lambda^E \supset H^1(\Gamma^\varepsilon)$ by [18]. The above definition is inspired by [24]. Note that the operator thus defined is negative the classical *DN* map of, e.g., [23].

On the vertex part Q_V, we consider the problem

$$\Delta u_\phi = 0, \quad u_\phi \in L^2(Q_V),$$
$$u_\phi|_{\Gamma^\varepsilon} = \phi, \quad \partial_n u_\phi|_{\partial Q} = 0,$$

and define Λ^V as the operator mapping the boundary values ϕ of u_ϕ to the negative traces of its normal derivative $-\partial_n u_\phi|_{\Gamma^\varepsilon}$, where $n = n_V$ is again the "edge-inward" normal. The self-adjointness of Λ^V on $\operatorname{dom} \Lambda^V \supset H^1(\Gamma^\varepsilon)$ follows by an unchanged argument.

Finally we introduce the operator Λ which on $\phi \in H^1(\Gamma^\varepsilon)$ is the sum $\Lambda\phi = \Lambda^V \phi + \Lambda^E \phi$. It is also a self-adjoint operator on $\operatorname{dom} \Lambda \supset H^1(\Gamma^\varepsilon)$. This can be ascertained either by the argument of [25], in which case it is defined as the inverse of a compact self-adjoint operator on the orthogonal complement to constants $L^2(\Gamma^\varepsilon) \ominus \{c\mathbf{1}\}$, extended to $\{c\mathbf{1}\}$ by zero or, alternatively, from its definition by a closed sesquilinear form.

The choice of $\Lambda^{E(V)}$ made above allows us to consider Γ_1 on the domain $\operatorname{dom} A_0 \dot{+} \Pi \operatorname{dom} \Lambda$. One then writes [24] the second Green identity in the following form:

$$\langle Au, v \rangle_{L^2(Q)} - \langle u, Av \rangle_{L^2(Q)} = \langle \Gamma_1 u, \Gamma_0 v \rangle_{L^2(\Gamma^\varepsilon)} - \langle \Gamma_0 u, \Gamma_1 v \rangle_{L^2(\Gamma^\varepsilon)}$$

for all $u, v \in \operatorname{dom} \Gamma_1 = \operatorname{dom} A_0 \dot{+} \Pi \operatorname{dom} \Lambda$, where the operator A is the null extension (see [34]) of the operator A_0 onto $\operatorname{dom} \Gamma_1$. Thus the triple $(\mathcal{H}, \Gamma_0, \Gamma_1)$ is closely related to a boundary quasi-triple of [26] (see also [35]) for the transmission problem considered; cf. [27] for an alternative approach.

The calculation of Π^* in [24] shows that $\Pi^* = \Gamma_1 A_0^{-1}$ and therefore Γ_1 as introduced above acts as follows:

$$\Gamma_1 : \; u = P_E u + P_V u \; \mapsto \; \partial_n P_e u|_{\Gamma^\varepsilon} - \partial_n P_V u|_{\Gamma^\varepsilon},$$

where P_E and P_V are the orthogonal projections of $L^2(Q)$ onto $L^2(Q_E), L^2(Q_V)$, respectively. Therefore, transmission problem at hand (at least, formally so far) corresponds to the interface (or "matching") condition $\Gamma_1 u = 0$.

Definition 1 ([24])**.** *The operator-valued function $M(z)$ defined on the domain* $\operatorname{dom} \Lambda$ *for $z \in \rho(A_0)$ (and in particular, for $z \in K_\sigma^\varepsilon$) by the formula*

$$M(z)\phi = \Gamma_1 S_z \phi = \Gamma_1 (1 - z A_0^{-1})^{-1} \Pi \phi \tag{3}$$

is referred to as the M-function of the problem (2).

The following result of [24] summarises the properties of the M−function which we will need in what follows.

Proposition 1 ([24], Theorem 3.3). *1. One has the following representation:*

$$M(z) = \Lambda + z\Pi^*(1 - zA_0^{-1})^{-1}\Pi, \quad z \in \rho(A_0). \tag{4}$$

2. $M(z)$ is an analytic operator-function with values in the set of closed operators in $L_2(\Gamma^\varepsilon)$ densely defined on the z-independent domain $\operatorname{dom}\Lambda$.

3. For $z, \zeta \in \rho(A_0)$ the operator $M(z) - M(\zeta)$ is bounded and

$$M(z) - M(\zeta) = (z - \zeta)S_{\bar{z}}^* S_\zeta$$

In particular, $\Im M(z) = (\Im z)S_{\bar{z}}^* S_{\bar{z}}$ and $(M(z))^* = M(\bar{z})$.

4. For $u_z \in \ker(A - zI) \cap \{\operatorname{dom} A_0 \dotplus \Pi \operatorname{dom}\Lambda\}$, the following formula holds:

$$M(z)\Gamma_0 u_z = \Gamma_1 u_z.$$

Alongside $M(z)$, we define $M_V(z)$ and $M_E(z)$, which pertain to the vertex Q_V and edge Q_E parts of the domain Q, respectively, by the formulae

$$\begin{aligned} M_V(z)\phi &= \Gamma_1^V S_z^V \phi = \Gamma_1^V \big(1 - z(A_0^V)^{-1}\big)^{-1}\Pi_V\phi, \\ M_E(z)\phi &= \Gamma_1^E S_z^E \phi = \Gamma_1^E \big(1 - z(A_0^E)^{-1}\big)^{-1}\Pi_E\phi. \end{aligned} \tag{5}$$

As before, in the notation we suppress the dependence on the parameter ε for brevity.

The value of the fact that for $\phi \in H^1(\Gamma^\varepsilon)$ one has $M(z)\phi = M_E(z)\phi + M_V(z)\phi$ is clear: in contrast to A_ε, which cannot be additively decomposed into "independent" terms pertaining to the vertex and edge parts of the medium Q owing to the transmission interface conditions on Γ^ε, the M-function is additive (see, e.g., [11,36], where this property was observed and exploited in the related settings of homogenisation and scattering, respectively). In what follows, we will observe that the resolvent $(A_\varepsilon - z)^{-1}$ can be expressed in terms of $M(z)$ via a version of the celebrated Kreĭn formula, thus reducing the asymptotic analysis of the resolvent to that of the corresponding M-function (see, e.g., [26,37] for alternative approaches to derivation of the Kreĭn formula in our setting).

Alongside the transmission problem (2), the boundary conditions of which can be now (so far, formally) represented as $u \in \operatorname{dom} A_0 \dotplus \Pi L^2(\Gamma^\varepsilon)$, $\Gamma_1 u = 0$, in what follows we will require a wider class of problems of this type. This class is formally given by the transmission conditions

$$u \in \operatorname{dom} A \equiv \operatorname{dom} A_0 \dotplus \Pi L^2(\Gamma^\varepsilon), \quad \beta_0\Gamma_0 u + \beta_1\Gamma_1 u = 0,$$

where β_1 is a bounded operator on $L^2(\Gamma^\varepsilon)$ and β_0 is a linear operator defined on the domain $\operatorname{dom}\beta_0 \supset \operatorname{dom}\Lambda$.

In general, the operator $\beta_0\Gamma_0 + \beta_1\Gamma_1$ is not defined on the domain $\operatorname{dom} A$. This problem is being taken care of by the following assumption, which will be satisfied throughout:

$$\beta_0 + \beta_1\Lambda \text{ defined on } \operatorname{dom}\Lambda \text{ is closable in } \mathcal{H}.$$

We remark that by Proposition 1 the operators $\beta_0 + \beta_1 M(z)$ are then closable for all $z \in \rho(A_0)$, and the domains of their closures coincide with $\operatorname{dom}\overline{\beta_0 + \beta_1\Lambda}$.

For any $f \in H$, $\phi \in \operatorname{dom}\Lambda$, the equality

$$(\beta_0\Gamma_0 + \beta_1\Gamma_1)(A_0^{-1}f + \Pi\phi) = \beta_1\Pi^* f + (\beta_0 + \beta_1\Lambda)\phi$$

shows that the operator $\beta_0\Gamma_0 + \beta_1\Gamma_1$ is correctly defined on $A_0^{-1}H \dotplus \Pi \operatorname{dom}\Lambda \subset \operatorname{dom} A$. Denoting $\mathcal{B} := \overline{\beta_0 + \beta_1\Lambda}$ with the domain $\operatorname{dom}\mathcal{B} \supset \operatorname{dom}\Lambda$, one checks that $H_\mathcal{B} := A_0^{-1}H \dotplus \Pi \operatorname{dom}\mathcal{B}$ is a Hilbert space with respect to the norm

$$\|u\|_\mathcal{B}^2 := \|f\|_H^2 + \|\phi\|_\mathcal{H}^2 + \|\mathcal{B}\phi\|_\mathcal{H}^2, \qquad u = A_0^{-1}f + \Pi\phi.$$

It is then proved [24] (Lemma 4.1) that $\beta_0\Gamma_0 + \beta_1\Gamma_1$ extends to a bounded operator from $H_\mathcal{B}$ to $\mathcal{H}$. For the sake of convenience, same notation $\beta_0\Gamma_0 + \beta_1\Gamma_1$ is preserved for this extension.

We will make use of the following version of the celebrated Kreĭn formula.

Proposition 2 ([24], Theorem 5.1). *Let $z \in \rho(A_0)$ be such that the operator $\overline{\beta_0 + \beta_1 M(z)}$ defined on $\operatorname{dom}\mathcal{B}$ is boundedly invertible. Then*

$$R_{\beta_0,\beta_1}(z) := (A_0 - z)^{-1} + S_z Q_{\beta_0,\beta_1}(z) S_{\bar{z}}^*, \text{ where } Q_{\beta_0,\beta_1} := -\left(\overline{\beta_0 + \beta_1 M(z)}\right)^{-1}\beta_1, \quad (6)$$

is the resolvent of a closed densely defined operator A_{β_0,β_1} with the domain

$$\operatorname{dom} A_{\beta_0,\beta_1} = \{u \in H_\mathcal{B} | (\beta_0\Gamma_0 + \beta_1\Gamma_1)u = 0\} = \ker(\beta_0\Gamma_0 + \beta_1\Gamma_1).$$

In particular, the (self-adjoint) operator of the transmission problem (2), which corresponds to the choice $\beta_0 = 0, \beta_1 = I$, admits the following characterisation in terms of its resolvent:

$$R_{0,I}(z) = (A_0 - z)^{-1} - S_z M^{-1}(z) S_{\bar{z}}^*. \quad (7)$$

In this case, one clearly has $H_\mathcal{B} = A_0^{-1}H \dotplus \Pi \operatorname{dom}\Lambda$ and $\operatorname{dom} A_{0,I} = \{u \in H_\mathcal{B} | \Gamma_1 u = 0\}$, which, together with the discussion at the beginning of this section, yields $A_{0,I} = A_\varepsilon$.

We remark that the operators β_0 and β_1 above can be assumed z-dependent, as this change does not impact the corresponding proofs of [24]. In this case, however, the corresponding operator-function $R_{\beta_0,\beta_1}(z)$ is shown to be the resolvent of a z-dependent operator family. Within the self-adjoint setup of the present paper, $R_{\beta_0,\beta_1}(z)$ is guaranteed to represent a generalised resolvent in the sense of [38–40].

3. Auxiliary Estimates

In this section, we collect a number of auxiliary statements required in our proof of the main result.

We start with the analysis of the operators Π_V and S_z^V introduced in Section 2. First, we note that each of these operators admits a decomposition into an orthogonal sum over N vertex domains $\{Q_v\}$ of Q. It therefore suffices to consider a single vertex domain Q_v (we recall for readers' convenience that the volume of this domain is assumed to be decaying with $\varepsilon \to 0$). Its boundary ∂Q_v contains a disjoint set of straight segments belonging to the internal boundary Γ^ε, which are, in line with what has been said above, denoted as Γ_{ev}^ε; the union of the latter is Γ_v^ε.

The decoupled operator A_0 has $L^2(Q_v)$ as its invariant subspace. We will denote by $A_0^{(v)}$ its self-adjoint restriction, $A_0^{(v)} := A_0|_{L^2(Q_v)}$. By construction, the operator $A_0^{(v)}$ is the Laplacian with the so-called Zaremba, or mixed Neumann–Dirichlet, boundary condition [17,41]. More precisely, it is subject to the Dirichlet boundary condition on Γ_v^ε and to Neumann boundary condition on its complement $\widetilde{\Gamma}_v^\varepsilon$. Clearly, this operator is boundedly invertible; moreover, the following statement holds.

Proposition 3 (see [31,42]). *There exists a constant $C > 0$ such that for all ε one has*

$$\|(A_0^{(v)})^{-1}\| \leq C|Q_v| \begin{cases} |\log\varepsilon|, & d = 2, \\ \varepsilon^{2-d}, & d \geq 3 \end{cases} \asymp \varepsilon|\log\varepsilon|^\gamma,$$

where, as before, $\gamma = 1$ for $d = 2$ and $\gamma = 0$ otherwise.

Remark 1. *The above proposition holds under more general conditions than those we impose. Namely, the domain Q_v is only required to be Lipschitz and no conditions whatsoever are imposed on the geometry of the set Γ_v^ε.*

Next, we turn our attention to the solution operator $S_z^{(v)} := S_z^V|_{\Gamma_v^\varepsilon}$ and the corresponding harmonic lift $\Pi^{(v)} := \Pi_V|_{\Gamma_v^\varepsilon}$. The two are clearly related by the formula

$$S_z^{(v)} = \big(1 - z(A_0^{(v)})^{-1}\big)^{-1}\Pi^{(v)}.$$

In order to bound the norm of $\Pi^{(v)}$, we can follow, e.g., the following approach. First, consider the corresponding Zaremba problem on Q_v^0. We proceed by relating the norm of the corresponding Poisson operator to the least Steklov eigenvalue of the bi-Laplacian, following the blueprint of [43], based in turn on Fichera's duality principle; see [44]. Since the boundary of Q_v^0 is non-smooth, in doing so we follow the generalisations developed in [45,46], with obvious modifications required when passing from the Dirichlet to Zaremba setup. The estimate for the said Steklov eigenvalue is then taken from the norm of the compact embedding of $H^2(Q_v^0)$ to the traces of normal derivatives on the contact plates; see, e.g., [47]. Rescaling back to Q_v, we obtain the following auxiliary result.

Lemma 1. *There exists $C > 0$ such that $\|\Pi^{(v)}\| \leq C$ for all ε.*

By Proposition 3, the above lemma yields the following estimate for the solution operator $S_z^{(v)}$.

Lemma 2. *For $\varepsilon \in (0,1)$, uniformly in $z \in K_\sigma^\varepsilon$ (and, in particular, for $z \in K_\sigma$) one has*

$$S_z^{(v)} = \big(1 + O(|z|\varepsilon|\log\varepsilon|^\gamma)\big)\Pi^{(v)} = \Pi^{(v)} + O(|z|\varepsilon|\log\varepsilon|^\gamma),$$

where the error bounds are understood in the uniform operator norm topology.

Our next step is the analysis of the "part" of the DN map Λ^V pertaining to the vertex domain Q_v. We will denote by Λ_v^V its self-adjoint restriction $\Lambda^V|_{L^2(\Gamma_v^\varepsilon)} =: \Lambda_v^V$.

First, we note that the spectrum of Λ_v^V (which can be termed as the Steklov spectrum of the sloshing problem pertaining to $A_0^{(v)}$; see [48]) is discrete and accumulates to negative infinity. The point $\lambda_1 = 0$ is the least (by absolute value) Steklov eigenvalue with $\psi_v = |\Gamma_v^\varepsilon|^{-1/2}\mathbf{1}|_{\Gamma_v^\varepsilon}$ being the corresponding eigenvector. For the second eigenvalue λ_2, one has the following estimate; see, e.g., [32] and references therein.

Lemma 3. *There exists $C > 0$ such that $|\lambda_2| \geq C\varepsilon^{-1}$.*

Introduce the N-dimensional orthogonal projection

$$\mathcal{P} := \sum_v \langle \cdot, \psi_v\rangle \psi_v,$$

define $\mathcal{P}_\perp := 1 - \mathcal{P}$ and consider the operator $\mathcal{P}_\perp M(z)\mathcal{P}_\perp$, which is well defined since $\mathcal{P}\,\mathrm{dom}\,\Lambda \subset \mathrm{dom}\,\Lambda$. By a straightforward estimate for sesquilinear forms, see [11] (Section 3.2), and taking into account (4) applied to $M_V(z)$ and combined with Proposition 3 as well as Lemmata 1, 3, one has the following statement.

Lemma 4. *There exists $C > 0$ such that for all $\varepsilon \in (0,1)$ one has*

$$\|(\mathcal{P}_\perp M(z)\mathcal{P}_\perp)^{-1}\| \leq C\varepsilon, \qquad z \in K_\sigma^\varepsilon,$$

where the operator $\mathcal{P}_\perp M(z)\mathcal{P}_\perp$ is considered as a linear (unbounded) operator in $\mathcal{P}_\perp\mathcal{H}$.

We conclude this section by noting that, alternatively, one can derive the bound on the Poisson operator $\Pi^{(v)}$ claimed in Lemma 1 by employing the scaling property of the Dirichlet-to-Neumann map Λ_v^V, similar to the argument of [32] referenced above, combined with a standard estimate on the solutions to the classical Neumann problem.

4. Norm-Resolvent Asymptotics

We will make use of the Kreĭn formula (7) to obtain a norm-resolvent asymptotics of the family A_ε. In doing so, we will compute the asymptotics of $M^{-1}(z)$ based on a Schur–Frobenius-type inversion formula, having first rewritten $M(z)$ as a 2×2 operator matrix relative to the orthogonal decomposition of the Hilbert space $\mathcal{H} = \mathcal{P}\mathcal{H}\oplus\mathcal{P}_\perp\mathcal{H}$. In the study of operator matrices, we rely upon the material of [49]; see also references therein.

The operator $M(z)$ admits the block matrix representation

$$M(z) = \begin{pmatrix} \mathbb{A} & B \\ \mathbb{E} & \mathbb{D} \end{pmatrix} \text{ with } \mathbb{A}, \mathbb{B}, \mathbb{E} \text{ bounded.}$$

For the inversion of $M(z)$, we then use the Schur–Frobenius inversion formula [49] (Theorem 2.3.3)

$$\overline{\begin{pmatrix} \mathbb{A} & \mathbb{B} \\ \mathbb{E} & \mathbb{D} \end{pmatrix}}^{-1} = \begin{pmatrix} \mathbb{A}^{-1} + \overline{\mathbb{A}^{-1}\mathbb{B}\overline{\mathcal{S}}^{-1}}\mathbb{E}\mathbb{A}^{-1} & -\overline{\mathbb{A}^{-1}\mathbb{B}\overline{\mathcal{S}}^{-1}} \\ -\overline{\mathcal{S}}^{-1}\mathbb{E}\mathbb{A}^{-1} & \overline{\mathcal{S}}^{-1} \end{pmatrix} \text{ with } \mathcal{S} := \mathbb{D} - \mathbb{E}\mathbb{A}^{-1}\mathbb{B}. \quad (8)$$

Note that by Proposition 1, one has $\Im M(z) = (\Im z) S_{\bar z}^* S_{\bar z}$. Moreover, since $S_z = (1 - zA_0^{-1})^{-1}\Pi$, one has

$$S_{\bar z}^* S_{\bar z} = \Pi^*(1 - zA_0^{-1})^{-1}(1 - \bar z A_0^{-1})^{-1}\Pi,$$

and therefore, for some constants $c_1, c_2 > 0$,

$$\langle S_{\bar z}^* S_{\bar z}\mathcal{P}\phi, \mathcal{P}\phi\rangle_{\mathcal{H}} = \left\|(1 - \bar z A_0^{-1})^{-1}\Pi\mathcal{P}\phi\right\|^2 \geq c_1\|\Pi\mathcal{P}\phi\|^2 \geq c_1\|\Pi_V\mathcal{P}\phi\|^2 \geq c_2\|\mathcal{P}\phi\|^2_{\mathcal{H}}$$

for all $\phi\in\mathcal{H}, z\in K_\sigma^\varepsilon$, where we have used the fact that the operator A_0 is bounded below by a positive constant. It follows that $\mathbb{A}^{-1} = (\mathcal{P}M(z)\mathcal{P})^{-1}$ is bounded.

Proceeding exactly as in [11] based on the estimate provided by Lemma 4 which now reads

$$\|\mathbb{D}^{-1}\| \leq C\varepsilon,$$

we use $\mathcal{S}^{-1} = (I - \mathbb{D}^{-1}\mathbb{E}\mathbb{A}^{-1}\mathbb{B})^{-1}\mathbb{D}^{-1}$ to obtain $\mathcal{S}^{-1} = O(\varepsilon)$.

Returning to (8), one obtains

$$M(z)^{-1} = \begin{pmatrix} \mathbb{A} & \mathbb{B} \\ \mathbb{E} & \mathbb{D} \end{pmatrix}^{-1} = \begin{pmatrix} \mathbb{A}^{-1} & 0 \\ 0 & 0 \end{pmatrix} + O(\varepsilon) \quad (9)$$

with a uniform estimate for the remainder term. Comparing our result with (6) of Proposition 2 with $\beta_0 := \mathcal{P}_\perp$ and $\beta_1 := \mathcal{P}$, one arrives at the following.

Theorem 1. *There exists $C > 0$ such that for all $\varepsilon \in (0,1)$ and $z \in K_\sigma^\varepsilon$ (in partucular, for all $z \in K_\sigma$) one has the estimate*

$$\left\|(A_\varepsilon - z)^{-1} - (A_{\beta_0,\beta_1} - z)^{-1}\right\| \leq C|z|^2\varepsilon$$

for a universal constant C and $\beta_0 = \mathcal{P}_\perp$, $\beta_1 = \mathcal{P}$, where the operator A_{β_0,β_1} is defined in Proposition 2.

Proof. The proof is identical to that of [11] (Theorem 3.1); we include it here for the sake of completeness. For the resolvent $(A_\varepsilon - z)^{-1}$, the formula (7) is applicable, in which for $M(z)^{-1}$ we use (9). As for the resolvent $\left(A_{\mathcal{P}_\perp,\mathcal{P}} - z\right)^{-1}$, Proposition 2 with $\beta_0 = \mathcal{P}_\perp$, $\beta_1 = \mathcal{P}$ is clearly applicable. Moreover, for this choice of β_0, β_1, the operator

$$Q_{\mathcal{P}_\perp,\mathcal{P}}(z) = -\left(\overline{\mathcal{P}_\perp + \mathcal{P}M(z)}\right)^{-1}\mathcal{P}$$

in (6) is easily computable (e.g., by the Schur–Frobenius inversion formula of [49], see (8)). The mentioned computation is facilitated by the fact that $\mathcal{P}_\perp + \mathcal{P}M(z)$ is triangular ($\mathbb{A} = \mathcal{P}M(z)\mathcal{P}$, $\mathbb{B} = \mathcal{P}M(z)\mathcal{P}_\perp$, $\mathbb{E} = 0$, $\mathbb{D} = I$ in (8)) with respect to the decomposition $\mathcal{H} = \mathcal{P}\mathcal{H} \oplus \mathcal{P}_\perp\mathcal{H}$, yielding

$$Q_{\mathcal{P}_\perp,\mathcal{P}}(z) = -\mathcal{P}\left(\mathcal{P}M(z)\mathcal{P}\right)^{-1}\mathcal{P}, \tag{10}$$

and the claim follows. □

Already the estimate of Theorem 1 establishes norm-resolvent convergence of the family A_ε to an operator which by (10) is a relative (i.e., with respect to the difference of resolvents) finite-rank perturbation of the decoupled operator A_0. However, in the case $z \in K_\sigma$, which we will assume henceforth, it is possible to obtain a further simplification of this answer, relating the leading-order asymptotic term to a self-adjoint operator on the limiting metric graph. This procedure follows the blueprint of our paper [11]. We next briefly outline the related argument. For the case of z not constrained to a compact, a similar argument yields a sequence of dimensionally reduced models, as mentioned in the Introduction.

Note first that $(A_0 - z)^{-1} = (A_0^V - z)^{-1} \oplus (A_0^E - z)^{-1}$ is easily analysed. Indeed, by Proposition 3 one has

$$(A_0^V - z)^{-1} = O(\varepsilon|\log\varepsilon|^\gamma). \tag{11}$$

Furthermore, the operator $(A_0^E - z)^{-1}$, by separation of variables, is $O(\varepsilon^2)$-close to the Dirichlet Laplacian on the space

$$H_G := \oplus_v L^2([0, l_e] \times \mathbf{1}_\varepsilon),$$

where $\mathbf{1}_e := \varepsilon^{-(d-1)/2}\mathbf{1}$ is the normalised constant function in the variable transverse to the edge e.

We remark that this is the only place where we use the assumption about the geometric shape of the edge parts Q_e of the thin structure. This can be generalised to the setup of [1], allowing for curvature and non-uniform thickness, leading to Laplace–Beltrami operators on the edges of the limiting graph.

The operator $(A_0 - z)^{-1}$ is therefore close, uniformly in $z \in K_\sigma$, to an operator that is unitary equivalent to the resolvent of A_0^G, where A_0^G is the Dirichlet-decoupled graph Laplacian pertaining to the graph G. The related error estimates are the same as in (11). The finite-dimensional second term on the right-hand side of (10) is therefore expected to encode the matching conditions at the vertices of the limiting graph G. In order to see this, one passes over to the generalised resolvent $R_\varepsilon(z) := P_E(A_\varepsilon - z)^{-1}P_E$, which is shown to admit the following asymptotics.

Theorem 2. *The operator family $R_\varepsilon(z)$ admits the following asymptotics in the operator-norm topology for $z \in K_\sigma$:*

$$R_\varepsilon(z) - R_{\text{eff}}(z) = O(\varepsilon),$$

where $R_{\text{eff}}(z)$ is the solution operator for the following spectral BVP on the edge domain Q_E:

$$\begin{aligned} -\Delta u - zu = f, \quad & f \in L^2(Q_E), \\ \beta_0(z)\Gamma_0^E u + \beta_1\Gamma_1^E u = 0, & \end{aligned} \tag{12}$$

with $\beta_0(z) = \mathcal{P}_\perp - \mathcal{P}B(z)\mathcal{P}$, $B(z) := -M_V(z)$ *and* $\beta_1 = \mathcal{P}$.

The boundary condition in (12) can be written in the more conventional form

$$\mathcal{P}_\perp u|_{\Gamma^\varepsilon} = 0, \quad \mathcal{P}\partial_n u = \mathcal{P}B(z)\mathcal{P}u\big|_{\Gamma^\varepsilon}.$$

Equivalently,

$$R_\varepsilon(z) - \left(A^E_{\mathcal{P}_\perp - \mathcal{P}B(z)\mathcal{P},\mathcal{P}} - z\right)^{-1} = O(\varepsilon),$$

where $A^E_{\mathcal{P}_\perp - \mathcal{P}B(z)\mathcal{P},\mathcal{P}}$, *for any fixed* z, *is the operator in* $L^2(Q_E)$ *defined by Proposition 2 relative to the triple* $(\mathcal{H}, \Pi_E, \Lambda^E)$, *where the term "triple" is understood in the sense of [24]. This operator is maximal anti-dissipative for* $z \in \mathbb{C}_+$ *and maximal dissipative for* $z \in \mathbb{C}_-$; *see [40].*

The *proof* of the theorem follows immediately from Theorem 1; see [11] (Theorem 3.6), together with the observation that

$$\mathcal{P}M(z)\mathcal{P} = \mathcal{P}M_E(z)\mathcal{P} + \mathcal{P}M_V(z)\mathcal{P}.$$

The next step of our argument is to introduce the truncated (reduced) boundary space $\breve{\mathcal{H}}$ in order to make all the ingredients finite-dimensional. In view of clarity, in what follows we consistently supply the (finite-dimensional) "truncated" spaces and operators pertaining to them by the breve overscript.

We put $\breve{\mathcal{H}} := \mathcal{P}\mathcal{H}$ (noting that in our setup $\breve{\mathcal{H}}$ is N-dimensional, where N is the number of vertices; see Section 2). Introduce the truncated Poisson operator on $\breve{\mathcal{H}}$ by $\breve{\Pi}_E := \Pi_E|_{\breve{\mathcal{H}}}$ and the truncated DN map $\breve{\Lambda}^E := \mathcal{P}\Lambda^E|_{\breve{\mathcal{H}}}$. Then, the following statement holds.

Proposition 4 ([11], Theorem 3.7). *1. The formula*

$$R_{\text{eff}}(z) = \left(A^E_0 - z\right)^{-1} - \breve{S}^E_z\left(\breve{M}_E(z) - \mathcal{P}B(z)\mathcal{P}\right)^{-1}(\breve{S}^E_{\bar{z}})^* \tag{13}$$

holds, where $\breve{S}^E_z$ *is the solution operator of the problem*

$$-\Delta u_\phi - z u_\phi = 0, \ \ u_\phi \in \operatorname{dom} A^E_0 \dotplus \operatorname{ran} \breve{\Pi}_E,$$
$$\Gamma^E_0 u_\phi = \phi, \quad \phi \in \breve{\mathcal{H}},$$

and $\breve{M}_E$ *is the M-operator defined in accordance with* (3), (5) *relative to the triple* $(\breve{\mathcal{H}}, \breve{\Pi}_E, \breve{\Lambda}^E)$.

2. The "effective" generalised resolvent $R_{\text{eff}}(z)$ *is represented as the generalised resolvent of the problem*

$$-\Delta u - zu = f, \quad f \in L^2(Q_E), \quad u \in \operatorname{dom} A^E_0 \dotplus \operatorname{ran} \breve{\Pi}_E,$$
$$\mathcal{P}\partial_n u\big|_{\Gamma^\varepsilon} = \mathcal{P}B(z)\mathcal{P}u\big|_{\Gamma^\varepsilon}.$$

3. The triple $(\breve{\mathcal{H}}, \breve{\Gamma}^{\varepsilon E}_0, \breve{\Gamma}^{\varepsilon E}_1)$ *is the classical boundary triple [50,51] for the operator* $A_{\max}$ *defined by the differential expression* $-\Delta$ *on the domain* $\operatorname{dom} A_{\max} = \operatorname{dom} A^E_0 \dotplus \operatorname{ran} \breve{\Pi}_E$. *Here,* $\breve{\Gamma}^{\varepsilon E}_0$ *and* $\breve{\Gamma}^{\varepsilon E}_1$ *are defined on* $\operatorname{dom} A_{\max}$ *as the operator of the boundary trace on* Γ *and* $\mathcal{P}\partial_n u$, *respectively.*

We now consider the operator $\mathcal{P}B(z)\mathcal{P}$ in (13); since $B = -M_V$ by definition, we invoke the estimates derived in Section 3 to obtain

$$\mathcal{P}B\mathcal{P} = -\mathcal{P}\Lambda^V\mathcal{P} - z\mathcal{P}\Pi^*_V\Pi_V\mathcal{P} + O(\varepsilon|\log\varepsilon|^\gamma) = -z\breve{\Pi}^*_V\breve{\Pi}_V + O(\varepsilon|\log\varepsilon|^\gamma),$$

with a uniform estimate for the remainder term. Here, the truncated Poisson operator $\breve{\Pi}_V$ is introduced as $\breve{\Pi}_V := \Pi_V|_{\breve{\mathcal{H}}}$ relative to the same truncated boundary space as above, $\breve{\mathcal{H}} = \mathcal{P}\mathcal{H}$. As a result, we obtain

$$R_{\text{eff}}(z) - R_{\text{hom}}(z) = O(\varepsilon|\log\varepsilon|^\gamma),$$

with

$$R_{\text{hom}}(z) := (A_0^E - z)^{-1} - \breve{S}_z^E\big(\breve{M}_E(z) + z\breve{\Pi}_V^*\breve{\Pi}_V\big)^{-1}(\breve{S}_{\bar{z}}^E)^*. \tag{14}$$

By a classical result of [40] (see also [38,39]), the operator $R_{\text{eff}}(z)$ is a generalised resolvent, so it defines a z-dependent family of closed densely defined operators in $L^2(Q_E)$, which are maximal anti-dissipative for $z \in \mathbb{C}_+$ and maximal dissipative for $z \in \mathbb{C}_-$. Writing the resolvent $(A_\varepsilon - z)^{-1}$ in the matrix form relative to the orthogonal decomposition $L^2(Q) = P_E L^2(Q) \oplus P_V L^2(Q) = L^2(Q_E) \oplus L^2(Q_V)$ then yields the following result.

Theorem 3. *The resolvent $(A_\varepsilon - z)^{-1}$ admits the following asymptotics in the uniform operator-norm topology:*

$$(A_\varepsilon - z)^{-1} = \mathcal{R}_{\text{eff}}(z) + O(\varepsilon|\log\varepsilon|^\gamma),$$

where the operator $\mathcal{R}_{\text{eff}}(z)$ has the following representation relative to the decomposition $L^2(Q_E) \oplus L^2(Q_V)$:

$$\mathcal{R}_{\text{eff}}(z) = \begin{pmatrix} R_{\text{eff}}(z) & \Big(\mathfrak{K}_{\bar{z}}\big[R_{\text{eff}}(\bar{z}) - (A_0^E - \bar{z})^{-1}\big]\Big)^*\breve{\Pi}_V^* \\ \breve{\Pi}_V\mathfrak{K}_z\big[R_{\text{eff}}(z) - (A_0^E - z)^{-1}\big] & \breve{\Pi}_V\mathfrak{K}_z\Big(\mathfrak{K}_{\bar{z}}\big[R_{\text{eff}}(\bar{z}) - (A_0^E - \bar{z})^{-1}\big]\Big)^*\breve{\Pi}_V^* \end{pmatrix}. \tag{15}$$

Here, $\mathfrak{K}_z := \Gamma_0^E|_{\mathfrak{N}_z}$ with $\mathfrak{N}_z := \operatorname{ran} S_z^E\mathcal{P}$, $z \in \mathbb{C}_\pm$, and the generalised resolvent $R_{\text{eff}}(z)$ is defined by (14).

The above theorem provides us with the simplest possible leading-order term $\mathcal{R}_{\text{eff}}(z)$ of the asymptotic expansion for $(A_\varepsilon - z)^{-1}$. However, it is not yet obvious whether it is the resolvent of some self-adjoint operator in the space $L^2(Q_E) \oplus \breve{\Pi}_V\breve{\mathcal{H}} \subset L^2(Q)$. It turns out that this is indeed so, which is seen via the following explicit construction.

Put $L^2(G) := \oplus_e L^2(0, l_e)$, $H^2(G) := \oplus_e H^2(0, l_e)$. For all $u \in H^2(G)$, denote by u_{ev} the limit of $u_e(x) := u|_e(x)$ at the vertex v. Let $H_{\text{eff}} := L^2(G) \oplus \mathbb{C}^N$, and set

$$\operatorname{dom}\mathcal{A}_{\text{hom}} = \big\{(u,\beta)^\top \in H_{\text{eff}} :\ u \in H^2(G),\ u_{ev} = u_{e'v} =: u_v \text{ for any } v \text{ and } e, e' \text{ incident to } v, \text{ and } \beta = \kappa u_V\big\}, \tag{16}$$

where u_V is the N-dimensional vector of $\{u_v\}_{v\in V}$ and κ is the diagonal matrix

$$\kappa := \operatorname{diag}\{|Q_v^0|^{1/2}\}. \tag{17}$$

The action of the operator is set by

$$\mathcal{A}_{\text{hom}}\begin{pmatrix} u \\ \beta \end{pmatrix} = \begin{pmatrix} -u'' \\ -\kappa^{-1}\partial_n u|_V \end{pmatrix}, \qquad \begin{pmatrix} u \\ \beta \end{pmatrix} \in \operatorname{dom}\mathcal{A}_{\text{hom}}, \tag{18}$$

where $\partial_n u|_V$ is the N-dimensional vector $\{\sum_{e\sim v}\partial_n u_e|_v\}_{v\in V}$, i.e., the vector for which each element is represented by the sum of edge-inward normal derivatives of the function u over all the edges incident to the vertex v. We write $e \sim v$ if and only if the edge e is incident to the vertex v.

The main result of the present work, which is obtained by computing explicitly the resolvent of (16)–(18) and comparing it with (15) (see details of a similar computation in [13]), is formulated next.

Theorem 4. *The resolvent $(A_\varepsilon - z)^{-1}$ admits the following estimate in the uniform operator norm topology, uniform in $z \in K_\sigma$:*

$$(A_\varepsilon - z)^{-1} - \Theta(\mathcal{A}_{\text{hom}} - z)^{-1}\Theta^* = O(\varepsilon|\log\varepsilon|^\gamma),$$

where Θ is a partial isometry from H_{eff} onto $L^2(Q)$, acting as follows:

- *For every edge $e \in G$, $e = [0, l_e]$, it embeds $u \in H^2(e)$ into $L^2(Q_e)$ as $u(x) \times \varepsilon^{-(d-1)/2}\mathbf{1}(y)$, where y is the variable in the direction transverse to that of x;*
- *For every vertex $v \in G$, it embeds the value u_v, i.e., the common value of $u \in H^2(G)$ at the vertex v, into $L^2(Q_v)$ as $\varepsilon^{-(d-1)/2}u_v\mathbf{1}$.*

5. Analysis of Vertex Matching Conditions

In the present section, we continue our study of the operator (16)–(18) associated with an arbitrary metric graph G, with a view to analyse its spectral structure. We will show that the matching conditions at graph vertices associated with the spectral problem for the mentioned operator, albeit closely resembling δ-type conditions with coupling constants linear in the spectral parameter z, can be in fact represented (up to a unitary gauge) by δ'-type matching conditions for all $z \neq 0$ (while for $z = 0$ they coincide with the classical Kirchhoff condition). We will assume throughout that this graph contains no loops, in line with the assumptions imposed on the thin network studied above. We will further assume without loss of generality that the graph G is connected and that the matrix κ is invertible.

Since the operator $\mathcal{A}_{\text{hom}}$ can be viewed as a self-adjoint out-of-space extension (see, e.g., [52] and references therein) of a symmetric differential operator on the metric graph G, it is amenable to the classical boundary triples theory; see [53,54]. We recall that for a closed and densely defined symmetric operator $\mathcal{A}$ on a separable Hilbert space H with domain $\operatorname{dom}\mathcal{A}$, a boundary triple is defined as follows.

Definition 2 ([55]). *A triple $(\mathcal{K}, \Gamma_0, \Gamma_1)$ consisting of an auxiliary Hilbert space $\mathcal{K}$ and linear mappings Γ_0, Γ_1 defined everywhere on* $\operatorname{dom}\mathcal{A}^*$ *is called a* boundary triple *for $\mathcal{A}^*$ if the following conditions are satisfied:*

(1) *The abstract Green's formula is valid*

$$(\mathcal{A}^*\vec{u}, \vec{v})_H - (\vec{u}, \mathcal{A}^*\vec{v})_H = (\Gamma_1\vec{u}, \Gamma_0\vec{v})_{\mathcal{K}} - (\Gamma_0\vec{u}, \Gamma_1\vec{v})_{\mathcal{K}}, \quad \vec{u}, \vec{v} \in \operatorname{dom}\mathcal{A}^* \tag{19}$$

(2) *For any $Y_0, Y_1 \in \mathcal{K}$ there exist $\vec{u} \in \operatorname{dom}\mathcal{A}^*$, such that $\Gamma_0\vec{u} = Y_0$, $\Gamma_1\vec{u} = Y_1$. In other words, the mapping $\vec{u} \mapsto \Gamma_0\vec{u} \oplus \Gamma_1\vec{u}$ from $\operatorname{dom}\mathcal{A}^*$ to $\mathcal{K} \oplus \mathcal{K}$ is surjective.*

It can be shown (see [55]) that a boundary triple for $\mathcal{A}^*$ exists, although it is not unique.

Definition 3. *Let $\mathcal{T} = (\mathcal{K}, \Gamma_0, \Gamma_1)$ be a boundary triple of $\mathcal{A}^*$.* The Weyl function *of $\mathcal{A}^*$ corresponding to $\mathcal{T}$ and denoted by $M(z)$, $z \in \mathbb{C} \setminus \mathbb{R}$, is an analytic operator-function with a positive imaginary part for $z \in \mathbb{C}_+$ (i.e., an operator R-function) with values in the algebra of bounded operators on $\mathcal{K}$ such that*

$$M(z)\Gamma_0\vec{u} = \Gamma_1\vec{u} \qquad \forall \vec{u} \in \ker(\mathcal{A}^* - zI).$$

For $z \in \mathbb{C} \setminus \mathbb{R}$ one has $(M(z))^* = (M(\bar{z}))$ and $\Im(z)\Im(M(z)) > 0$.

A comparison with the assertion 4 of Proposition 1 shows that the Weyl function $M(z)$ in the context of the boundary triples theory is intimately related to the object introduced in Definition 1. The overall setup leading to its construction is, however, different and is based on the explicit choice of the boundary operators Γ_0 and Γ_1.

Definition 4. *An extension $\mathscr{A}$ of a closed densely defined symmetric operator $\mathcal{A}$ is called* almost solvable *and is denoted by $\mathscr{A} = A_B$, if there exist a boundary triple $(\mathcal{K}, \Gamma_0, \Gamma_1)$ for $\mathcal{A}^*$ and a bounded operator $B : \mathcal{K} \to \mathcal{K}$ defined everywhere in $\mathcal{K}$ such that*

$$\vec{u} \in \operatorname{dom} A_B \iff \Gamma_1\vec{u} = B\Gamma_0\vec{u}.$$

This definition implies that $\operatorname{dom} A_B \subset \operatorname{dom}\mathcal{A}^*$ and A_B is a restriction of $\mathcal{A}^*$ to the linear set $\operatorname{dom} A_B := \{\vec{u} \in \operatorname{dom}\mathcal{A}^* : \Gamma_1\vec{u} = B\Gamma_0\vec{u}\}$. In this context, the operator B plays the rôle of a parameter for the family of extensions $\{A_B \mid B : \mathcal{K} \to \mathcal{K}\}$. It can be shown (see [36] for references) that the resolvent set of A_B is non-empty (i.e., A_B is maximal), both A_B and $(A_B)^* = A_{B^*}$ are restrictions of $\mathcal{A}^*$ to their domains, and A_B and B are self-adjont or dissipative simultaneously.

Under the additional assumption that $\mathcal{A}$ is *simple* (or, in other words, completely non-self-adjoint), that is, it has no reducing self-adjoint "parts", the spectrum of A_B coincides, counting multiplicities, with the set of points $z_0 \in \mathbb{C}$ into which $(M(z_0) - B)^{-1}$ does not admit analytic continuation. In the general case, however, the spectrum is a union of the "zeroes" of the operator-valued function $M(z) - B$ introduced above and the spectrum of the self-adjoint "part" of the symmetric operator $\mathcal{A}$ in its Wold decomposition [56].

Our immediate aim is to construct a convenient boundary triple for the operator $\mathcal{A}_{\text{eff}}$. In doing so, we rely upon the framework (in a particular case of a loop-graph with exactly one vertex) of the paper [57].

We define the symmetric operator $\mathcal{A}$ as follows (cf. (16)):

$$\operatorname{dom}\mathcal{A} = \Big\{(u,\beta)^\top \in H_{\text{eff}} : u \in H^2(G),\ u_{ev} = u_{e'v} =: u_v \text{ for any } v \\ \text{and } e, e' \text{ incident to } v,\quad \partial_n u|_V = 0, \text{ and } \beta = \kappa u_V\Big\},$$

where u_V is, as above, the N-dimensional vector of $\{u_v\}_{v\in V}$, κ is the diagonal matrix (17), and $\partial_n u|_V$ is the N-dimensional vector of $\{\sum_{e\sim v}\partial_n u_e|_v\}_{v\in V}$. The action of the operator $\mathcal{A}$ is set by (18). The operator thus defined is clearly symmetric in $L^2(G)\oplus\mathbb{C}^N$; its adjoint $\mathcal{A}^*$ is defined by the same expression (18) on the domain

$$\operatorname{dom}\mathcal{A}^* = \Big\{(u,\beta)^\top \in H_{\text{eff}} :\ u \in H^2(G),\ u_{ev} = u_{e'v} =: u_v \text{ for any } v \\ \text{and } e, e' \text{ incident to } v\Big\}.$$

We have the following lemma.

Lemma 5. *Let $\mathcal{K} = \mathbb{C}^N$, $\Gamma_0(u,\beta)^\top := \partial_n u|_V$, and $\Gamma_1(u,\beta)^\top := \kappa^{-1}\beta - u_V$. The triple $(\mathcal{K},\Gamma_0,\Gamma_1)$ is a boundary triple for the operator $\mathcal{A}^*$. The operator $\mathcal{A}_{\text{eff}}$ is a self-adjoint almost solvable extension of $\mathcal{A}$, corresponding to the matrix $B = 0$ with respect to this boundary triple.*

The *proof* of the above lemma is obtained via integration by parts; for details, see [57].

We shall further require two ordinary differential operators on the metric graph G together with their boundary triples. Consider $A^\delta_{\max}$ to be the operator generated by the negative Laplacian $-\Delta$ on $L^2(G)$, defined on the domain

$$\operatorname{dom} A^\delta_{\max} = \{u \in H^2(G) : u_v := u_{ev} = u_{e'v} \text{ for any } v \text{ and } e, e' \text{ incident to } v\}.$$

In the paper [58], it is shown that this is a natural choice of a maximal operator if one seeks to consider the so-called δ-type matching conditions at the graph vertices, i.e., matching conditions of the type

$$u \text{ continuous at every vertex } v \text{ and } \partial_n u|_V = \eta^2 u_V, \tag{20}$$

where η is a diagonal matrix. The conditions (20) reduce to Kirchhoff, or standard, matching conditions under the choice $\eta = 0$. The natural boundary triple for $A^\delta_{\max}$ is $(\mathcal{K},\Gamma^\delta_0,\Gamma^\delta_1)$, where $\mathcal{K} = \mathcal{C}^N$, $\Gamma^\delta_0 u = u_V$, and $\Gamma^\delta_1 u = \partial_n u|_V$. The corresponding Weyl function ("M-matrix") admits the form [58] $M_\delta(z) := \{m^\delta_{vv'}\}_{v,v'\in V}$, where

$$m^\delta_{vv'}(k) = \begin{cases} -k\sum_{e\sim v}\cot kl_e, & v = v', \\ k\sum_{e\sim v, e\sim v'}(\sin kl_e)^{-1}, & v \neq v'. \end{cases} \tag{21}$$

Here, $k := \sqrt{z}$ is such that $\Im k \geq 0$.

Next, we consider the *magnetic* Laplacian on the graph G subject to δ'-type matching at the vertices. Namely, we assume that for all edges e the action of the operator is described as

$$-\left(\frac{d}{dx} + \mathrm{i}\tau_e\right)^2, \tag{22}$$

where τ_e is an edgewise-constant magnetic potential. In order to introduce δ'-type matching, consider the co-normal derivatives

$$\partial_n^\tau u_e|_v = \begin{cases} u_e'|_v + \mathrm{i}\tau_e u_e|_v & \text{if } v \text{ is the left endpoint of } e, \\ -(u_e'|_v + \mathrm{i}\tau_e u_e|_v) & \text{otherwise.} \end{cases}$$

We say that the magnetic Laplacian on G is subject to δ'-type matching at the vertices if

$$\partial_n^\tau u|_v := \partial_n^\tau u_e|_v = \partial_n^\tau u_{e'}|_v \text{ for any } e, e' \sim v, \text{ and } \Sigma u|_V = \eta^2 \partial_n^\tau u|_V \quad \forall\, v \in V, \tag{23}$$

where $\partial_n^\tau u|_V$ and $\Sigma u|_V$ denote the vectors $\{\partial_n^\tau u|_v\}_{v\in V}$ and $\{\sum_{e\sim v} u_e|_v\}_{v\in V}$, respectively.

By an argument similar to that of [58], see also [13], one easily checks that $(\mathcal{K}, \hat{\Gamma}_0, \hat{\Gamma}_1)$ is a boundary triple for the magnetic Laplacian $A^{\delta'}_{\max}$ defined on

$$\operatorname{dom} A^{\delta'}_{\max} = \{u \in H^2(G) : \partial_n^\tau u|_v := \partial_n^\tau u_e|_v = \partial_n^\tau u_{e'}|_v \text{ for any } v \text{ and } e, e' \text{ incident to } v\},$$

if $\mathcal{K} = \mathbb{C}^N$, $\hat{\Gamma}_0 u := \{\partial_n^\tau u|_v\}_{v\in V}$, and, finally, $\hat{\Gamma}_1 u := -\{\sum_{e\sim v} u_e|_v\}_{v\in V}$. The corresponding M-matrix $\hat{M}(z) := \{\hat{m}_{vv'}\}_{v,v'\in V}$ admits the form [58]

$$\hat{m}_{vv'}(k) = \begin{cases} -k^{-1}\sum_{e\sim v} \cot kl_e, & v = v', \\ -k^{-1}\sum_{e\sim v, e\sim v'} \exp(\mathrm{i}\sigma_e(v,v')\tau_e l_e)(\sin kl_e)^{-1}, & v \neq v'. \end{cases}$$

Here $k = \sqrt{z}$ such that $\Im k \geq 0$ and $\sigma_e(v, v') = 1$ if e is directed from v to v', $\sigma_e(v, v') = -1$ otherwise.

We will now fix the values of the magnetic potential as follows: $\tau_e := \pi/l_e$. The operator $A^{\delta'}_{\max}$ corresponding to this choice will be henceforth denoted by $\hat{A}_{\max}$. Its M-matrix $\hat{M}$ relative to the triple $(\mathcal{K}, \hat{\Gamma}_0, \hat{\Gamma}_1)$ admits the form

$$\hat{m}_{vv'}(k) = \begin{cases} -k^{-1}\sum_{e\sim v} \cot kl_e, & v = v', \\ k^{-1}\sum_{e\sim v, e\sim v'} (\sin kl_e)^{-1}, & v \neq v', \end{cases}$$

which coincides with (21) up to the factor z^{-1}. We remark that the operator $\hat{A}_{\max}$ and any of its self-adjoint restrictions can be unitary transformed into a regular (non-magnetic) Laplacian on the same graph G by a standard gauge transform; this will, however, be reflected in the corresponding change of matching conditions at the vertices. Motivated by applications to electromagnetic wave propagation [16], here we prefer to proceed with the magnetic setup.

Returning to the analysis of the operator $\mathcal{A}_{\text{eff}}$, we now have the following lemma.

Lemma 6. *Relative to the boundary triple of Lemma 5, the Weyl M-matrix of the operator $\mathcal{A}_{\text{eff}}$ admits the form*

$$M(z) = -M_\delta^{-1}(z) - \frac{1}{z}\kappa^{-2} = -\frac{1}{z}(\hat{M}^{-1}(z) + \kappa^{-2}).$$

Proof. In view of Lemma 5, consider the vector $(u, \beta)^\top \in \operatorname{dom} \mathcal{A}^*$ such that

$$\mathcal{A}^*\begin{pmatrix} u \\ \beta \end{pmatrix} = z\begin{pmatrix} u \\ \beta \end{pmatrix}.$$

By the definition of $\mathcal{A}^*$, one equivalently has $-u'' = zu$ on G and $-\kappa^{-1}\partial_n u|_V = z\beta$. Abbreviating $\Gamma_0(u,\beta)^\top =: \alpha$, one therefore has, in view of the definition of Γ_0 (see Lemma 5):

$$\beta = -\frac{1}{z}\kappa^{-1}\alpha.$$

Furthermore, taking now into account the definition of the second boundary operator Γ_1, one has

$$\Gamma_1(u,\beta)^\top = \kappa^{-1}\beta - u_V = -\frac{1}{z}\kappa^{-2}\alpha - u_V.$$

Note that the function u, by the definition of $\mathcal{A}^*$, must be continuous at every $v \in V$ and therefore belongs to the domain of the operator $A^\delta_{\max}$. Therefore, $u \in \ker(A^\delta_{\max} - z)$ and thus one has $\Gamma^\delta_1 u = M_\delta(z)\Gamma^\delta_0 u$. By construction, one has $\Gamma^\delta_0 u = u|_V$ and $\Gamma^\delta_1 u = \partial_n u|_V$, whence $u_V = M_\delta^{-1}(z)\alpha$. Ultimately,

$$M(z)\alpha = \Gamma_1(u,\beta)^\top = -\frac{1}{z}\kappa^{-2}\alpha - M_\delta^{-1}(z)\alpha,$$

as claimed. □

Let $\mathcal{E}(\cdot)$ denote the orthogonal operator spectral measure of the self-adjoint operator $\mathcal{A}_{\text{eff}}$ and $E(\cdot)$ the orthogonal spectral measure of the self-adjoint operator $\hat{A}_{-\kappa^2}$, where the latter is defined as the almost solvable extension of $(A^{\delta'}_{\max})^*$ corresponding to the parameterising operator $B = -\kappa^2$. In other words, it is the magnetic Laplacian on the graph G subject to the condition $\tau_e = \pi/l_e$ for all $e \in E$ and δ'-type matching conditions (23) at the graph vertices with $\eta = \kappa$. We have the following statement.

Theorem 5. *For any $c_0 > 0$, the operators $\mathcal{A}_{\text{eff}}\mathcal{E}(c_0,\infty)$ and $\hat{A}_{-\kappa^2}E(c_0,\infty)$ are unitary equivalent.*

Proof. In the generic case when $\mathcal{A}$ is simple (in particular when all l_e are rationally independent), the claim follows immediately from Lemma 6. Indeed, in this case no positive z can be an eigenvalue of the operator $\mathcal{A}$ and the same applies to the operator $(A^{\delta'}_{\max})^*$. Therefore, any reducing self-adjoint "part" of either symmetric operator can only be zero. In the general case, if $z_0 > 0$ is an eigenvalue of $\mathcal{A}$, then the corresponding eigenfunction solves $-u'' = z_0 u$ on G subject to $-\kappa^{-1}\partial_n u|_V = z_0\beta$, $u_V = \kappa^{-1}\beta$, $\partial_n u|_V = 0$ and therefore $u_V = \partial_n u|_V = 0$. Thus, any eigenfunction must be of the form $c_e \sin\sqrt{z_0}x$ on each e; moreover, one must also have $\sin\sqrt{z_0}l_e = 0$ for all e such that $c_e \neq 0$ and $\partial_n u|_V = 0$. For each of these, the function that is edgewise transformed as $c_e \sin\sqrt{z_0}x \mapsto c_e\exp(-i\tau_e x)\cos\sqrt{z_0}x$ is shown to be an eigenfunction of $(A^{\delta'}_{\max})^*$ corresponding to the eigenvalue z_0. The same argument applied in the opposite direction completes the proof. □

Remark 2. *The analysis of the unitary equivalence linking $\mathcal{A}_{\text{eff}}\mathcal{E}(c_0,\infty)$ and $\hat{A}_{-\kappa^2}E(c_0,\infty)$ is an exciting possible development from the point of view of classical functional analysis, since it appears to be a natural graph-based generalisation of the classical Hilbert transform. This can be seen, in particular, from the explicit calculation in the case of an infinite chain graph [14].*

The above theorem shows that the spectral analysis of the operator $\mathcal{A}_{\text{eff}}$ in relation to its non-zero spectrum reduces to that of the magnetic graph Laplacian with (non-trivial) δ'-type matching condition at the graph vertices, which could come as a surprise given that the eigenvalue problem for $\mathcal{A}_{\text{eff}}$ yields, for the first component of the eigenvector $(u,\beta)^\top$, the equation $-u'' = zu$ on G subject to $\partial_n u|_V = -z\kappa^2 u|_V$, which on the face of it is a δ-type matching condition, albeit with coupling constants proportional to the spectral parameter z. It would seem natural, therefore, for the self-adjoint operator $\mathcal{A}_{\text{eff}}$ to be a H^{-1} singular perturbation of the graph Laplacian with standard boundary conditions. Instead, our result shows that it is in fact a more singular H^{-2} perturbation. The underlying reasons of this peculiar behaviour are discussed in detail in [59], where the relationship of operators of the class considered with those in the area of zero-range potentials with internal structure, as introduced by B. S. Pavlov [60,61], is explained. We also point out that the assertion of Theorem 5 has been observed in a particular case of the cycle graph with one vertex (the loop) in [14] in the context of a high-contrast homogenisation problem on the real line.

We conclude the spectral analysis of the operator $\mathcal{A}_{\text{eff}}$ by a brief discussion of its kernel and the comparison of the latter with that of the operator $\hat{A}_{-\kappa^2}$. It turns out that, unlike what happens with its non-zero spectrum, the kernel $\ker \mathcal{A}_{\text{eff}}$ of the operator $\mathcal{A}_{\text{eff}}$ is that of the Kirchhoff graph Laplacian.

Note first that $\ker \mathcal{A}_{\text{eff}}$ necessarily belongs to the non-simple (i.e., self-adjoint) part of $\mathcal{A}$. Indeed, for $(u, \beta)^\top \in \ker \mathcal{A}_{\text{eff}}$ it has to satisfy

$$-u'' = 0 \text{ on } G, \qquad \partial_n u|_V = 0, \quad u_V = \kappa^{-1}\beta,$$

where the boundary conditions are equivalent to $\Gamma_0(u,\beta)^\top = 0$, $\Gamma_1(u,\beta)^\top = 0$. On the other hand, this is precisely the condition for u to be in the kernel of a graph Laplacian with Kirchhoff matching conditions. One therefore infers from [5] that $\dim\ker \mathcal{A}_{\text{eff}} = 1$ (in our case of connected graphs), and the elements of $\ker \mathcal{A}_{\text{eff}}$ are constants on G.

The kernel of $\hat{A}_{-\kappa^2}$ is spanned by functions u such that

$$-u'' = 0 \text{ on } G \text{ and } \Sigma u|_V = \kappa^2 \partial_n^\tau u|_V.$$

It is easily checked that a non-trivial solution to this problem could exist only if $\partial_n^\tau u|_V = 0$, $\Sigma u|_V = 0$ or, in other words, if $\hat{\Gamma}_0 u = \hat{\Gamma}_1 u = 0$. This means that, precisely as in the case of $\mathcal{A}_{\text{eff}}$, the kernel of $\hat{A}_{-\kappa^2}$ necessarily belongs to the non-simple (i.e., self-adjoint) part of the symmetric operator $(A^{\delta'}_{\max})^*$. The question of its existence and dimension admits a simple answer in terms of the graph topology. It is clear that it is trivial in the case when G is a tree; in general, its dimension is shown to be equal to the cyclomatic number χ of the graph G. In particular, this yields unitary equivalence of $\mathcal{A}_{\text{eff}}$ and $\hat{A}_{-\kappa^2}$ in the case where G contains exactly one cycle.

Remark 3. *The result of the present section seems to have been overlooked in a number of now-classical papers dealing with Sturm–Liouville problems on an interval with boundary conditions depending on a spectral parameter; see, e.g., [62–67].*

Conjecture 1. *The above discussion raises the question of which definition of δ'-type interaction on a graph is motivated physically, i.e., whether it is the one emerging from the analysis of thin networks as the operator $\mathcal{A}_{\text{hom}}$, see* (16)–(18)*, or the traditional (see [5]) definition* (22)–(23)*. At first sight, the difference between the two operators is insignificant: it is only in their kernels. However, it can happen to be of paramount importance if, e.g., one considers an ε-periodic graph with δ'-type matching conditions, in which case the homogenisation procedure [11,12] will lead to drastically different outcomes for the two related setups, as it relies upon a "threshold effect" [68] in the behaviour of the least eigenvalue of the operator on the fundamental cell for small quasimomenta.*

Author Contributions: Writing—original draft, K.D.C., Y.Y.E. and A.V.K.; Writing—review & editing, K.D.C., Y.Y.E. and A.V.K. All authors have read and agreed to the published version of the manuscript.

Funding: K.D.C. and Y.Y.E. are grateful for the financial support of EPSRC Grant EP/L018802/2. KDC and AVK are grateful for the financial support of EPSRC Grant EP/V013025/1. Y.Y.E. and A.V.K. are grateful to IIMAS–UNAM for the hospitality and financial support during the research visit when part of this work was carried out.

Data Availability Statement: No new data were generated or analysed during this study.

Acknowledgments: The authors acknowledge valuable comments by the referees, which have helped eliminate a number of inaccuracies.

Conflicts of Interest: There are no conflicts of interest to be declared.

References

1. Post, O. *Spectral Analysis on Graph-Like Spaces;* Lecture Notes in Mathematics 2039; Springer: Berlin/Heidelberg, Germany, 2012.
2. Exner, P.; Post, O. Convergence of spectra of graph-like thin manifolds. *J. Geom. Phys.* **2005**, *54*, 77–115. [CrossRef]
3. Kuchment, P.; Zeng, H. Convergence of spectra of mesoscopic systems collapsing onto a graph. *J. Math. Anal. Appl.* **2001**, *258*, 671–700. [CrossRef]
4. Kuchment, P.; Zeng, H. Asymptotics of spectra of Neumann Laplacians in thin domains. *Contemp. Math.* **2003**, *327*, 199–213.
5. Berkolaiko, G.; Kuchment, P. Introduction to Quantum Graphs; In *Mathematical Surveys and Monographs;* American Mathematical Society: Providence, RI, USA, 2012.
6. Mikhailova, A.B.; Pavlov, B.S.; Prokhorov, L.V. Intermediate Hamiltonian via Glazman's splitting and analytic perturbation for meromorphic matrix-functions. *Math. Nachr.* **2007**, *280*, 1376–1416. [CrossRef]
7. Pavlov, B.S. Krein formula with compensated singularities for the ND-mapping and the generalized Kirchhoff condition at the Neumann Schrödinger junction. *Russ. J. Math. Phys.* **2008**, *15*, 364–388. [CrossRef]
8. Kadic, M.; Bückmann, T.; Stenger, N.; Thiel, M.; Wegener, M. On the practicability of pentamode mechanical metamaterials. *Appl. Phys. Lett.* **2012**, *100*, 191901. [CrossRef]
9. Milton, G.W.; Cherkaev, A.V. Which Elasticity Tensors are Realizable? *J. Eng. Mater. Technol.* **1995**, *117*, 483. [CrossRef]
10. Cherednichenko, K.; Kiselev, A.V.; Velčić, I.; Žubrinić, J. Effective behaviour of critical-contrast PDEs: Micro-resonances, frequency conversion, and time dispersive properties. II. *arXiv* **2023**, arXiv:2307.01125.
11. Cherednichenko, K.D.; Ershova, Y.Y.; Kiselev, A.V. Effective behaviour of critical-contrast PDEs: Micro-resonances, frequency conversion, and time dispersive properties. I. *Commun. Math. Phys.* **2020**, *375*, 1833–1884. [CrossRef]
12. Cherednichenko, K.D.; Ershova, Y.Y.; Kiselev, A.V. Time-dispersive behaviour as a feature of critical contrast media. *SIAM J. Appl. Math.* **2019**, *79*, 690–715. [CrossRef]
13. Cherednichenko, K.; Ershova, Y.; Kiselev, A.; Naboko, S. Unified approach to critical-contrast homogenisation with explicit links to time-dispersive media. *Trans. Mosc. Math. Soc.* **2019**, *80*, 295–342. [CrossRef]
14. Cherednichenko, K.D.; Kiselev, A.V. Norm-resolvent convergence of one-dimensional high-contrast periodic problems to a Kronig-Penney dipole-type model. *Commun. Math. Phys.* **2017**, *349*, 441–480. [CrossRef]
15. Rubinstein, J.; Schatzman, M. Variational Problems on multiply connected thin strips I: Basic estimates and convergence of the Laplacian spectrum. *Arch. Ration. Mech. Anal.* **2001**, *160*, 271–308. [CrossRef]
16. Kiselev, A.V.; Ryadovkin, K.S. Phase transition in a periodic tubular structure. *arXiv* **2023**, arXiv:2303.00872; to appear in: SIAM J. Appl. Math. (2024).
17. Zaremba, S. Sur un probléme mixte relatif à l'équation de Laplace (French). In *Bulletin International de l'Académie des Sciences de Cracovie. Classe des Sciences Mathématiques et Naturelles, Serie A: Sciences Mathématiques;* Uniwersytet Jagielloński: Krakow, Poland, 1910; pp. 313–344.
18. Brown. R.M. The mixed problem for Laplace's equation in a class of Lipschitz domains. *Commun. Partial Diff. Eqns.* **1994**, *19*, 1217–1233. [CrossRef]
19. Taylor, J.L.; Ott, K.A.; Brown, R.M. The mixed problem in Lipschitz domains with general decompositions of the boundary. *Trans. Am. Math. Soc.* **2013**, *365*, 2895–2930. [CrossRef]
20. Grieser, D. Spectra of graph neighborhoods and scattering. *Proc. Lond. Math. Soc.* **2008**, *97*, 718–752. [CrossRef]
21. Birman, M.S. On the averaging procedure for periodic operators in a neighborhood of an edge of an internal gap. *St. Petersburg Math. J.* **2004**, *15*, 507–513. [CrossRef]
22. Craster, R.V.; Kaplunov, J.; Pichugin, A.V. High-frequency homogenization for periodic media. *Proc. R. Soc. Lond. Ser. A* **2010**, *466*, 2341–2362. [CrossRef]
23. Friedlander, L. On the density of states of periodic media in the large coupling limit. *Commun. Part. Diff. Eq.* **2002**, *27*, 355–380. [CrossRef]
24. Ryzhov, V. Linear operators and operator functions associated with spectral boundary value problems. *Oper. Adv. Appl.* **2020**, *276*, 576–626.
25. Schechter, M. A generalization of the problem of transmission. *Ann. Scuola Norm. Sup. Pisa* **1960**, *14*, 207–236.
26. Behrndt, J.; Langer, M. Boundary value problems for elliptic partial differential operators on bounded domains. *J. Func. Anal.* **2007**, *243*, 536–565. [CrossRef]
27. Brown, M.; Marletta, M.; Naboko, S.; Wood, I. Boundary triples and *M*-functions for non-selfadjoint operators with applications to elliptic PDEs and block operator matrices. *J. Lond. Math. Soc.* **2008**, *77*, 700–718. [CrossRef]
28. Birman, M.S. On the self-adjoint extensions of positive definite operators. *Math. Sb.* **1956**, *38*, 431–450.
29. Kreĭn, M.G. Theory of self-adjoint extensions of semibounded hermitian operators and applications II. *Mat. Sb.* **1947**, *21*, 365–404.
30. Višik, M.I. On general boundary problems for elliptic differential equations (Russian). *Trudy Moskov. Mat. ObŠc.* **1952**, *1*, 187–246.
31. Denzler, J. Bounds for the heat diffusion through windows of given area. *J. Math. Anal. Appl.* **1998**, *217*, 405–422. [CrossRef]
32. Agranovich, M.S. On a mixed Poincaré-Steklov type spectral problem in a Lipschitz domain. *Russ. J. Math. Phys.* **2006**, *13*, 239–244. [CrossRef]
33. Grubb, G. The mixed boundary value problem, Krein resolvent formulas and spectral asymptotic estimates. *J. Math. Anal. Appl.* **2011**, *382*, 339–363. [CrossRef]

34. Ryzhov, V. Weyl-Titchmarsh function of an abstract boundary value problem, operator colligations, and linear systems with boundary control. *Complex Anal. Oper. Theory* **2009**, *3*, 289–322. [CrossRef]
35. Behrndt, J.; Rohleder, J. Spectral analysis of selfadjoint elliptic differential operators, Dirichlet-to-Neumann maps, and abstract Weyl functions. *Adv. Math.* **2015**, *285*, 1301–1338. [CrossRef]
36. Cherednichenko, K.D.; Kiselev, A.V.; Silva, L.O. Scattering theory for non-selfadjoint extensions of symmetric operators. *Oper. Theory Adv. Appl.* **2020**, *276*, 194–230.
37. Abels, H.; Grubb, G.; Wood, I.G. Extension theory and Kreĭn-type resolvent formulas for non-smooth boundary value problems. *J. Func. Anal.* **2014**, *266*, 4037–4100. [CrossRef]
38. Neumark, M. Spectral functions of a symmetric operator (Russian). *Bull. Acad. Sci. URSS. Sér. Math.* **1940**, *4*, 277–318.
39. Neumark, M. Positive definite operator functions on a commutative group (Russian). *Bull. Acad. Sci. URSS Sér. Math.* **1943**, *7*, 237–244.
40. Strauss, A.V. Generalised resolvents of symmetric operators (Russian). *Izv. Akad. Nauk SSSR Ser. Mat.* **1954**, *18*, 51–86.
41. Girouard, A.; Polterovich, I. Spectral geometry of the Steklov problem (survey article). *J. Spec. Theory* **2017**, *7*, 321–359. [CrossRef] [PubMed]
42. Denzler, J. Windows of given area with minimal heat diffusion. *Trans. Am. Math. Soc.* **1999**, *351*, 569–580. [CrossRef]
43. Kuttler, J.R. Remarks on a Stekloff eigenvalue problem. *SIAM J. Numer. Anal.* **1972**, *9*, 1–5. [CrossRef]
44. Fichera, G. Su un principio di dualità per talune formole di maggiorazione relative alle equazioni differenziali (Italian). *Atti Accad. Naz. Lincei Rend. Cl. Sci. Fis. Mat. Nat.* **1956**, *19*, 411–418.
45. Auchmuty, G. The S.V.D. of the Poisson kernel. *J. Fourier Anal. Appl.* **2017**, *23*, 1517–1536. [CrossRef]
46. Bucur, D.; Ferrero, A.; Gazzola, F. On the first eigenvalue of a fourth order Steklov problem. *Calc. Var.* **2009**, *35*, 103–131. [CrossRef]
47. Ferrero, A.; Gazzola, F.; Weth, T. On a fourth order Steklov eigenvalue problem. *Analysis* **2005**, *25*, 315–332. [CrossRef]
48. Levitin M.; Parnovski, L.; Polterovich, I.; Scher, D. Sloshing, Steklov and corners: Asymptotics of sloshing eigenvalues. *J. Anal. Math.* **2022**, *146*, 65–125. [CrossRef]
49. Tretter, C. *Spectral Theory of Block Operator Matrices and Applications*; Imperial College Press: London, UK, 2008.
50. Derkach, V.A.; Malamud M.M. Generalised resolvents and the boundary value problems for Hermitian operators with gaps. *J. Funct. Anal.* **1991**, *95*, 1–95. [CrossRef]
51. Gorbachuk, V.I.; Gorbachuk, M.L. *Boundary Value Problems for Operator Differential Equations*; Kluwer Academic Publishers: Dordrecht, The Netherlands, 1991.
52. Štraus, A.V. Functional models and generalized spectral functions of symmetric operators. *St. Petersburg Math. J.* **1999**, *10*, 733–784.
53. Behrndt, J.; Hassi, S.; de Snoo, H. *Boundary Value Problems, Weyl Functions, and Differential Operators*; Springer: Berlin/Heidelberg, Germany, 2020.
54. Schmüdgen, K. *Unbounded Self-Adjoint Operators on Hilbert Space*; Graduate Texts in Mathematics 265; Springer: Dordrecht, The Netherlands, 2012.
55. Kočubeĭ, A.N. Extensions of symmetric operators and of symmetric binary relations. *Math. Notes* **1975**, *17*, 25–28. [CrossRef]
56. Gorbachuk, M.I.; Gorbachuk, V.I. *M.G. Krein's Lectures on Entire Operators*; Springer: Dordrecht, The Netherlands, 2012.
57. Cherednichenko, K.; Ershova, Y.; Naboko, S. Functional model for generalised resolvents and its application to time-dispersive media. *arXiv* **2021**, arXiv:2111.05387.
58. Ershova, Y.; Karpenko, I.I.; Kiselev, A.V. Isospectrality for graph Laplacians under the change of coupling at graph vertices. *J. Spectr. Theory* **2016**, *6*, 43–66. [CrossRef]
59. Cherednichenko, K.D.; Ershova, Y.Y.; Kiselev, A.V.; Ryzhov, V.A.; Silva, L.O. Asymptotic analysis of operator families and applications to resonant media. *Oper. Theory Adv. Appl.* **2023**, *291*, 239–311.
60. Pavlov, B.S. A model of zero-radius potential with internal structure (Russian). *Teoret. Mat. Fiz.* **1984**, *59*, 345–353.
61. Pavlov, B.S. The theory of extensions, and explicitly soluble models. *Russ. Math. Surv.* **1987**, *42*, 127–168. [CrossRef]
62. Evans, W.D. A non-selfadjoint differential operator in $L^2_\sigma[a,b)$. *Quart. J. Math.* **1970**, *21*, 371–383. [CrossRef]
63. Fulton, C.T. Two-point boundary value problems with eigenvalue parameter contained in the boundary conditions. *Proc. R. Soc. Edinb A.* **1977**, *77*, 293–308. [CrossRef]
64. Hinton, D.B. An expansion theorem for an eigenvalue problem in the boundary conditions. *Quart. J. Math.* **1977**, *30*, 33–42. [CrossRef]
65. Schneider, A. A note on eigenvalue problems with eigenvalue parameter in the boundary conditions. *Math. Z.* **1974**, *136*, 163–167. [CrossRef]
66. Shkalikov, A.A. Boundary problems for ordinary differential equations with parameter in the boundary conditions. *J. Soviet. Math.* **1983**, *33*, 1311–1342. [CrossRef]
67. Walter, J. Regular eigenvalue problems with eigenvalue parameter in the boundary condition. *Math. Z.* **1973**, *133*, 301–312. [CrossRef]
68. Birman, M.S.; Suslina, T.A. Second order periodic differential operators. Threshold properties and homogenisation. *St. Petersburg Math. J.* **2004**, *15*, 639–714. [CrossRef]

mathematics

MDPI

Article

Utilization of a Genetic Algorithm to Identify Optimal Geometric Shapes for a Seismic Protective Barrier

Vladimir Bratov [1,*], Andrey Murachev [2] and Sergey V. Kuznetsov [3]

1 School of Computing, Engineering & The Built Environment, Edinburgh Napier University, Edinburgh EH10 5DT, UK
2 Department of Theoretical Mechanics, Peter the Great St. Petersburg Polytechnic University, St. Petersburg 195251, Russia; andrey.murachev@spbstu.ru
3 Ishlinsky Institute for Problems in Mechanics, Moscow 119526, Russia; kuzn-sergey@yandex.ru
* Correspondence: v.bratov@napier.ac.uk

Abstract: The utilization of seismic barriers for protection against the hazardous impact of natural or technogenic waves is an extremely promising emerging technology to secure buildings, structures and entire areas against earthquake-generated seismic waves, high-speed-transport-induced vibrations, etc. The current research is targeted at studying the effect of seismic-barrier shape on the reduction of seismic-wave magnitudes within the protected region. The analytical solution of Lamb's problem was used to verify the adopted numerical approach. It was demonstrated that the addition of complementary geometric features to a simple barrier shape provides the possibility of significantly increasing the resulting seismic protection. A simple genetic algorithm was employed to evaluate the nontrivial but extremely effective geometry of the seismic barrier. The developed approach can be used in various problems requiring optimization of non-parameterizable geometric shapes. The applicability of genetic algorithms and other generative algorithms to discover optimal (or close to optimal) geometric configurations for the essentially multiscale problems of the interaction of mechanical waves with inclusions is discussed.

Keywords: seismic barrier; earthquake; vibration; complex geometry; genetic algorithm; optimization; multiscale modeling; heterogeneous media and structures

MSC: 74P10; 74J25; 74H15

1. Introduction

The use of seismic barriers to protect buildings and structures from the potentially harmful energy of surface waves of natural or technogenic origin is a relatively new emerging approach in modern seismic protection developments. In contrast to the traditional approaches, which, in most cases, rely on the installation of damping between the structure basement and the structure itself, the considered seismic barriers provide the possibility of ensuring territorial protection, significantly reducing the amplitude of displacements and accelerations in the area encircled by the barrier and embracing protected buildings and structures. Installation of a seismic barrier makes it possible to ensure the required levels of protection for buildings and structures inside the protected area, and it possesses a number of significant advantages over more traditional seismic protection methods. The major advantages of the considered territorial seismic protection include the following:

- Barriers protect the area on which the structure is placed, along with the structure itself;
- Seismic barriers are not a part of the protected structure; thus, degradation and fracture of barriers do not directly affect the strength and stability of the structure;
- The functionality of the barrier is not significantly affected by the frequency spectrum of oncoming seismic waves (unlike more traditional seismic protection systems that are designed to operate in a certain frequency range);

- Basement slabs are also protected;
- Barriers can be used to protect structures placed on weak soils that potentially can be exposed to the earthquake-induced liquefaction (as seismic barriers prevent the propagation of seismic waves into the protected area);
- Barriers can be retrofitted to protect the existing structures;
- In many cases, the creation of seismic barriers is economically much more efficient compared with the more traditional seismic protection systems.

The potential of various seismic protective structures based on specially designed metamaterials [1–7], seismic barriers [1,7–16], or even seismic barriers filled with metamaterials [2,4,5,8] was explicitly revealed in a large number of previous works. Earlier research clearly demonstrates the possibility of using seismic barriers for effective protection against seismic waves of different origins. A seismic barrier represents a vertical inclusion made from a material having physical properties different from the ambient medium. The interaction of the incident wave with this inclusion (seismic barrier) can result in the diffraction, reflection, refraction and scattering of waves. In addition to this, part of the wave energy can be captured inside the barrier. These effects can result in a decrease in wave amplitude behind the barrier. A schematic representation of the vertical barrier is given in Figure 1: A time-dependent load $P(t)$ is initiating a wave travelling along the surface of a half-plane simulating the effect of an earthquake at areas not directly adjacent to the earthquake epicentre; the seismic barrier (black rectangular area in Figure 1) interacts with a seismic wave caused by the applied dynamic surface load, which generates surface Rayleigh waves along with bulk waves emanating from the point force load; the wave amplitudes are measured by a sensor located on the free surface behind the barrier.

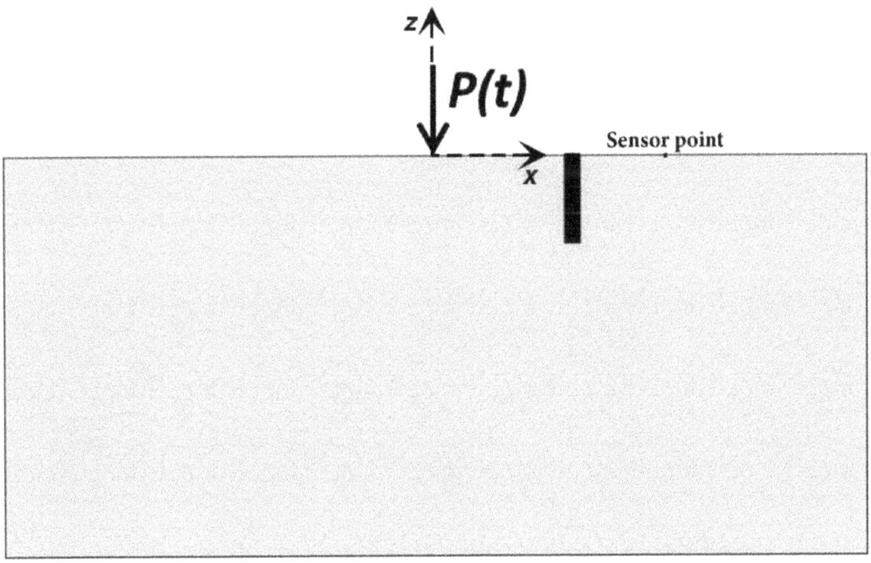

Figure 1. Vertical seismic barrier for protection against vibrations induced by load $P(t)$.

The efficiency of a seismic barrier (i.e., its capability of decreasing the amplitudes of displacements and accelerations) is determined by the properties of materials forming the barrier and the barrier geometry. The effect of material properties has been studied before, e.g., [17]. The present study is one of the first attempts to address the effect of the barrier shape. Previously, only barriers of the simplest (mostly rectangular) cross-sections were considered (e.g., [14–16]). In [18], Brennan et al. present a review of the state of the art in the utilization of soft barriers for protection against seismic hazards. The most sophisticated cases of seismic protection geometry reported by the authors are a V-shaped protective barrier numerically studied by Flora et al. [3] and a bowl-shaped barrier with an analytical solution reported by Le et al. [19]. Some authors study periodic protective elements as part of the structure basement (e.g., [20]) or that are erected in a vicinity of the structure to be protected (e.g., [15]). In addition to the fact that the available literature about the utilization of barriers for vibration protection is limited to the simplest barrier shapes (i.e., the shape-optimization problem was previously only solved in terms of finding optimal barrier dimensions), there is no practical possibility for a direct quantitative comparison of the observed protective properties to previously published results. The reason for this is that the quantitative data are highly specific to the problem geometry, boundary conditions and loading conditions. At the same time, the protective properties observed below for barriers of simple shape (i.e., a rectangular barrier filled with a single material) are an indication of the protective properties reported in the available literature. It should also be noted that though there are almost no published attempts to optimize the shape of seismic protective barriers, the problem of shape optimization for acoustic protective barriers is much more explored in literature. For example, Duhamel [21] utilizes a genetic algorithm for the shape optimization of noise barriers.

2. Numerical Model

Consider the 2D generalized plane-strain problem (Figure 1) for a halfplane with a barrier of an arbitrary shape. The motion within the halfplane is given by the Lame equation for displacements u_i in the directions x_i:

$$\rho u_{i,tt} = (\lambda + \mu) u_{j,ji} + \mu u_{i,jj}, \tag{1}$$

where μ and ν are the material Lame parameters and ρ is the material mass density. Stresses are coupled with strains by Hooke's law:

$$\sigma_{ij} = \lambda \delta_{ij} u_{k,k} + \mu \left(u_{i,j+} u_{j,i} \right), \tag{2}$$

At $t = 0$, the halfplane is stress-free, and the velocity field is zero everywhere in the plane:

$$\sigma_{ij}\big|_{t=0} = u_{,t}\big|_{t=0} = 0. \tag{3}$$

The load $P(t)$ is applied as a concentrated force at the surface at $x = 0$. The time dependence of the force is given graphically in Figure 2. The force amplitude $P_{peak} = 1000$ N.

$$\sigma_{xz}\big|_{z=0} = 0; \ \sigma_{zz}\big|_{z=0} = P(t)\delta(x). \tag{4}$$

This problem (1–4) is solved numerically by utilizing the finite-element method (FEM) [17]. To study the interaction of the incident waves with a barrier, a transient dynamic solution accounting for inertia effects should be employed. Herein, the LS-DYNA commercial FEM software [22] was used to solve the dynamic linear-elastic problem of surface wave propagation within the studied domain, taking advantage of explicit time integration.

Solutions for various shapes of a protective barrier are compared with the reference solution, which corresponds to the case when no protective barrier between the source and the sensor point is installed. The reference configuration represents the well-studied 2D Lamb's outer problem [23,24]. The existence of the exact analytical solution for Lamb's outer problem provides the possibility of verifying the accuracy of the numerical FEM approach.

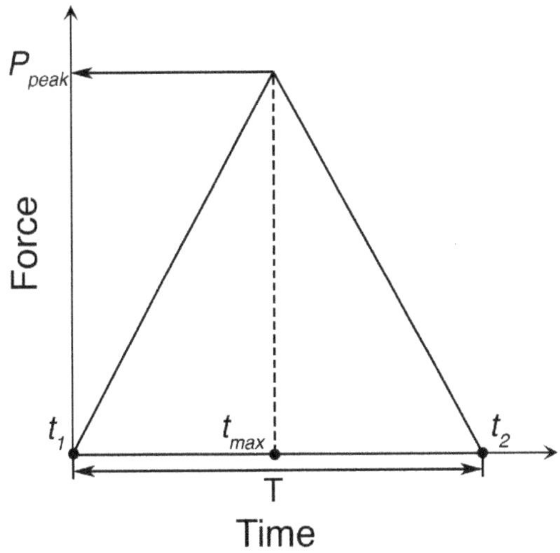

Figure 2. Temporal shape of $P(t)$ used in simulations.

The analytical solution for Lamb's problem with the $P(t)$ given in Figure 2 can be found as a convolution of the Lamb's solution [23–25] for the Dirac delta function with $P(t)$ [26,27]. For a point force applied at the surface and given by:

$$F_x(x,t) = P_h\delta(x)\delta(t);\; F_z(x,t) = P_v\delta(x)\delta(t), \tag{5}$$

where F_x and F_y are the horizontal and vertical components of the force (see Figure 1), P_h and P_v are the corresponding force amplitudes in the horizontal and vertical directions and $\delta()$ being the Dirac delta function, the solution for displacements at the plane surface can be obtained; for example, following Kausel [24]. In (6)–(9), the solution is given separately for the vertical (u_{zz}, u_{zx}) and horizontal (u_{xx}, u_{xz}) components of total displacement caused by the horizontal and the vertical component of the surface load:

$$u_{xx} = \frac{P_h\beta}{\pi\mu|x|}\begin{cases} 0,\; for\; \tau < a \\ \frac{4\tau^2(1-\tau^2)\sqrt{\tau^2-a^2}}{(2\tau^2-1)^4+16\tau^4(\tau^2-a^2)(1-\tau^2)},\; for\; a \le \tau \le 1 \\ \frac{-\sqrt{\tau^2-1}}{(2\tau^2-1)^2+4\tau^2\sqrt{\tau^2-a^2}\sqrt{\tau^2-1}},\; for\; \tau > 1 \end{cases} \tag{6}$$

for displacements in the horizontal (or x) direction caused by the horizontal component of the surface-loading force. Displacements in the vertical (or z) direction caused by the force applied in the vertical direction are:

$$u_{zz} = \frac{P_v\beta}{\pi\mu|x|}\begin{cases} 0,\; for\; \tau < a \\ \frac{-(2\tau^2-1)^2\sqrt{\tau^2-a^2}}{(2\tau^2-1)^4+16\tau^4(\tau^2-a^2)(1-\tau^2)},\; for\; a \le \tau \le 1 \\ \frac{-\sqrt{\tau^2-a^2}}{(2\tau^2-1)^2+4\tau^2\sqrt{\tau^2-a^2}\sqrt{\tau^2-1}},\; for\; \tau > 1 \end{cases} \tag{7}$$

Displacements in the horizontal direction induced by the vertical component of the force are:

$$u_{xz} = \frac{P_v\beta}{\pi\mu|x|}\begin{cases} \frac{2\tau(2\tau^2-1)\sqrt{\tau^2-a^2}\sqrt{1-\tau^2}}{(2\tau^2-1)^4+16\tau^4(\tau^2-a^2)(1-\tau^2)},\; for\; a \le \tau \le 1 \\ \frac{\pi(2\tau_R^2-1)^3}{4(1-4\tau_R^2+8\tau_R^6(1-a^2))}\,\delta(\tau-\tau_R),\; else \end{cases} \tag{8}$$

Displacements in the vertical direction induced by the horizontal component of the force are:

$$u_{zx} = -\frac{P_h\beta}{\pi\mu|x|}\begin{cases} \frac{2\tau(2\tau^2-1)\sqrt{\tau^2-a^2}\sqrt{1-\tau^2}}{(2\tau^2-1)^4+16\tau^4(\tau^2-a^2)(1-\tau^2)}, & for\ a \le \tau \le 1 \\ \frac{\pi(2\tau_R^2-1)^3}{4(1-4\tau_R^2+8\tau_R^6(1-a^2))}\delta(\tau-\tau_R), & else \end{cases} \tag{9}$$

In the equations above, $\tau = \frac{t\beta}{|x|}$, $\tau_R = \frac{\beta}{C_R}$, $a = \frac{\beta}{\alpha} = \frac{1-2\nu}{2(1-\nu)}$, where α and β are the longitudinal and transversal wave speeds and C_R is the Rayleigh wave speed. The resulting displacement for any point on the halfspace surface at any time can be obtained as the sum of the above displacements (i.e., $u = u_{zz} + u_{zx} + u_{xx} + u_{xz}$).

Obviously, making use of the above explicit solution for displacements and by performing convolutions for time and the x-coordinate along the surface, one can obtain surface displacements for an arbitrary system of forces, each having an arbitrary time profile, applied on the surface of the studied halfspace. Performing the convolution with the time profile of the point load (given in Figure 2), an exact analytical solution for the reference configuration of the problem (no barrier) will be obtained. Figure 3 plots the comparison between the exact analytical solution and the corresponding numerical solution of the same problem. The observed remarkable coincidence between the exact and the numerical result confirms the applicability of the developed numerical approach to accurately solve the problem of the propagation of elastic waves within the halfplane. The same numerical model was utilized to calculate the amplitudes of displacements and accelerations at a sensor point in the case of a halfspace with a barrier.

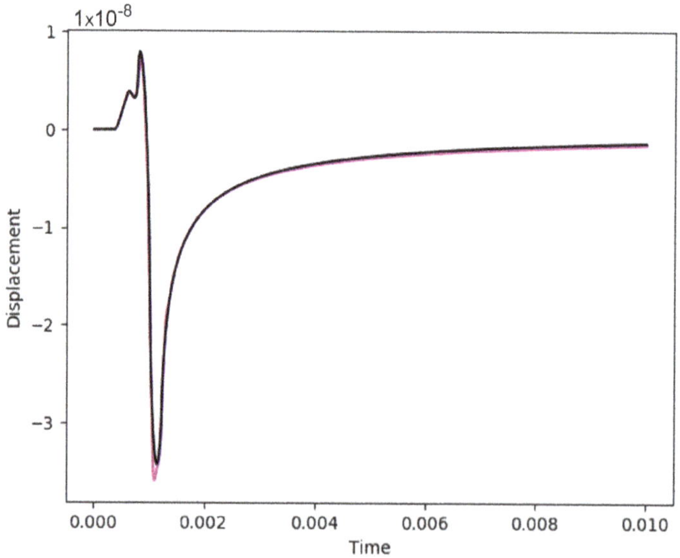

Figure 3. Comparison of displacement history at the sensor point for the exact analytical solution (black line) and the results of FEM numerical simulations (red line) for the reference configuration (no barrier).

Unfortunately, there is no possibility of developing an exact analytical solution for the more complicated problem of the interaction between a wave and an inclusion within a halfplane. The developed numerical model was validated in comparison to the exact analytical solution in the case of no seismic barrier (or, equally, a seismic barrier with properties similar to the ones in the surrounding media). It is demonstrated that the numerical approach can accurately simulate the propagation of dynamic waves, including bulk waves and surface waves (i.e., Rayleigh waves). It is expected that the same model, for which the properties of the material are altered for some locations, should still accurately simulate the propagation of elastic waves in a media with a seismic barrier.

The load duration T is chosen to be equal to the time needed for the fastest (longitudinal) wave to travel a distance equal to the barrier depth in the material representing the halfspace. There is no doubt that the load duration will significantly affect the wave interaction with a barrier of a prescribed geometry and, hence, will have a substantial impact on the effective barrier protective properties. It is apparent that barriers with sizes significantly smaller than the oncoming wave wavelength cannot have any notable influence on the passing wave. Waves with wavelengths comparable to or smaller than the barrier size can be significantly transformed in the process of interacting with the barrier. Credibly, the waves with wavelengths substantially smaller than the barrier dimensions can be effectively filtered by the barrier with no pronounced dependency on the wavelength. Thereby, the most interesting, from both the theoretical and the practical points of view, is the case of waves with wavelengths comparable to the dimensions of the protective barrier itself.

For each of the tested barrier configurations, the maximum overtime displacement induced by the load $P(t)$ at the sensor point (see Figure 1) is scaled with the magnitude of displacement at the same point for the case with no seismic barrier to obtain the protection factor for the corresponding barrier configuration. Namely:

$$PF = \frac{\max_t \left(u\left(x_{sp}, t\right)\right)}{u_{max}^{ref}}, \tag{10}$$

where x_{sp} gives the location of the sensor point, $u(x,t)$ gives the time history of the total displacement at a point with location x and u_{max}^{ref} gives the magnitude of displacement at the sensor point for a reference configuration with no barrier.

For example, a protection factor equal to 2.0 means that the magnitude of displacements in the sensor point is reduced by a factor of 2. The aim of this study is to identify barrier shapes that provide the maximum possible protection, i.e., find the barrier's geometric shape that maximizes the protection factor. Within the framework of the 2D plane-strain problem, it is assumed that the barrier can occupy an area of 1.0 m × 2.0 m. Finite-element mesh consists of quadratic elements, with the element side equal to 1/3 dm, which gives a domain of 30 × 60 elements for the seismic barrier area. The domain is divided into clusters of 2 × 2 elements each, and the barrier area is represented by a 15 × 30 array of clusters. Each of the 450 clusters can be occupied by either a material with the same properties as the halfplane or by another material (barrier material)—see Figure 4. This means that a barrier can be represented as a binary vector in 450-dimensional space. The medium is simulated as a linear elastic material with the properties typical of soil: elastic modulus, E = 30 MPa; density, ρ = 1750 kg/m^3; Poisson's ratio, ν = 0.3. The material representing the barrier has properties typical of concrete: E = 30 GPa; ρ = 2400 kg/m^3; Poisson's ratio, ν = 0.35.

Initially, an attempt to "guess" a good (providing maximized protection factor) barrier shape was performed. The hypothesis was that adding "scatters" to a simple rectangular-shaped seismic barrier can significantly improve its efficiency. Also, several periodic configurations were tested. Table 1 shows some of the tested barrier shapes with their corresponding protection factors.

Case 1 in Table 1 corresponds to a simple rectangular shape of a seismic barrier (securing a protection factor of 2.2). Case 11 corresponds to the domain completely filled with concrete, with a protection factor of 2.64, being the maximum PF observed at that stage. It is concluded that, within the framework of the problem solved, there is most probably impossible to design a seismic barrier with a PF exceeding the one for the entire domain filled with concrete.

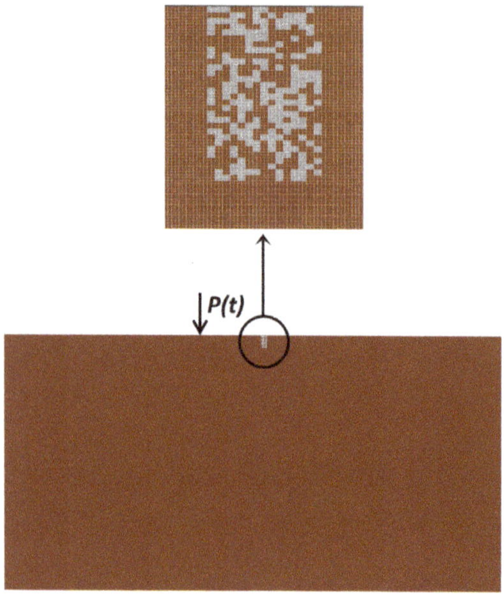

Figure 4. Finite-element discretization and randomly generated barrier geometry.

Table 1. Several of the tested seismic barrier shapes and the computed corresponding protection factors (PFs). Larger protection factors correspond to better protective properties. The brown material corresponds to soil (with the same properties as the surrounding medium) and the gray material corresponds to concrete.

	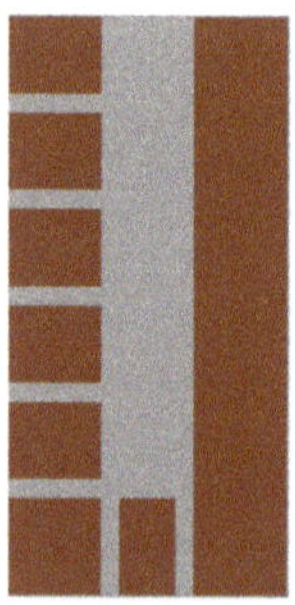	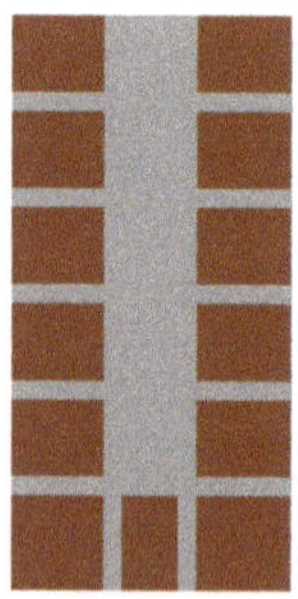	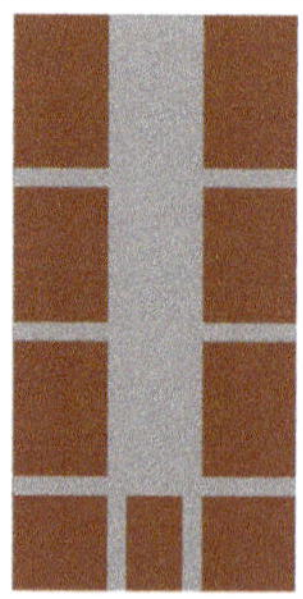		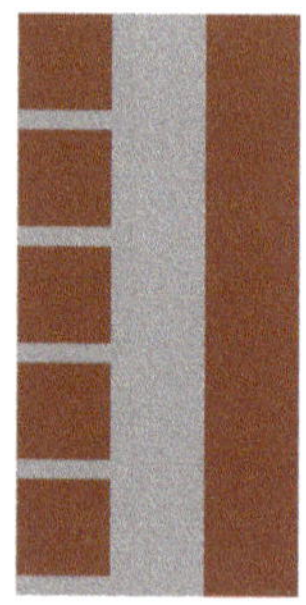
1. PF = 2.19	2. PF = 2.0	3. PF = 1.82	4. PF = 2.03	5. PF = 2.14	6. PF = 2.23

7. PF = 2.24	8. PF = 1.46	9. PF = 1.98	10. PF = 2.22	11. PF = 2.64	

3. Genetic Algorithm

To reaffirm the conclusion of the previous section and to develop a method for finding an optimal geometric configuration for similar problems (for example, for the case of several different materials that could be used to form a barrier), some method to evaluate the optimal or close-to-optimal configuration of a seismic barrier is needed. This paper is not studying the case of more than one material utilized to form the barrier (in addition to the material of the surrounding media), but this can be explored in the following work. Calculation of the PF for a given geometric configuration of a barrier requires around 5 min of processor time on a high-performance computer system. Obviously, there is no practical possibility of calculating the PF for every possible configuration of a barrier (2^{450} possible configurations). To find "good" geometric shapes for a seismic barrier, it was decided to employ a genetic algorithm (see, e.g., [21,28]) to initiate a process for the evolutionary development of barrier geometric configurations that converges to a configuration with the maximum PF possible. The idea of a genetic algorithm is to mimic the principles of natural selection and genetic inheritance to optimize barrier configurations with respect to protective properties over successive generations. Figure 5 gives the basic scheme of a genetic algorithm.

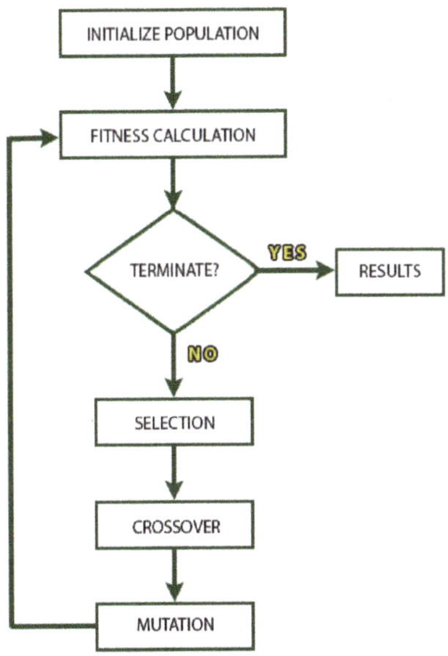

Figure 5. The basic scheme of a genetic algorithm.

Initially, a population of diverse geometric configurations is randomly generated. Alternatively, if there are assumptions about how a "good" seismic barrier configuration should look, these can be used for the generation of the initial population. Every species (barrier configuration) in the population is evaluated using the previously described numerical method to evaluate the PF. For the problem being solved, the PF serves as a fitness function that quantifies the performance of a given barrier shape.

Once the protection factors are calculated, through the process of selection, configurations with higher PFs are chosen as parents for the next generation. The selected barrier configurations undergo genetic operations, including crossovers and mutations, mimicking genetic recombination and variation. Crossover involves combining the geometries of two parent configurations to create a species for the new generation.

Mutation introduces small random changes into the geometry to explore new possibilities. This is implemented by switching materials in a random position inside a barrier.

Combining the operations of crossover and mutation, a new population with the same number of species is generated. The species (barrier geometry) with the highest PF in the previous generation is taken to the next generation unchanged to guarantee that the maximum PF is a non-decreasing function of evolution time.

The process of selection, crossover, and mutation is repeated iteratively. As the generations progress, the population tends to converge toward more optimal configurations, driven by the improvement in the PF. By iteratively applying these genetic operators and evaluating the PF of the resulting configurations, the genetic algorithm explores and refines the space of possible geometric configurations, ultimately converging toward those that exhibit maximum possible protective properties.

4. Implementation of the Genetic Algorithm

Initial population. There are no constraints imposed on the possible shape of the barrier. This means that there are no "prohibited" configurations for the barrier, and any shape can be potentially achieved in the process of evolution. To receive the results presented below, the population (i.e., barrier shapes) for the first generation is generated randomly. This means that every point within a barrier (see Figure 4) is assigned a material with either the properties of concrete or soil with equal probability. This results in approximately equal fractions of soil and concrete within the initial barrier configurations. There is also the possibility of using different fractions of concrete and soil within the initial population. To investigate the convergence of the developed algorithm by demonstrating the independence of the evolution results on the variation of the initial population, computations were performed for various initial fractions of concrete (see Table 2). It was found that the final result of the GA evolution for this problem is not significantly affected by the initial fraction of concrete. For equal fractions of concrete and soil, the GA convergence is slightly faster. The following results were obtained for an initial population with equal fractions of concrete and soil.

Table 2. Examples of initial configurations of barriers with various fractions of concrete. The brown material corresponds to soil (with the same properties as the surrounding medium) and the gray material corresponds to concrete.

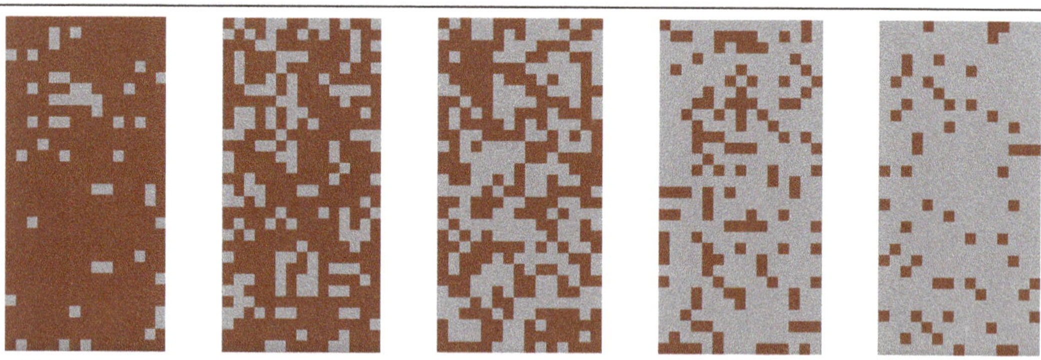

10% of concrete	30% of concrete	50% of concrete	70% of concrete	90% of concrete

Population size. Calculations were performed for different numbers (4, 8, 12, 24, 40) of species (barrier geometries) in a population. It was found empirically that the final result of the GA evolution for this problem is not significantly affected by the population size. It was found empirically that for the problem solved and the computational system utilized for these simulations, having 8–12 species in a population provides the fastest possible convergence. The below results were obtained for a population of 12 species.

Calculation of fitness. For each individual species, fitness is calculated utilizing (10) and with the results of the described above FEM solution (Figure 4), solving the problem of (1)–(4). Fitness is determined by the protection factor, which reflects the ability of the definite barrier geometry to decrease the amplitudes of vibrations at the surface behind the barrier.

Selection. Following the calculation of fitness for each species in the generation, half of the species with the lowest fitness is dismissed. The other half of the species is used to produce the next generation via crossover and mutation. The species with the highest fitness in the generation is taken to the next generation unchanged to guarantee that the fitness is never decreasing through the evolutionary process.

Crossover. A very simple approach to performing a crossover between two species was adopted: the child's geometry takes the upper half of the geometry from one parent and the lower half of geometry from the other parent. Random crossover between the higher-fitness species of the previous generation was used to create the species of the new generation (except for the species with the highest fitness from the previous generation, which is always copied to the next generation).

Mutation. Mutation involves switching the material (soil→concrete or concrete→soil) at a random location of a barrier. It was found empirically that switching materials at more than one point at a time does not have any benefit and impedes the convergence. The probability of mutation is ½.

Genetic algorithm. Once the new generation is produced following selection, crossover and mutation, the fitness is calculated for this new generation, and then the process is looped until the condition for termination is fulfilled. For the computations presented below, the process was terminated once the predefined number of generations was reached (typically 1000 generations). As seen from the results presented below, the fitness is still improving at the typical termination time, but the observed changes to the barrier geometry are minor and there is no major benefit to performing computations for additional generations. Alternatively, the termination condition can be the achievement of a definite fitness value or a significant decrease in the convergence rate.

5. Results

Figure 6 plots the observed evolution of fitness (maximum protection factor) for the genetic algorithm evolution that was started from a randomly generated initial population. The same figure shows that the final (biggest PF) configuration of seismic barrier geometry gives a protection factor of 2.84. This is significantly higher than the protection factor for a barrier with the entire domain filled with concrete (PF = 2.64, see Table 1), which was previously believed to provide the biggest PF possible.

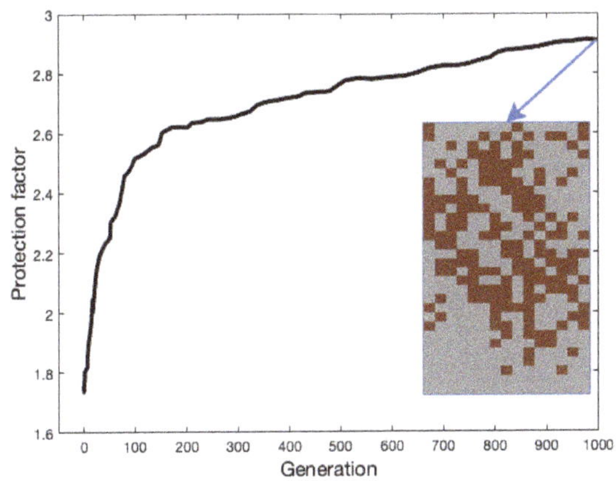

Figure 6. Evolution of fitness (PF) during the genetic algorithm evolution. The picture shows the final barrier shape securing a protection factor of 2.84.

In order to study the convergence and repeatability, the developed GA evolution was executed several times, starting from a randomly generated initial population. A total of more than 50 GA runs were executed, all resulting in very similar optimal barrier geometries after 1000 generations. Some of the final shapes obtained and their corresponding protection factors are presented in Table 3.

Table 3. Final shapes and their corresponding protection factors for three GA evolutions with a randomly generated initial population. Comparison with the protection factor for a barrier with the entire domain filled with concrete.

GA run 1. PF = 2.84	GA run 2. PF = 2.75	GA run 3. PF = 2.82	100% concrete. PF = 2.64

An attempt was made to initiate a new GA evolution using final barrier shapes from previous evolutions as an initial population. This attempt is also important for evaluating the optimal shape as, typically, there is still an increase in the fitness at the termination time of a GA starting from a random initial population (Figure 6). Figure 7 shows the resulting evolution of the PF and the final barrier shape obtained, with the PF = 2.92.

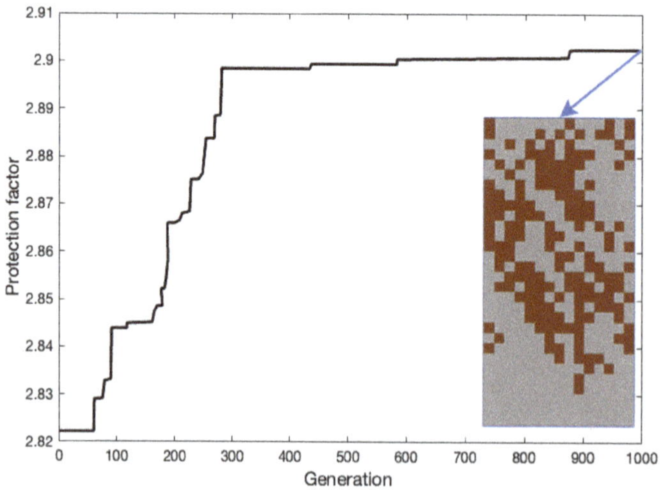

Figure 7. Evolution of fitness (PF) during the genetic algorithm evolution using barrier shapes from Table 2 as the initial population. The picture shows the final barrier shape securing a protection factor of 2.92.

6. Conclusions and Discussion

Utilization of a GA provided the possibility of revealing barrier shapes with protection factors significantly exceeding the protection factor for a homogeneous barrier entirely filled with concrete, which was previously believed to give the highest possible PF. All of the obtained "good" barrier shapes are very similar, which is an indication of the GA convergence and repeatability. The barrier shapes with high PFs have concrete concentrated at the lower end of the barrier from the direction of the wave arrival (left). The rest of the barrier resembles a layered structure inclined at 45°. It is not yet clear what diffraction processes lead to this kind of structure. To understand these processes, presumably a finer discretization of the barrier domain is required. The observed asymmetry of the optimized barrier is the result of the essential asymmetry of the studied problem. For most practical cases of seismic protection, the direction from which the seismic wave is expected to arrive is known. However, if this direction is not known, the constraint on the barrier symmetry should be imposed before the shape optimization is initiated. This present work does not explore this case.

Unfortunately, increasing the barrier discretization density implies a significant increase in both the time needed to obtain the numerical solution for a given barrier shape and the number of PF calculations for convergence of the GA. For example, the calculations presented in Figure 6 (single run of a GA evolution for 1000 generations) required a week of computational time on a high-performance computer system. For this reason, any significant complication of the existing problem will result in the practical impossibility of obtaining the GA convergence.

At the same time, it is believed that there exists a possibility to significantly reduce the number of PF computations needed for the evaluation of "good" barrier geometric shapes. This will require a substantial modification of the GA or its substitution by another generative algorithm (e.g., indirect evolutionary algorithm [29]).

A crucial aspect of the interaction between mechanical waves and inclusions is the inherent multiscale character of this phenomenon. The dimensions of the geometric features within these inclusions are closely tied to the frequency of the incident waves with which the inclusions can effectively interact. Therefore, methods that presuppose the possible existence of certain geometric patterns hold the potential for both significant acceleration of the algorithm convergence and uncovering highly efficient, intricate geometric configurations. The compositional pattern-producing network (CPPN) (see, e.g., [30]) is one of the generative approaches that can be used to efficiently construct the optimal geometric shapes for inclusions interacting with waves. At this stage, it is not possible to directly compare the performance of the developed genetic algorithm to other available generative algorithms as this will require the implementation of these algorithms to solve the same problem and the following long-term simulations.

It should also be noted that a wide variety of fundamental and engineering problems require the optimization of shape in the case when it is not possible to parametrize this shape with several parameters. For example, this kind of problem appears in the construction of various protective structures, in the design of metamaterials, in materials engineering, etc. In most cases, the evaluation of optimal geometric shape results in an attempt to predict what the optimal shape should be. As demonstrated in this research, even a very simple GA can be efficiently applied to evaluate extremely nontrivial geometric shapes that cannot be parameterized.

Author Contributions: Conceptualization, V.B.; Methodology, V.B.; Software, A.M.; Validation, A.M.; Formal analysis, S.V.K.; Investigation, V.B. and A.M.; Writing—original draft, V.B.; Writing—review & editing, V.B.; Supervision, S.V.K. All authors have read and agreed to the published version of the manuscript.

Funding: This research was partly funded by the Ministry of Science and Higher Education of the Russian Federation as part of the World-class Research Center program: Advanced Digital Technologies (contract No. 075-15-2022-311 dated 20 April 2022).

Data Availability Statement: The data presented in this study are available on request from the corresponding author.

Conflicts of Interest: The authors declare no conflict of interest.

References

1. Adam, M.; Estorff, O. Reduction of train-induced building vibrations by using open and filled trenches. *Comput. Struct.* **2005**, *83*, 11–24. [CrossRef]
2. Li, T.; Su, Q.; Kaewunruen, S. Seismic metamaterial barriers for ground vibration mitigation in railways considering the train-track-soil dynamic interactions. *Constr. Build. Mater.* **2020**, *260*, 119936. [CrossRef]
3. Flora, A.; Lombardi, D.; Nappa, V.; Bilotta, E. Numerical analyses of the effectiveness of soft barriers into the soil for the mitigation of seismic risk. *J. Earthq. Eng.* **2018**, *22*, 63–93. [CrossRef]
4. Mu, D.; Shu, H.; Zhao, L.; An, S. A Review of Research on Seismic Metamaterials. *Adv. Eng. Mater.* **2020**, *22*, 1901148. [CrossRef]
5. Brûlé, S.; Enoch, S.; Guenneau, S. Emergence of seismic metamaterials: Current state and future perspectives. *Phys. Lett. A* **2020**, *384*, 126034. [CrossRef]
6. Geng, Q.; Zhu, S.; Chong, K.P. Issues in design of one-dimensional metamaterials for seismic protection. *Soil Dyn. Earthq. Eng.* **2018**, *107*, 264–278. [CrossRef]
7. Kacin, S.; Ozturk, M.; Sevim, U.K.; Mert, B.A.; Ozer, Z.; Akgol, O.; Unal, E.; Karaaslan, M. Seismic metamaterials for low-frequency mechanical wave attenuation. *Nat. Hazards* **2021**, *107*, 213–229. [CrossRef]
8. Albino, C.; Godinho, L.; Amado-Mendes, P.; Alves-Costa, P.; Dias-da-Costa, D.; Soares, D. 3D FEM analysis of the effect of buried phononic crystal barriers on vibration mitigation. *Eng. Struct.* **2019**, *196*, 109340. [CrossRef]
9. Sun, F.; Xiao, L.; Bursi, O.S. Optimal design and novel configuration of a locally resonant periodic foundation (LRPF) for seismic protection of fuel storage tanks. *Eng. Struct.* **2019**, *189*, 147–156. [CrossRef]
10. Wenzel, M.; Bursi, O.S.; Antoniadis, I. Optimal finite locally resonant metafoundations enhanced with nonlinear negative stiffness elements for seismic protection of large storage tanks. *J. Sound Vib.* **2020**, *483*, 115488. [CrossRef]
11. Rezaie, A.; Rafiee-Dehkharghani, R.; Dolatshahi, K.M.; Mirghaderi, S.R. Soil-buried wave barriers for vibration control of structures subjected to vertically incident shear waves. *Soil Dyn. Earthq. Eng.* **2018**, *109*, 312–323. [CrossRef]
12. Jesmani, M.; Fallahil, M.A.; Kashani, H.F. Effects of geometrical properties of rectangular trenches intended for passive isolation in sandy soils. *Earth Sci. Res.* **2012**, *1*, 137–151. [CrossRef]
13. Kim, S.H.; Das, M.P. Artificial seismic shadow zone by acoustic metamaterials. *Mod. Phys. Lett. B* **2013**, *27*, 1350140. [CrossRef]
14. Palermo, A.; Krödel, S.; Marzani, A.; Daraio, C. Engineered metabarrier as shield from seismic surface waves. *Sci. Rep.* **2016**, *6*, 39356. [CrossRef] [PubMed]
15. Gupta, A.; Sharma, R.; Thakur, A.; Gulia, P. Metamaterial foundation for seismic wave attenuation for low and wide frequency band. *Sci. Rep.* **2023**, *13*, 2293. [CrossRef] [PubMed]
16. Bratov, V.A.; Ilyashenko, A.V.; Kuznetsov, S.V.; Lin, T.K.; Morozov, N.F. Homogeneous horizontal and vertical seismic barriers: Mathematical foundations and dimensional analysis. *Mater. Phys. Mech.* **2020**, *44*, 61–65.
17. Bratov, V.; Kuznetsov, S.; Morozov, N. Seismic barriers filled with solid elastic and granular materials. *Comp. Anal. Math. Mech. Solids* **2022**, *27*, 1761–1770. [CrossRef]
18. Brennan, A.J.; Klar, A.; Madabhushi, S. Mitigation of Seismic Accelerations by Soft Materials Embedded in the Ground. *Proc. Inst. Civ. Eng.-Ground Improv.* **2019**, *3*, 117–140.
19. Le, T.; Lee, V.W.; Luo, H. Out of plane (SH) soil-structure interaction: A shear wall with rigid and flexible ring foundation. *Earth Sci.* **2016**, *29*, 44–55. [CrossRef]
20. Martakis, P.; Aguzzi, G.; Dertimanis, V.K.; Chatzi, E.N.; Colombi, A. Nonlinear periodic foundations for seismic protection: Practical design, realistic evaluation and stability considerations. *Soil Dyn. Earthq. Eng.* **2021**, *150*, 106934. [CrossRef]
21. Duhamel, D. Shape optimization of noise barriers using genetic algorithms. *J. Sound Vib.* **2006**, *297*, 432–443. [CrossRef]
22. ANSYS Company. *LS-DYNA Manual R12.0*; Livermore Software Technology (LST); ANSYS Company: Canonsburg, PA, USA, 2020.
23. Lamb, H. On the propagation of tremors over the surface of an elastic solid. *Philos. Trans. R. Soc. Lond. A* **1904**, *203*, 1–42.
24. Kausel, E. Lamb's problem at its simplest. *Proc. R. Soc. A Math. Phys. Eng. Sci.* **2012**, *469*, 20120462. [CrossRef]
25. Eringen, A.C.; Suhubi, E.S. *Elastodynamics, Volume II: Linear Theory*; Academic Press: New York, NY, USA, 1975.
26. Bratov, V.A.; Kuznetsov, S.V.; Morozov, N.F. Lamb's problems and related problems of dynamics. *J. Appl. Math. Mech.* **2022**, *86*, 451–469.
27. Morozov, N.F.; Bratov, V.A.; Kuznetsov, S.V. Seismic barriers for protection against surface and head waves: Multiple scatters and metamaterials. *Mech. Solids* **2021**, *56*, 911–921. [CrossRef]
28. Mirjalili, S. Genetic Algorithm. In *Evolutionary Algorithms and Neural Networks. Studies in Computational Intelligence*; Springer: Cham, Switzerland, 2019; Volume 780.

29. Hart, E.; Le Goff, L.K. Artificial evolution of robot bodies and control: On the interaction between evolution, learning and culture. *Philos. Trans. R. Soc. B* **2022**, *377*, 20210117. [CrossRef] [PubMed]
30. Stanley, K.O.; D'Ambrosio, D.B.; Gauci, J. A hypercube-based encoding for evolving large-scale neural networks. *Artif. Life* **2009**, *15*, 185–212. [CrossRef]

Σ mathematics

MDPI

Article

A General Case of a Line Contact Lubricated by a Non-Newtonian Giesekus Fluid

Ilya I. Kudish [1,*] and Sergei S. Volkov [2]

[1] ILRIMA Consulting, Inc., 19396 Warbler Ln., Millersburg, MI 49759, USA
[2] Research and Education Center "Materials", Don State Technical University, 1 Gagarina sq., 344001 Rostov-on-Don, Russia; fenix_rsu@mail.ru
* Correspondence: ilyakudish@gmail.com; Tel.: +1-586-3603-704

Abstract: A steady plane hydrodynamic problem of lubrication of a lightly loaded contact of two parallel cylinders lubricated by a non-Newtonian fluid with Giesekus rheology is considered. The advantage of this non-Newtonian rheology is its ability to properly describe the real behavior of formulated lubricants at high and low shear stresses. The problem is solved by using a modification of the regular perturbation method with respect to the small parameter α, characterizing the degree to which the polymeric molecules of the additive to the lubricant follow the streamlines of the lubricant flow. It is assumed that the lubricant relaxation time and the value of α are of the order of the magnitude of the ratio of the characteristic gap between the contact surfaces and the contact length. The obtained analytical solution of the problem is analyzed numerically for the dependencies of the problem characteristics such as contact pressure, fluid flux, lubrication film thickness, friction force, energy loss in the lubricated contact, etc., on the problem input parameters.

Keywords: Giesekus lubricant rheology; hydrodynamic lubrication problem; method of regular perturbations; lubricant mobility factor; lubricant relaxation time

MSC: 76D08

1. Introduction

Most machine elements work in lubricated environments. Generally, the usage of lubrication allows a decrease in frictional losses in contacts and, thus, an increase in their durability. When such lubricated contacts are mathematically modeled, the rheology of the lubricant is considered to be Newtonian or non-Newtonian. The former case is significantly simpler and a lot of studies for the cases of homogeneous material and coated solids [1–10] were dedicated to it. The latter case is more complex, but it better corresponds to the real behavior of lubricants because polymeric additives make their behavior significantly non-Newtonian, which allows for a distinctly different lubricant behavior at low and high shear stresses. Different rheological models for lubricant behavior were used to catch various aspects of real lubricant behavior such as lubricant relaxation time, etc. An overview and analysis of different fluid rheological models is given in [11,12]. One of the simplest rheological models of non-Newtonian lubricant behavior among generalized Newtonian fluids [13] is the Ree–Eyring model [14–16]. Some studies of fluids with complex non-Newtonian rheology were presented in [17–22].

Experimental data show that lubricant viscosity depends not only on pressure but also on lubricant shear stress and velocity [23]. The actual behavior of polymer-formulated lubricant flowing through narrow gaps is most accurately described by the Giesekus rheological model [12,24,25]. The Giesekus model differs from the Maxwell, Jeffris, Oldroyd A and B, etc., rheological models by the presence of a nonlinear term determined by the mobility factor α and stress state of the fluid. It provides the opportunity to take into account the degree to which the fluid viscosity is dependent on the shear stress in the fluid. In particular, this model allows for proper accounting of relatively high viscosity for low and relatively low viscosity for high shear stresses in the fluid. The model takes into account four parameters of the fluid: the relaxation time, the viscosities of base oil and polymeric additive and the mobility factor. If the mobility factor is set to zero, then the Giesekus model turns into the convected Maxwell model [11,12]. The Giesekus model is nonlinear and, therefore, more complex for analysis. A relatively simple case of a fluid with Giesekus rheology flowing between two parallel surfaces is considered in [26,27]. In [28], the analysis of a lightly loaded trust bearing lubricated by a fluid with the Giesekus rheology was performed by the perturbation methods. However, the presence of convective terms and the base oil were not taken into account.

The present paper is the continuation of the series of the author's studies of hydro- and elastohydrodynamic problems for Newtonian [29,30] and non-Newtonian [13] lubricating fluids including the study of a lightly loaded trust bearing lubricated with a fluid with the Giesekus rheology [31]. Also, the authors previously considered a similar but more simple problem under the condition when the mobility factor α is significantly greater than the ratio of the characteristic gap between the surfaces and length of the contact [32]. In this case, in a two-term asymptotic solution, the inertia terms can be neglected. The first two terms of asymptotic representation for contact pressure with respect to the small parameter α are obtained in the analytical form. Numerical analysis of the obtained relationships for contact pressure, lubrication film thickness, lubricant flux, friction, contact energy loss, etc., as functions of the input parameters of the Giesekus model is performed. Some applications of perturbation methods to steady problems similar to the one used are described in [33,34].

2. Formulation of the Lubrication Problem and General Description of the Analysis Applied

Let us consider a steady plane problem for a lubricated contact of two parallel infinite cylinders of the radius R (see Figure 1), both of which are made of rigid materials (the case of cylinders of different radii is considered at the end of the paper). It is assumed that the lubricant is described by the Giesekus model [12] with a constant viscosity μ and relaxation time λ_1. The x-axis of the coordinate system is directed along the contact and perpendicular to the cylinder axes, the y-axis is directed along the cylinder axes, and the z-axis is perpendicular to both x- and y-axes. A continuous lubricant layer separates the cylinders which steadily roll and slide with the surface linear speeds $\overline{u_1}$ and $\overline{u_2}$ in the direction of the x-axis. The upper cylinder is subjected to the normal load P along the z-axis. The lubricant velocity components are represented by functions $u(x,y,z)$, $v(x,y,z)$, and $w(x,y,z)$. Due to the problem geometry, $v(x,y,z) = \frac{\partial v(x,y,z)}{\partial y} = 0$. Therefore, the problem parameters are independent of y and the motion equations of such a fluid are as follows [13,29].

$$
\begin{gathered}
u\frac{\partial u}{\partial x} + w\frac{\partial u}{\partial z} = \frac{1}{\rho}\left(\frac{\partial p_{xx}}{\partial x} + \frac{\partial p_{zx}}{\partial z}\right),\ \frac{\partial p_{xy}}{\partial x} + \frac{\partial p_{zy}}{\partial z} = 0, \\
u\frac{\partial w}{\partial x} + w\frac{\partial w}{\partial z} = \frac{1}{\rho}\left(\frac{\partial p_{zx}}{\partial x} + \frac{\partial p_{zz}}{\partial z}\right).
\end{gathered} \tag{1}
$$

In addition to that, one needs to consider the continuity equation. It is assumed that the fluid is incompressible, i.e., $\rho(x,z) = constant$. That leads to the continuity equation

$$
\frac{\partial u}{\partial x} + \frac{\partial w}{\partial z} = 0. \tag{2}
$$

In this case, the stress tensor components are as follows

$$
p_{xx} = -p + \tau_{xx},\ p_{xy} = \tau_{xy} = 0,\ \ p_{zx} = \tau_{zx},\ p_{zz} = -p + \tau_{zz},\ p_{zy} = \tau_{zy} = 0, \tag{3}
$$

where p is the pressure and τ_{xx}, τ_{xy}, τ_{zx}, τ_{zz}, and τ_{zy} are the additional stress components acting in the corresponding directions. These tensor components satisfy the Giesekus fluid model which is a certain generalization of the Maxwell model and takes into account the degree to which the additive polymeric molecules follow the lubricant flow. The rheological equations are as follows [12]

$$\tau = \tau_s + \tau_p,\ \mu = \mu_s + \mu_p,\ \tau_s = \mu_s \dot{\gamma},$$
$$\tau_p + \lambda_1 \tau_{p(1)} - \alpha \frac{\lambda_1}{\mu_p}\{\tau_p \cdot \tau_p\} = \mu_p \dot{\gamma}, \quad (4)$$

where τ is the full stress tensor while τ_s and τ_p are the solvent and polymer stress tensors, respectively, μ_s and μ_p are the constant solvent and polymer dynamic viscosities, $\dot{\gamma}$ is the deformation tensor, λ_1 is the constant relaxation time, and α is the dimensionless constant mobility factor describing the degree of the alignment of polymeric molecules with the lubricant flow, $0 \leq \alpha \leq 1$.

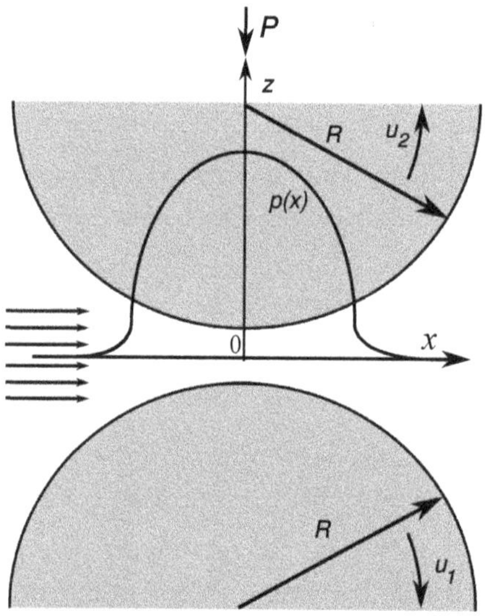

Figure 1. The general view of a lubricated contact.

In (4), the following definitions of the tensor operators $\tau_{p(1)}$ and $\{\tau_p \cdot \tau_p\}$ [12] are used

$$\tau_{p(1)xx} = u\frac{\partial \tau_{pxx}}{\partial x} + w\frac{\partial \tau_{pxx}}{\partial z} - 2\frac{\partial u}{\partial z}\tau_{pxz} - 2\frac{\partial u}{\partial x}\tau_{pxx},$$
$$\tau_{p(1)xz} = u\frac{\partial \tau_{pxz}}{\partial x} + w\frac{\partial \tau_{pxz}}{\partial z} - \frac{\partial u}{\partial x}\tau_{pzx} - \frac{\partial u}{\partial z}\tau_{pzz} - \frac{\partial w}{\partial x}\tau_{pxx} - \frac{\partial w}{\partial z}\tau_{pzx},$$
$$\tau_{p(1)zz} = u\frac{\partial \tau_{pzz}}{\partial x} + w\frac{\partial \tau_{pzz}}{\partial z} - 2\frac{\partial w}{\partial z}\tau_{pzz} - 2\frac{\partial w}{\partial x}\tau_{pzx}, \quad (5)$$
$$\{\tau_p \cdot \tau_p\}_{xx} = \tau_{pxx}^2 + \tau_{pxz}^2,$$
$$\{\tau_p \cdot \tau_p\}_{xz} = \tau_{pxx}\tau_{pxz} + \tau_{pxz}\tau_{pzz},$$
$$\{\tau_p \cdot \tau_p\}_{zz} = \tau_{pxz}^2 + \tau_{pzz}^2.$$

and [12]

$$\dot{\gamma}_{xx} = 2\frac{\partial u}{\partial x},\ \dot{\gamma}_{xz} = \frac{\partial u}{\partial z} + \frac{\partial w}{\partial x},\ \dot{\gamma}_{zz} = 2\frac{\partial w}{\partial z}. \quad (6)$$

It is necessary to impose some boundary conditions on the components u and w of the lubricant velocity on the surfaces of the contacting solids. It is assumed that the no-slip and no-penetration conditions are valid on the solid boundaries. The contact is considered to be concentrated, i.e., $L_z/L_x \ll 1$ (where L_x and L_z are the characteristic contact sizes along the x- and z-axes) or equivalently $dh(x)/dx \ll 1$, where $h(x)$ is the gap between the cylinder surfaces at point x. Using the above boundary conditions will take the form

$$u(x, -\frac{h(x)}{2}) = \overline{u_1},\ u(x, \frac{h(x)}{2}) = \overline{u_2}, \quad (7)$$

$$w(x, -\frac{h(x)}{2}) = -\frac{\overline{u_1}}{2}\frac{dh(x)}{dx},\ w(x, \frac{h(x)}{2}) = \frac{\overline{u_2}}{2}\frac{dh(x)}{dx}. \quad (8)$$

It will be shown later that in the simplified formulation the function of pressure p is independent of the variable z. Therefore, under pressure, one can impose the boundary conditions stemming from the fact that at the contact boundaries the fluid pressure p is equal to the atmospheric one and, therefore, can be neglected. To avoid cavitation at the exit from the contact [13], the boundary conditions on p have the form

$$p(x_i) = p(x_e) = \frac{dp(x_e)}{dx} = 0, \quad (9)$$

where x_i and x_e are the coordinates of the inlet and exit points of the contact.

In addition to that, it is required that the stress $-p_{zz}$ from (3) supports the normal load P applied to the cylinder, i.e.,

$$\int\limits_{x_i}^{x_e}[p(x) - \tau_{zz}(x, h(x)/2)]dx = P. \quad (10)$$

Being interested in a two-term asymptotic problem solution in α, in (9) and (10), it is assumed that the two-term asymptotic expansion of pressure $p(x,z)$ in α (i.e., a linear function of α) is independent of z. That assumption is confirmed later for small α up to the order of $O(\alpha)$, $\alpha \ll 1$, i.e., in the asymptotic expansion of p in $\alpha \ll 1$, terms p_0 and p_1 are independent of z, while terms p_2, $p_3, \ldots$ depend on both x and z.

The goal is to determine the contact pressure $p(x)$, the gap between the two cylinders $h(x)$, the coordinate of the exit point from the lubricated contact x_e, the components of the tensor τ in the fluid, etc. Typically, the point of entrance in the lubricated contact, i.e., the coordinate of the inlet point x_i, is known, and it is determined by the amount of lubricant supplied to the contact. The region where the problem solution is searched is bounded by $x = x_i$ and $x = x_e$ as well as by $z = -h(x)/2$ and $z = h(x)/2$. Out of these four boundaries of the region, three are unknown, i.e., $x = x_e$, $z = -h(x)/2$, and $z = h(x)/2$ are unknown. One needs to find a perturbation solution to the above-determined problem in the case when $\alpha \ll 1$. In this case, x_e and $h(x)$ are just slightly perturbed values of the exit point x_{e0} and gap $h_0(x)$ realized in a lubricated contact with Newtonian fluid, i.e., when $\alpha = 0$.

The general case of

$$\alpha = \alpha_0\epsilon,\ \alpha_0 = O(1),\ \epsilon = \frac{L_z}{L_x} \ll 1, \quad (11)$$

will be considered. Here, α_0 is a nonnegative constant. This case allows us to take into account some of the convective as well as major and minor dissipative terms of the equations. Also, it allows us to consider the limiting cases $\alpha \gg \epsilon$ and $\alpha \ll \epsilon$ by correspondingly taking $\alpha_0 \gg 1$ and $\alpha_0 \ll 1$.

3. Simplification of the Rheological Equations and the Equations of Lubricant Motion

It was shown that the problem solution should be searched for not in the original (x, z) independent variables but in slightly modified ones (x_0, z_0). It is due to the fact that the exit $x = x_e$ and upper $z = h(x)/2$ and lower $z = -h(x)/2$ boundaries of the lubricated contact are perturbed ones of the corresponding boundaries $x = x_{e0}$ and $z = \pm h_0(x_0)/2$ of the contact lubricated by a Newtonian fluid (for $\alpha = 0$). Therefore, let us assume that

$$x_e = x_{e0} + \alpha x_{e1} + \ldots,\ h(x) = h_0(x_0) + \alpha h_1(x_0) + \ldots, \quad (12)$$

where x_{e1} and $h_1(x_0)$ will be determined as a part of the problem solution. Then, the problem can be projected on the region $[x_i, x_{e0}] \times [-h_0(x_0)/2, h_0(x_0)/2]$ by introducing the following new independent variables

$$x = x_0 + \alpha x_{e1}\frac{x_0 - x_i}{x_{e0} - x_i},\ z = \frac{h(x)}{h_0(x_0)} z_0. \tag{13}$$

When $x_0 = x_i$ and $x_0 = x_{e0}$ with the precision of $O(\alpha^2)$, $\alpha \ll 1$, one obtains $x = x_i$ and $x = x_e$, respectively, and vice versa (see (12) and (13)). The actual definition of the new variable x_0 can be taken as $x_0 = (x_i + x_{e0})/2 + \xi(x_{e0} - x_i)/2,\ -1 \leq \xi \leq 1$, which varies between x_i and x_{e0}.

As it was shown earlier, this projection of the original problem onto the region $[x_i, x_{e0}] \times [-h_0(x_0)/2, h_0(x_0)/2]$ is also dictated by the asymptotic expansion of the boundary conditions on pressure $p(x)$ at the exit point $x = x_e$ [31,32].

Let us consider the solution form which should be used in this case. One needs to modify integration and differentiation operators while converting the problem from variables (x, z) to variables (x_0, z_0) as follows (see (12) and (13))

$$\begin{gathered} dx = [1 + \alpha\frac{x_{e1}}{x_{e0} - x_i} + \ldots]dx_0,\ dz = [1 + \alpha\frac{h_1(x_0)}{h_0(x_0)} + \ldots]dz_0, \\ \frac{\partial}{\partial x} = [1 - \alpha\frac{x_{e1}}{x_{e0} - x_i} + \ldots]\frac{\partial}{\partial x_0},\ \frac{\partial}{\partial z} = [1 - \alpha\frac{h_1(x_0)}{h_0(x_0)} + \ldots]\frac{\partial}{\partial z_0}. \end{gathered} \tag{14}$$

Let us consider an arbitrary sufficiently smooth function $f(\alpha, x, z)$ which has to be expanded asymptotically in $\alpha \ll 1$. Then, based on (12) and (13), one has [31,32]

$$\begin{aligned} f(\alpha, x, z) &= f(0, x, z) + \alpha f_{1*}(x, z) + \ldots \\ &= f(0, x_0 + \frac{\alpha x_{e1}(x_0 - x_i)}{x_{e0} - x_i} + \ldots, \frac{(h_0(x_0) + \alpha h_1(x_0) + \ldots)z_0}{h_0(x_0)}) \\ &+ \alpha f_{1*}(x_0 + \frac{\alpha x_{e1}(x_0 - x_i)}{x_{e0} - x_i} + \ldots, \frac{(h_0(x_0) + \alpha h_1(x_0) + \ldots)z_0}{h_0(x_0)}) + \ldots \\ &= f(0, x_0, z_0) + \alpha[f_1(x_0, z_0) + \frac{x_{e1}(x_0 - x_i)}{x_{e0} - x_i}\frac{\partial f(0, x_0, z_0)}{\partial x_0} + \frac{z_0 h_1(x_0)}{h_0(x_0)}\frac{\partial f(0, x_0, z_0)}{\partial z_0}] + \ldots, \end{aligned} \tag{15}$$

where $f_{1*}(x, z) = \frac{\partial f(0, x, z)}{\partial \alpha}$ and the derivative of f with respect to x_0 and z_0 are also taken for $\alpha = 0$. Therefore, the solution of the problem one will search using the Taylor expansions of the unknown functions such as $p(x)$, $u(x, z)$, $\tau_{xx}(x, z)$, etc., about x_0 and (x_0, z_0), retaining just the first two terms of the expansions as well as taking into account the sample expansion (15) and the perturbed expressions for x_e and $h(x)$ from (12) and (13).

In the form the problem is formulated above it is very complex for any analytical analysis except the perturbation method. Therefore, the first goal is to simplify Equations (1)–(8) and to reduce them to much simpler equations which would allow for an analytical solution and derivation of analogs of the Reynolds equation. One needs to retain only the terms of higher orders of magnitude. As it was mentioned earlier, due to the fact that the thickness of the lubrication layer is much smaller than the extent of the lubricated contact, one has a small parameter $\epsilon = L_z/L_x \ll 1$. Also, let U_x and U_z be the characteristic velocities of the lubricating fluid in the directions of the x- and z-axes. Then, the following scaling of the equations can be introduced

$$\begin{gathered} \lambda_1' = \frac{U_x}{L_x}\lambda_1,\ x' = \frac{x}{L_x},\ z' = \frac{z}{L_z},\ p' = p\frac{L_z^2}{\mu_* U_x L_x}, \\ u' = \frac{u}{U_x},\ w' = \frac{w}{U_z},\ s_0 = 2\frac{\overline{u_2} - \overline{u_1}}{\overline{u_1} + \overline{u_2}},\ \{\mu', \mu_s', \mu_p'\} = \frac{1}{\mu_*}\{\mu, \mu_s, \mu_p\}, \\ \{\tau_{xx}', \tau_{sxx}', \tau_{pxx}', \tau_{zx}', \tau_{szx}', \tau_{pzx}', \tau_{zz}', \tau_{szz}', \tau_{pzz}'\} = \frac{L_z\{\tau_{xx}, \tau_{sxx}, \tau_{pxx}, \tau_{zx}, \tau_{szx}, \tau_{pzx}, \tau_{zz}, \tau_{szz}, \tau_{pzz}\}}{\mu_* U_x}, \end{gathered} \tag{16}$$

where μ_* is the characteristic value of the lubricant viscosity. For simplicity, in the further analysis, the primes at the dimensionless variables are dropped.

By introducing the above scaling in the continuity Equation (2), one obtains

$$\frac{\partial u}{\partial x}+\frac{U_z}{U_x}\frac{L_x}{L_z}\frac{\partial w}{\partial z}=0. \quad (17)$$

To retain both terms in Equation (17), one needs to assume that

$$\frac{U_z}{U_x}\frac{L_x}{L_z}=1. \quad (18)$$

Then, Equation (17) becomes

$$\frac{\partial u}{\partial x}+\frac{\partial w}{\partial z}=0. \quad (19)$$

Introducing scaling (16) in Equations (1) and (3)–(6) leads to equations

$$\epsilon Re_0\{u\frac{\partial u}{\partial x}+w\frac{\partial u}{\partial z}\}=-\frac{\partial p}{\partial x}+\epsilon\frac{\partial \tau_{xx}}{\partial x}+\frac{\partial \tau_{zx}}{\partial z},$$
$$\epsilon^3 Re_0\{u\frac{\partial w}{\partial x}+w\frac{\partial w}{\partial z}\}=\epsilon^2\frac{\partial \tau_{zx}}{\partial x}-\frac{\partial p}{\partial z}+\epsilon\frac{\partial \tau_{zz}}{\partial z}, \quad (20)$$

$$\tau_{xx}=\tau_{sxx}+\tau_{pxx},\ \tau_{zx}=\tau_{szx}+\tau_{pzx},\ \tau_{zz}=\tau_{szz}+\tau_{pzz},\tau_{sxx}=\epsilon 2\mu_s\frac{\partial u}{\partial x},$$
$$\tau_{szx}=\mu_s(\epsilon^2\frac{\partial w}{\partial x}+\frac{\partial u}{\partial z}),\tau_{szz}=\epsilon 2\mu_s\frac{\partial w}{\partial z}=-\epsilon 2\mu_s\frac{\partial u}{\partial x}, \quad (21)$$

$$\tau_{pxx}+\lambda_1[u\frac{\partial \tau_{pxx}}{\partial x}+w\frac{\partial \tau_{pxx}}{\partial z}-\frac{2}{\epsilon}\frac{\partial u}{\partial z}\tau_{pzx}-2\frac{\partial u}{\partial x}\tau_{pxx}]-\alpha\frac{\lambda_1}{\epsilon\mu_p}(\tau_{pxx}^2+\tau_{pzx}^2)=2\epsilon\mu_p\frac{\partial u}{\partial x},$$

$$\tau_{pzx}+\lambda_1[u\frac{\partial \tau_{pzx}}{\partial x}+w\frac{\partial \tau_{pzx}}{\partial z}-\frac{1}{\epsilon}\frac{\partial u}{\partial z}\tau_{pzz}-\frac{\partial u}{\partial x}\tau_{pzx}-\frac{\partial w}{\partial z}\tau_{pzx}-\epsilon\frac{\partial w}{\partial x}\tau_{pxx}]$$
$$-\alpha\frac{\lambda_1}{\epsilon\mu_p}(\tau_{pxx}+\tau_{pzz})\tau_{pzx}=\mu_p(\frac{\partial u}{\partial z}+\epsilon^2\frac{\partial w}{\partial x}), \quad (22)$$

$$\tau_{pzz}+\lambda_1[u\frac{\partial \tau_{pzz}}{\partial x}+w\frac{\partial \tau_{pzz}}{\partial z}-2\frac{\partial w}{\partial z}\tau_{pzz}-2\epsilon\frac{\partial w}{\partial x}\tau_{pzx}]-\alpha\frac{\lambda_1}{\epsilon\mu_p}(\tau_{pzx}^2+\tau_{pzz}^2)=2\epsilon\mu_p\frac{\partial w}{\partial z},$$

where $Re_0=\frac{\rho U_x L_z}{\mu_*}$ is the local Reynolds number.

Let us assume that not only $\epsilon \ll 1$ but also $Re_0=O(1)$, $\alpha \ll 1$, and

$$\lambda_1=\lambda\epsilon,\ \lambda=O(1),\ \epsilon\ll 1. \quad (23)$$

The latter assumption makes the rheological equations solvable. Otherwise, for $\lambda_1=O(1)$, $\epsilon \ll 1$, these equations do not have a reasonable solution.

Then, Equations (20)–(22) can be simplified. Based on the example of function $f(\alpha,x,z)$ expansion for $\alpha \ll 1$, one needs to search the problem solution in the form

$$\{\tau_{sxx}(x,z),\tau_{sxz}(x,z),\tau_{szz}(x,z)\}=\{\tau_{sxx0}(x_0,z_0),\tau_{sxz0}(x_0,z_0),\tau_{szz0}(x_0,z_0)\}$$
$$+\alpha\{\tau_{sxx1*}(x_0,z_0),\tau_{sxz1*}(x_0,z_0),\tau_{szz1*}(x_0,z_0)\}+\ldots,$$
$$\{\tau_{sxx1*}(x_0,z_0),\tau_{sxz1*}(x_0,z_0),\tau_{szz1*}(x_0,z_0)\}$$
$$=\{\tau_{sxx1}(x_0,z_0),\tau_{sxz1}(x_0,z_0),\tau_{szz1}(x_0,z_0)\} \quad (24)$$
$$+x_{e1}\frac{x_0-x_i}{x_{e0}-x_i}\frac{\partial}{\partial x_0}\{\tau_{sxx0}(x_0,z_0),\tau_{sxz0}(x_0,z_0),\tau_{szz0}(x_0,z_0)\}$$
$$+z_0\frac{h_1(x_0)}{h_0(x_0)}\frac{\partial}{\partial z_0}\{\tau_{sxx0}(x_0,z_0),\tau_{sxz0}(x_0,z_0),\tau_{szz0}(x_0,z_0)\},$$

$$\{\tau_{pxx}(x,z),\tau_{pxz}(x,z),\tau_{pzz}(x,z)\}=\{\tau_{pxx0}(x_0,z_0),\tau_{pxz0}(x_0,z_0),\tau_{pzz0}(x_0,z_0)\}$$
$$+\alpha\{\tau_{pxx1*}(x_0,z_0),\tau_{pxz1*}(x_0,z_0),\tau_{pzz1*}(x_0,z_0)\}+\ldots,$$
$$\{\tau_{pxx1*}(x_0,z_0),\tau_{pxz1*}(x_0,z_0),\tau_{pzz1*}(x_0,z_0)\}$$
$$=\{\tau_{pxx1}(x_0,z_0),\tau_{pxz1}(x_0,z_0),\tau_{pzz1}(x_0,z_0)\} \tag{25}$$
$$+x_{e1}\frac{x_0-x_i}{x_{e0}-x_i}\frac{\partial}{\partial x_0}\{\tau_{pxx0}(x_0,z_0),\tau_{pxz0}(x_0,z_0),\tau_{pzz0}(x_0,z_0)\}$$
$$+z_0\frac{h_1(x_0)}{h_0(x_0)}\frac{\partial}{\partial z_0}\{\tau_{pxx0}(x_0,z_0),\tau_{pxz0}(x_0,z_0),\tau_{pzz0}(x_0,z_0)\},$$

$$p(x)=p_0(x_0)+\alpha p_{1*}(x_0)+\ldots,$$
$$p_{1*}(x_0)=p_1(x_0)+x_{e1}\frac{x_0-x_i}{x_{e0}-x_i}\frac{dp_0(x_0)}{dx_0},$$
$$u(x,z)=u_0(x_0,z_0)+\alpha u_{1*}(x_0,z_0)+\ldots,$$
$$u_{1*}(x_0,z_0)=u_1(x_0,z_0)+x_{e1}\frac{x_0-x_i}{x_{e0}-x_i}\frac{\partial u_0(x_0,z_0)}{\partial x_0}+z_0\frac{h_1(x_0)}{h_0(x_0)}\frac{\partial u_0(x_0,z_0)}{\partial z_0}, \tag{26}$$
$$w(x,z)=w_0(x_0,z_0)+\alpha w_{1*}(x_0,z_0)+\ldots,$$
$$w_{1*}(x_0,z_0)=w_1(x_0,z_0)+x_{e1}\frac{x_0-x_i}{x_{e0}-x_i}\frac{\partial w_0(x_0,z_0)}{\partial x_0}+z_0\frac{h_1(x_0)}{h_0(x_0)}\frac{\partial w_0(x_0,z_0)}{\partial z_0},$$

where $p_0(x_0)$, $u_0(x_0,z_0)$, $w_0(x_0,z_0)$, $\tau_{sxx0}(x_0,z_0)$, $\tau_{sxz0}(x_0,z_0)$, $\tau_{szz0}(x_0,z_0)$, $\tau_{pxx0}(x_0,z_0)$, $\tau_{pxz0}(x_0,z_0)$, $\tau_{pzz0}(x_0,z_0)$ and $u_1(x_0,z_0)$, $w_1(x_0,z_0)$, $\tau_{sxx1}(x_0,z_0)$, $\tau_{sxz1}(x_0,z_0)$, $\tau_{szz1}(x_0,z_0)$, $\tau_{pxx1}(x_0,z_0)$, $\tau_{pxz1}(x_0,z_0)$, $\tau_{pzz1}(x_0,z_0)$ are the unknown main and first-order approximations of the corresponding functions while the gap $h(x)$ between two contact rigid solids and functions $h_0(x_0)$ and $h_1(x_0)$ are described by the equations (see (12))

$$h(x)=h_e+\frac{x^2-x_e^2}{\gamma_0},$$
$$h(x)=h_0(x_0)+\alpha h_1(x_0)+\ldots,\ h_e=h_{e0}+\alpha h_{e1}+\ldots, \tag{27}$$

$$h_0(x_0)=h_{e0}+\frac{x_0^2-x_{e0}^2}{\gamma_0},\ h_1(x_0)=h_{e1}+\frac{2x_{e1}}{\gamma_0}\Big[\frac{x_0(x_0-x_i)}{x_{e0}-x_i}-x_{e0}\Big], \tag{28}$$

where $\gamma_0=\frac{2L_zR_e}{L_x^2}$, h_e is the a priori unknown lubrication film thickness at the exit point $x=x_e$, R_e is the effective curvature radius of the contacting solids, $R_e=R/2$ and h_{e0}, h_{e1}, x_{e0}, and x_{e1} are unknown components of the perturbed solution.

Substituting (12) and (15) and using (14) in Equations (19)–(22), taking into account (11), and expanding all equations in $\alpha\ll 1$ for the first two terms of each of the above expansions, one obtains equations

$$-\frac{\partial p_0}{\partial x_0}+\frac{\partial\tau_{zx0}}{\partial z_0}=0,\ \frac{\partial p_0}{\partial z_0}=0,$$
$$\frac{Re_0}{\alpha_0}\Big\{u_0\frac{\partial u_0}{\partial x_0}+w_0\frac{\partial u_0}{\partial z_0}\Big\}=-\frac{\partial p_{1*}}{\partial x_0}+\frac{x_{e1}}{x_{e0}-x_i}\frac{\partial p_0}{\partial x_0}+\frac{1}{\alpha_0}\frac{\partial\tau_{xx0}}{\partial x_0}+\frac{\partial\tau_{zx1*}}{\partial z_0}-\frac{h_1}{h_0}\frac{\partial\tau_{zx0}}{\partial z_0}, \tag{29}$$
$$-\frac{\partial p_{1*}}{\partial z_0}+\frac{h_1}{h_0}\frac{\partial p_0}{\partial z_0}+\frac{1}{\alpha_0}\frac{\partial\tau_{zz0}}{\partial z_0}=0,$$

as well as the solutions for the tensor components as follows

$$\tau_{xx} = \tau_{sxx} + \tau_{pxx},\ \tau_{zx} = \tau_{szx} + \tau_{pzx},\ \tau_{zz} = \tau_{szz} + \tau_{pzz},$$

$$\tau_{sxx0} = 0,\ \tau_{szz0} = 0, \tau_{sxx1*} = \frac{2\mu_s}{\alpha_0}\frac{\partial u_0}{\partial x_0},$$
$$\tau_{szx0} = \mu_s \frac{\partial u_0}{\partial z_0}, \tau_{szx1*} = \mu_s[\frac{\partial u_{1*}}{\partial z_0} - \frac{h_1}{h_0}\frac{\partial u_0}{\partial z_0}], \quad (30)$$

$$\tau_{szz1*} = \frac{2\mu_s}{\alpha_0}\frac{\partial w_0}{\partial z_0},$$
$$\tau_{pxx0} = 2\lambda\mu_p(\frac{\partial u_0}{\partial z_0})^2,\ \tau_{pzx0} = \mu_p\frac{\partial u_0}{\partial z_0},\ \tau_{pzz0} = 0. \quad (31)$$

$$\tau_{pxx1*} = 6\lambda^3\mu_p(\frac{\partial u_0}{\partial z_0})^4 + 4\lambda\mu_p\frac{\partial u_0}{\partial z_0}[\frac{\partial u_{1*}}{\partial z_0} - \frac{h_1}{h_0}\frac{\partial u_0}{\partial z_0}]$$
$$+\lambda\mu_p(\frac{\partial u_0}{\partial z_0})^2[4\lambda^2(\frac{\partial u_0}{\partial z_0})^2 + 1] + \frac{2\mu_p}{\alpha_0}\{\frac{\partial u_0}{\partial x_0} - 3\lambda^2\frac{\partial u_0}{\partial z_0}[u_0\frac{\partial^2 u_0}{\partial x_0\partial z_0} + w_0\frac{\partial^2 u_0}{\partial z_0^2}]\},$$
$$\tau_{pzx1*} = 3\lambda^2\mu_p(\frac{\partial u_0}{\partial z_0})^3 + \mu_p[\frac{\partial u_{1*}}{\partial z_0} - \frac{h_1}{h_0}\frac{\partial u_0}{\partial z_0}] - \frac{\lambda\mu_p}{\alpha_0}\{2\frac{\partial u_0}{\partial z_0}\frac{\partial u_0}{\partial x_0} + u_0\frac{\partial^2 u_0}{\partial x_0\partial z_0} + w_0\frac{\partial^2 u_0}{\partial z_0^2}\}, \quad (32)$$
$$\tau_{pzz1*} = \lambda\mu_p(\frac{\partial u_0}{\partial z_0})^2 - \frac{2\mu_p}{\alpha_0}\frac{\partial u_0}{\partial\partial x_0}.$$

Here, the approximation of the continuity Equation (19) for the main terms of u and w is used in the form

$$\frac{\partial u_0}{\partial x_0} + \frac{\partial w_0}{\partial z_0} = 0. \quad (33)$$

Equations (29)–(31) show that functions p_0 and p_{1*} are independent of z_0, i.e., $p_0 = p_0(x_0)$ and $p_{1*} = p_{1*}(x_0)$. That supports the form (26) in which $p(x)$ will be searched as well as the choice of the boundary and additional conditions imposed on pressure p in the form (9) and (10).

4. Derivation of the Lubricant Velocity Components u and w

Before solving the continuity Equation (19) and equations of motion (29), one needs to determine the boundary conditions imposed on u_0, u_1, w_0, and w_1. To do that, let us take conditions (7) and (8) and substitute into them the expansions from (26). Expanding the obtained boundary conditions in $\alpha \ll 1$, one obtains the boundary conditions for the main and first-order approximations of u and w in the form

$$u_0(x_0, -\frac{h_0(x_0)}{2}) = \overline{u_1},\ u_0(x_0, \frac{h_0(x_0)}{2}) = \overline{u_2},$$
$$u_1(x_0, -\frac{h_0(x_0)}{2}) = -x_{e1}\frac{x_0 - x_i}{x_{e0} - x_i}\frac{\partial u_0(x_0, -h_0(x_0)/2)}{\partial x_0} + \frac{h_1(x_0)}{2}\frac{\partial u_0(x_0, -h_0(x_0)/2)}{\partial z_0}, \quad (34)$$
$$u_1(x_0, \frac{h_0(x_0)}{2}) = -x_{e1}\frac{x_0 - x_i}{x_{e0} - x_i}\frac{\partial u_0(x_0, h_0(x_0)/2)}{\partial x_0} - \frac{h_1(x_0)}{2}\frac{\partial u_0(x_0, h_0(x_0)/2)}{\partial z_0},$$

$$w_0(x_0, -\frac{h_0(x_0)}{2}) = -\frac{\overline{u_1}}{2}\frac{dh_0(x_0)}{dx_0},$$
$$w_0(x_0, \frac{h_0(x_0)}{2}) = \frac{\overline{u_2}}{2}\frac{dh_0(x_0)}{dx_0},$$
$$w_1(x_0, -\frac{h_0(x_0)}{2}) = -x_{e1}\frac{x_0 - x_i}{x_{e0} - x_i}\frac{\partial w_0(x_0, -h_0(x_0)/2)}{\partial x_0}$$
$$+\frac{h_1(x_0)}{2}\frac{\partial w_0(x_0, -h_0(x_0)/2)}{\partial z_0} + \frac{\overline{u_1}}{2}\{\frac{x_{e1}}{x_{e0} - x_i}\frac{dh_0(x_0)}{dx_0} - \frac{dh_1(x_0)}{dx_0}\}, \quad (35)$$
$$w_1(x_0, \frac{h_0(x_0)}{2}) = -x_{e1}\frac{x_0 - x_i}{x_{e0} - x_i}\frac{\partial w_0(x_0, h_0(x_0)/2)}{\partial x_0}$$
$$-\frac{h_1(x_0)}{2}\frac{\partial w_0(x_0, h_0(x_0)/2)}{\partial z_0} - \frac{\overline{u_2}}{2}\{\frac{x_{e1}}{x_{e0} - x_i}\frac{dh_0(x_0)}{dx_0} - \frac{dh_1(x_0)}{dx_0}\}.$$

Now, one can turn to solving Equations (19) and (29). Solving the first equation in (29) with boundary conditions from (34) and taking into account that p_0 is independent of z_0, one finds that

$$u_0(x_0, z_0) = \overline{u_1} + (\frac{z_0}{h_0} + \frac{1}{2})(\overline{u_2} - \overline{u_1}) + (z_0^2 - \frac{h_0^2}{4})\frac{1}{2\mu}\frac{dp_0}{dx_0}. \tag{36}$$

An asymptotic expansion of the continuity equation in α produces Equation (33) and a similar equation $\frac{\partial u_{1*}}{\partial x_0} - \frac{x_{e1}}{x_{e0}-x_i}\frac{\partial u_0}{\partial x_0} + \frac{\partial w_{1*}}{\partial z_0} - \frac{h_1}{h_0}\frac{\partial w_0}{\partial z_0} = 0$ for u_{1*} and w_{1*}. Let us integrate Equation (33) with respect to z_0 from $-h_0/2$ to z_0 which leads to

$$w_0(x_0, z_0) = -\frac{\overline{u_1}}{2}\frac{dh_0}{dx_0} - \int\limits_{-h_0/2}^{z_0} \frac{\partial u_0}{\partial x_0} dz_0. \tag{37}$$

A similar analysis for w_{1*} produces the expression

$$w_{1*}(x_0, z_0) = -\frac{\overline{u_1}}{2}(\frac{dh_1}{dx_0} - \frac{x_{e1}}{x_{e0}-x_i}\frac{dh_0}{dx_0}) - \int\limits_{-h_0/2}^{z_0} \{\frac{\partial u_{1*}}{\partial x_0} + [\frac{h_1}{h_0} - \frac{x_{e1}}{x_{e0}-x_i}]\frac{\partial u_0}{\partial x_0}\} dz_0. \tag{38}$$

However, the solution to the latter equation does not represent a significant interest as w_{1*} is not involved in any relationships for the first-order approximations of functions p, u, etc., and it will not be considered in detail.

Using (36) in (37), one obtains

$$\begin{aligned} w_0(x_0, z_0) = &-\frac{\overline{u_1}}{2}\frac{dh_0}{dx_0} + (z_0^2 - \frac{h_0^2}{4})\frac{\overline{u_2}-\overline{u_1}}{2h_0^2}\frac{dh_0}{dx_0} + (z_0^3 + \frac{h_0^3}{8})\frac{1}{3\mu h_0}\frac{dh_0}{dx_0}\frac{dp_0}{dx_0} \\ &-(z_0^3 - \frac{3h_0^2 z_0}{4} - \frac{h_0^3}{4})\frac{1}{6\mu h_0^2}\frac{d}{dx_0}(h_0^2\frac{dp_0}{dx_0}). \end{aligned} \tag{39}$$

To show that this expression for w_0 satisfies the second boundary condition from (35) at $z_0 = h_0/2$, it is sufficient to use the Reynolds equation of order zero (see below Equation (44)).

Now, let us determine u_1. Keeping in mind that p_0 and p_{1*} are independent of z_0 and using the third equation in (29) and the third and fourth boundary conditions in (34) as well as the representation of u_{1*} from (26), one obtains

$$\begin{aligned} u_1(x_0, z_0) = &\, x_{e1}\frac{x_0-x_i}{x_{e0}-x_i}\frac{dh_0}{dx_0}(z_0\frac{\overline{u_2}-\overline{u_1}}{h_0^2} + \frac{h_0}{4\mu}\frac{dp_0}{dx_0}) - \frac{h_0 h_1}{4\mu}\frac{dp_0}{dx_0} + (z_0^2 - \frac{h_0^2}{4})\frac{1}{2\mu}\frac{dp_1}{dx_0} \\ &-\frac{3\lambda^2\mu_p}{\mu^2}\frac{dp_0}{dx_0}\{\frac{3}{2}(z_0^2 - \frac{h_0^2}{4})(\frac{\overline{u_2}-\overline{u_1}}{h_0})^2 + (z_0^3 + \frac{h_0^3}{8})\frac{\overline{u_2}-\overline{u_1}}{h_0}\frac{1}{\mu}\frac{dp_0}{dx_0} - (z_0 + \frac{h_0}{2})\frac{\overline{u_2}-\overline{u_1}}{h_0}\frac{h_0^2}{4\mu}\frac{dp_0}{dx_0} \\ &+(z_0^4 - \frac{h_0^4}{16})\frac{1}{4\mu^2}(\frac{dp_0}{dx_0})^2\} - z_0(\overline{u_2} - \overline{u_1})\frac{h_1}{h_0^2} + \frac{\lambda\mu_p}{2\alpha_0\mu}(z_0^2 - \frac{h_0^2}{4})\{(\overline{u_2} - \overline{u_1})^2\frac{1}{h_0^3}\frac{dh_0}{dx_0} \\ &+\frac{\overline{u_1}+\overline{u_2}}{2}\frac{1}{\mu}\frac{d^2p_0}{dx_0^2} - \frac{h_0^2}{4\mu^2}\frac{dp_0}{dx_0}\frac{d^2p_0}{dx_0^2} - \frac{h_0}{4\mu^2}\frac{dh_0}{dx_0}(\frac{dp_0}{dx_0})^2\} + \frac{Re_0}{\alpha_0\mu}\{-\frac{\overline{u_2}-\overline{u_1}}{4h_0}[(z_0^4 - \frac{h_0^4}{16})\frac{\overline{u_2}-\overline{u_1}}{6h_0^2} \\ &+(z_0^3 + \frac{h_0^3}{8})\frac{\overline{u_1}+\overline{u_2}}{3h_0} + (z_0^2 - \frac{h_0^2}{4})\frac{3\overline{u_1}+\overline{u_2}}{4} - (z_0 + \frac{h_0}{2})h_0\frac{\overline{u_1}+\overline{u_2}}{12})]\frac{dh_0}{dx_0} + [(z_0^5 + \frac{h_0^5}{32})\frac{\overline{u_2}-\overline{u_1}}{15h_0} \\ &+(z_0^4 - \frac{h_0^4}{16})\frac{\overline{u_1}+\overline{u_2}}{12} - (z_0^2 - \frac{h_0^2}{4})h_0^2\frac{5\overline{u_1}+\overline{u_2}}{24} - (z_0 + \frac{h_0}{2})h_0^3\frac{\overline{u_2}-\overline{u_1}}{240}]\frac{1}{4\mu}\frac{d^2p_0}{dx_0^2} \\ &+[-z_0^3 - \frac{3z_0^2 h_0}{2} + \frac{z_0 h_0^2}{4} + \frac{3h_0^3}{8}]\frac{\overline{u_1}}{12}\frac{1}{\mu}\frac{dh_0}{dx_0}\frac{dp_0}{dx_0} + [\frac{z_0^6}{45} + \frac{z_0^3 h_0^3}{18} + \frac{z_0^2 h_0^4}{16} - \frac{z_0 h_0^5}{72} - \frac{23h_0^6}{1440}]\frac{1}{8\mu^2}\frac{dp_0}{dx_0}\frac{d^2p_0}{dx_0^2} \\ &+[(z_0 + \frac{h_0}{2})^4 - h_0^3(z_0 + \frac{h_0}{2})]\frac{h_0}{96\mu^2}\frac{dh_0}{dx_0}(\frac{dp_0}{dx_0})^2\}. \end{aligned} \tag{40}$$

5. Reynolds Equations and Their Analysis

Based on the above analysis, let us derive the zero- and first-order Reynolds equations. Integrating the continuity Equation (2) with respect to z from $-h(x)/2$ to $h(x)/2$, using the boundary conditions (7) and (8) as well as changing the order of integration and differentiation leads to the equation

$$\frac{d}{dx}\int_{-h/2}^{h/2} u(x,z)dz = 0, \tag{41}$$

which will lead to the necessary Reynolds equations. Using (14) and the asymptotic representations of u from (26) and h from (27) and (28) and expanding (41) in $\alpha \ll 1$, one derives the zero-order Reynolds equation

$$\frac{d}{dx_0}\int_{-h_0/2}^{h_0/2} u_0(x_0,z_0)dz_0 = 0, \tag{42}$$

and the first-order Reynolds equation

$$\frac{d}{dx_0}\{\int_{-h_0/2}^{h_0/2} u_{1*}(x_0,z_0)dz_0 + \frac{h_1}{h_0}\int_{-h_0/2}^{h_0/2} u_0(x_0,z_0)dz_0\} = 0. \tag{43}$$

Substituting the expression for u_0 from (36) into (42) leads to the zero-order Reynolds equation which is just the traditional Reynolds equation for a Newtonian fluid

$$\frac{d}{dx_0}\{\frac{h_0^3}{12\mu}\frac{dp_0}{dx_0} - \frac{\overline{u_1}+\overline{u_2}}{2}h_0\} = 0. \tag{44}$$

Similarly, substituting u_0, u_1, and u_{1*} from (36), (40), and (26) into (42), one derives the first-order Reynolds equation

$$\begin{aligned}
&\frac{d}{dx_0}\{\frac{h_0^3}{12\mu}\frac{dp_1}{dx_0} - \frac{\overline{u_1}+\overline{u_2}}{2}h_1 + x_{e1}\frac{x_0-x_i}{x_{e0}-x_i}\frac{h_0^3}{12\mu}\frac{d^2p_0}{dx_0^2} + \frac{h_1h_0^2}{4\mu}\frac{dp_0}{dx_0} - \frac{3\lambda^2\mu_p}{4\mu^2}h_0\frac{dp_0}{dx_0}[(\overline{u_2}-\overline{u_1})^2 \\
&+\frac{h_0^4}{20\mu^2}(\frac{dp_0}{dx_0})^2] + \frac{\lambda\mu_p h_0^3}{12\alpha_0\mu}[\frac{(\overline{u_2}-\overline{u_1})^2}{h_0^3}\frac{dh_0}{dx_0} + \frac{\overline{u_1}+\overline{u_2}}{2}\frac{1}{\mu}\frac{d^2p_0}{dx_0^2} - \frac{h_0^2}{4\mu^2}\frac{dp_0}{dx_0}\frac{d^2p_0}{dx_0^2} - \frac{h_0}{4\mu^2}\frac{dh_0}{dx_0}(\frac{dp_0}{dx_0})^2] \\
&-\frac{Re_0}{16\alpha_0\mu}[\frac{(\overline{u_2}-\overline{u_1})(3\overline{u_2}+7\overline{u_1})}{15}h_0^2\frac{dh_0}{dx_0} + \frac{\overline{u_2}+11\overline{u_1}}{90}\frac{h_0^5}{\mu}\frac{d^2p_0}{dx_0^2} \\
&+\frac{\overline{u_1}}{3}\frac{h_0^4}{\mu}\frac{dh_0}{dx_0}\frac{dp_0}{dx_0} - \frac{3h_0^7}{140\mu^2}\frac{dp_0}{dx_0}\frac{d^2p_0}{dx_0^2} - \frac{h_0^6}{20\mu^2}\frac{dh_0}{dx_0}(\frac{dp_0}{dx_0})^2]\} = 0,
\end{aligned} \tag{45}$$

which takes into account the presence of polymeric additive in the lubricant.

Now, let us determine the boundary and additional conditions for p_0 and p_1. Substituting the representation for $p(x)$ from (26) in (9) and expanding them in $\alpha \ll 1$ leads to the boundary conditions on p_0 and p_1 in the form

$$p_0(x_i) = p_0(x_{e0}) = 0,\ \frac{dp_0(x_{e0})}{dx_0} = 0, \tag{46}$$

$$p_1(x_i) = p_1(x_{e0}) = 0,\ \frac{dp_1(x_{e0})}{dx_0} = -x_{e1}\frac{d^2p_0(x_{e0})}{dx_0^2}. \tag{47}$$

Let us introduce dimensionless variables (16) in (10) which will result in the equation

$$\int_{x_i}^{x_e}\{p(x) - \epsilon\tau_{zz}(x,\frac{h(x)}{2})\}dx = P. \tag{48}$$

Taking into account that $\alpha \ll 1$ and that $\tau_{zz0} = 0$ (see (30) and (31)), and taking into account (14), (24) and (25), and expanding this integral condition in $\alpha \ll 1$ and integrating by parts produces equations

$$\int\limits_{x_i}^{x_{e0}} p_0(x_0)dx_0 = P, \tag{49}$$

$$\int\limits_{x_i}^{x_{e0}} p_1(x_0)dx_0 = 0. \tag{50}$$

Now, one can finally formulate the specific boundary value problems for the zero-order Reynolds equation, i.e., for the given values of R_e, $\overline{u_1}$, $\overline{u_2}$, P, and x_i, one needs to solve Equations (44), (46) and (49), and the first equation in (28) for $p_0(x_0)$, $h_0(x_0)$, h_{e0}, and x_{e0} while for the first-order Reynolds equation for the given values of R_e, $\overline{u_1}$, $\overline{u_2}$, P, x_i and main solution $p_0(x_0)$, $h_0(x_0)$, h_{e0}, and x_{e0}, one needs to solve Equations (45), (47) and (50), and the second equation in (28) for $p_1(x_0)$, $h_1(x_0)$, h_{e1}, and x_{e1}. Let us use the dimensionless variables (16) more convenient for the analysis of hydrodynamic lubrication problems [29] by assuming that

$$\{a, c_0, c_1\} = \{x_i, x_{e0}, x_{e1}\},\ L_x = \frac{2R_e}{\theta},\ L_z = h_{0e},$$
$$U_x = \frac{\overline{u_1}+\overline{u_2}}{2},\ U_z = \frac{\overline{u_1}+\overline{u_2}}{2}\frac{h_{0e}\theta}{2R_e},\ \overline{u_1'} = 1 - \frac{s_0}{2},\ \overline{u_2'} = 1 + \frac{s_0}{2} \tag{51}$$

and

$$\{F'_{fr-}, F'_{fr+}\} = \frac{1}{P}\{F_{fr-}, F_{fr+}\},\ E' = \frac{1}{P(\overline{u_1}+\overline{u_2})}E,$$
$$Q' = \frac{2}{(\overline{u_1}+\overline{u_2})h_{0e}}Q,\ N_1' = \frac{2h_{0e}}{\mu_*(\overline{u_1}+\overline{u_2})}N_1,\ \theta^2 = \frac{P}{3\pi\mu_*(\overline{u_1}+\overline{u_2})},\ \gamma_0 = \theta^2\frac{h_{e0}}{2R_e}, \tag{52}$$

Here, θ is a dimensionless constant dependent on some of the problem input data, a and c are the dimensionless coordinates of the contact inlet and exit points, respectively, $c = c_0 + \alpha c_1 + \ldots$, s_0 is the slide-to-roll ratio, γ_0 is the main term of the dimensionless lubrication film thickness at the exit from the contact, Q is the lubricant volume flux through the gap between the contact solids, N_1 is the additional pressure which in the dimensional form is given by $N_1 = \tau_{xx} - \tau_{zz}$, F_{fr-} and F_{fr+} which are the friction forces acting on the surfaces of the two cylinders, respectively, which in the dimensional form are equal to $F_{fr\pm} = \int\limits_{x_i}^{x_e} \tau_{xz}(x, \pm h(x)/2)dx$, while E is the loss of energy in the contact, which in the dimensional form is $E = \int\limits_{x_i}^{x_e} dx \int\limits_{-h/2}^{h/2} \tau_{xz}(x,z)\frac{\partial u(x,z)}{\partial z}dz$. In the dimensionless form, these formulas have the following form (primes are omitted)

$$F_{fr\pm} = \frac{1}{6\pi\gamma_0\theta}\{\int\limits_a^{c_0} \tau_{xz0}(x_0, \pm\frac{h_0(x_0)}{2})dx_0$$
$$+\alpha\int\limits_a^{c_0}[\tau_{xz1*}(x_0, \pm\frac{h_0(x_0)}{2}) + \frac{c_1}{c_0-a}\tau_{xz0}(x_0, \pm\frac{h_0(x_0)}{2})]dx_0\} + \ldots, \tag{53}$$

$$E = \frac{1}{12\pi\gamma_0\theta}\{\int\limits_a^{c_0} dx_0 \int\limits_{-h_0(x_0)/2}^{h_0(x_0)/2} \tau_{xz0}(x_0, z_0)\frac{\partial u_0(x_0,z_0)}{\partial z_0}dz_0$$
$$+\alpha\int\limits_a^{c_0} dx_0 \int\limits_{-h_0(x_0)/2}^{h_0(x_0)/2} [\tau_{xz0}(x_0, z_0)\frac{\partial u_{1*}(x_0,z_0)}{\partial z_0} \tag{54}$$
$$+\tau_{xz1*}(x_0, z_0)\frac{\partial u_0(x_0,z_0)}{\partial z_0} + \frac{c_1}{c_0-a}\tau_{xz0}(x_0, z_0)\frac{\partial u_0(x_0,z_0)}{\partial z_0}]dz_0\} + \ldots,$$

$$Q(x_0) = h_0(x_0) \int\limits_{-1/2}^{1/2} u_0(x_0, z)dz + \alpha\{h_1(x_0) \int\limits_{-1/2}^{1/2} u_0(x_0, z)dz \\ + h_0(x_0) \int\limits_{-1/2}^{1/2} u_1(x_0, z)dz\} + \ldots, \tag{55}$$

$$N_1 = \tau_{xx}(x, z) - \tau_{zz}(x, z) = \tau_{xx0}(x_0, z_0) + \alpha[\tau_{xx1*}(x_0, z_0) - \tau_{zz1*}(x_0, z_0)] + \ldots, \tag{56}$$

$$h_e = 1 + \alpha h_{e1} + \ldots \tag{57}$$

Obviously, $Q(x_0) = const$, i.e., it is independent of x_0.

In these dimensionless variables, the small parameter ϵ involved in (23) is determined as follows $\epsilon = \frac{\gamma_0}{\theta}$ (see (51) and (52)).

By introducing the dimensionless variables from (52), one obtains the following dimensionless problems for the Reynolds equation of the zero-order

$$\frac{h_0^3}{12\mu}\frac{dp_0}{dx_0} = h_0 - 1, \; p_0(a) = p_0(c_0) = 0, \\ \gamma_0(h_0 - 1) = x_0^2 - c_0^2, \int\limits_a^{c_0} p_0(x_0)dx_0 = 6\pi\gamma_0^2. \tag{58}$$

and the following dimensionless problem for the Reynolds equation of the first-order

$$\frac{d}{dx_0}\{\frac{h_0^3}{12\mu}\frac{dp_1}{dx_0} - h_1 + c_1\frac{x_0-a}{c_0-a}\frac{h_0^3}{12\mu}\frac{d^2p_0}{dx_0^2} + \frac{h_1h_0^2}{4\mu}\frac{dp_0}{dx_0} - \frac{3\lambda^2\mu_p}{4\mu^2}h_0\frac{dp_0}{dx_0}[s_0^2 + \frac{h_0^4}{20\mu^2}(\frac{dp_0}{dx_0})^2] \\ + \frac{\lambda\mu_p h_0^3}{12\alpha_0\mu}[\frac{s_0^2}{h_0^3}\frac{dh_0}{dx_0} + \frac{1}{\mu}\frac{d^2p_0}{dx_0^2} - \frac{h_0^2}{4\mu^2}\frac{dp_0}{dx_0}\frac{d^2p_0}{dx_0^2} - \frac{h_0}{4\mu^2}\frac{dh_0}{dx_0}(\frac{dp_0}{dx_0})^2] - \frac{Re_0}{16\alpha_0\mu}[\frac{2s_0(5-s_0)}{15}h_0^2\frac{dh_0}{dx_0} \\ + \frac{12-5s_0}{90}\frac{h_0^5}{\mu}\frac{d^2p_0}{dx_0^2} + \frac{2-s_0}{6}\frac{h_0^4}{\mu}\frac{dh_0}{dx_0}\frac{dp_0}{dx_0} - \frac{3h_0^7}{140\mu^2}\frac{dp_0}{dx_0}\frac{d^2p_0}{dx_0^2} - \frac{h_0^6}{20\mu^2}\frac{dh_0}{dx_0}(\frac{dp_0}{dx_0})^2]\} = 0, \\ h_1(x_0) = h_{e1} + \frac{2c_1}{c_0-a}[h_0 - 1 + \frac{a(c_0-x_0)}{\gamma_0}], \\ p_1(a) = p_1(c_0) = 0, \; \frac{dp_1(c_0)}{dx_0} = -c_1\frac{d^2p_0(c_0)}{dx_0^2}. \tag{59}$$

Equation (58) describe the lubrication process by a Newtonian fluid [13,29] and their solution has the form

$$p_0(x_0) = 12\mu\gamma_0^2\{K_2(x_0) - \gamma_0K_3(x_0)\}, \tag{60}$$

where the unknown constants γ_0 and c_0 satisfy the following system of algebraic equations [13,29]

$$K_2(c_0) - \gamma_0K_3(c_0) = 0, \mu\{\overline{K_2(c_0)} - \gamma_0\overline{K_3(c_0)}\} = \frac{\pi}{2}. \tag{61}$$

Double integration of the Reynolds equation from (59) with its boundary and additional condition leads to the solution

$$p_1(x_0) = 12\mu\{3\gamma_0^3[\gamma_0K_4(x_0) - K_3(x_0)]h_{e1} - 2c_0c_1\gamma_0^2(3\gamma_0K_4(x_0) - 2K_3(x_0)) \\ + \frac{9\lambda^2\mu_p}{5\mu}[(5s_0^2 + 36)\gamma_0^4K_4(x_0) - \gamma_0^5(5s_0^2 + 108)K_5(x_0) + 108\gamma_0^6K_6(x_0) - 36\gamma_0^7K_7(x_0)] \\ + \frac{\lambda\mu_p}{12\alpha_0\mu}[2c_0\gamma_0^2(s_0^2 + 12)K_3(x_0) - (s_0^2 + 12)\gamma_0^2L_3(x_0) + 72\gamma_0^3(L_4(x_0) - \gamma_0L_5(x_0))] \\ - \frac{Re_0}{16\mu\alpha_0}[\frac{4c_0(-s_0^2+12)}{15}\gamma_0^2K_3(x_0) + \frac{2(s_0^2+\frac{12}{7})}{15}L_1(x_0) + \frac{8}{35}\gamma_0L_2(x_0) - \frac{72\gamma_0^2}{35}L_3(x_0)]\}, \tag{62}$$

where the unknowns c_1 and h_{e1} are determined from the system of two linear algebraic equations

$$12\mu\{3\gamma_0^3[\gamma_0K_4(c_0)-K_3(c_0)]h_{e1}-2c_0c_1\gamma_0^2(3\gamma_0K_4(c_0)-2K_3(c_0))+\frac{9\lambda^2\mu_p}{5\mu}[(5s_0^2$$
$$+36)\gamma_0^4K_4(c_0)-\gamma_0^5(5s_0^2+108)K_5(c_0)+108\gamma_0^6K_6(c_0)-36\gamma_0^7K_7(c_0)]$$
$$+\frac{\lambda\mu_p}{12\alpha_0\mu}[2c_0\gamma_0^2(s_0^2+12)K_3(c_0)-(s_0^2+12)\gamma_0^2L_3(c_0)+72\gamma_0^3(L_4(c_0)-\gamma_0L_5(c_0))]$$
$$-\frac{Re_0}{16\mu\alpha_0}[\frac{4c_0(-s_0^2+12)}{15}\gamma_0^2K_3(c_0)+\frac{2(s_0^2+\frac{12}{7})}{15}L_1(c_0)+\frac{8}{35}\gamma_0L_2(c_0)-\frac{72\gamma_0^2}{35}L_3(c_0)]\}=0, \quad (63)$$

$$12\mu\{3\gamma_0^3[\gamma_0\overline{K_4(c_0)}-\overline{K_3(c_0)}]h_{e1}-2c_0c_1\gamma_0^2(3\gamma_0\overline{K_4(c_0)}-2\overline{K_3(c_0)})$$
$$+\frac{9\lambda^2\mu_p}{5\mu}[(5s_0^2+36)\gamma_0^4\overline{K_4(c_0)}-\gamma_0^5(5s_0^2+108)\overline{K_5(c_0)}+108\gamma_0^6\overline{K_6(c_0)}-36\gamma_0^7\overline{K_7(c_0)}]$$
$$+\frac{\lambda\mu_p}{12\alpha_0\mu}[2c_0\gamma_0^2(s_0^2+12)\overline{K_3(c_0)}-(s_0^2+12)\gamma_0^2\overline{L_3(c_0)}+72\gamma_0^3(\overline{L_4(c_0)}-\gamma_0\overline{L_5(c_0)})]$$
$$-\frac{Re_0}{16\mu\alpha_0}[\frac{4c_0(-s_0^2+12)}{15}\gamma_0^2\overline{K_3(c_0)}+\frac{2(s_0^2+\frac{12}{7})}{15}\overline{L_1(c_0)}+\frac{8}{35}\gamma_0\overline{L_2(c_0)}-\frac{72\gamma_0^2}{35}\overline{L_3(c_0)}]\}=0, \quad (64)$$

In (60)–(64), the following integrals—$K_n(x_0)$, $L_n(x_0)$, and $\overline{K_n(c_0)}$, $\overline{L_n(c_0)}$, $n \geq 1$—are used

$$K_n(x)=\int_a^x\frac{dx_0}{\gamma_0^nh_0^n(x_0)}=\frac{1}{2(n-1)(\gamma_0-c_0^2)}[\frac{x}{(\gamma_0-c_0^2+x^2)^{n-1}}-\frac{a}{(\gamma_0-c_0^2+a^2)^{n-1}}]$$
$$+\frac{2n-3}{2(n-1)(\gamma_0-c_0^2)}K_{n-1}(x),\ n>1,$$
$$\overline{K_n(c_0)}=\int_a^{c_0}K_n(x_0)dx_0=c_0K_n(c_0)-\frac{1}{2}L_n(c_0),$$
$$L_n(x)=\int_a^x\frac{2x_0dx_0}{\gamma_0^nh_0^n(x_0)}=\frac{1}{1-n}[\frac{1}{(\gamma_0-c+0^2+x^2)^{n-1}}-\frac{1}{(\gamma_0-c_0^2+a^2)^{n-1}}],\ n>1,$$
$$\overline{L_n(c_0)}=\int_a^{c_0}L_n(x_0)dx_0=\frac{1}{1-n}[K_{n-1}(c_0)-\frac{c_0-a}{(\gamma_0-c_0^2+a^2)^{n-1}}],\ n>1. \quad (65)$$

The solution of the nonlinear system (61) for c_0 and γ_0 can be found iteratively using Newton's method. To start the iteration process, an initial approximation of the solution should be chosen according to [13,29].

Therefore, the approach to solution of the problem is first to solve the system (61) for γ_0 and c_0 and then to solve the linear system (63) and (64) for h_{e1} and c_1. Having the values of γ_0, h_{e1}, c_0, and c_1 allows to analytically calculate the functions of pressure $p(x_0)$ and gap $h(x_0)$ as well as the exit lubrication film thickness $h_e = 1+\alpha h_{e1}+\ldots$ and the exit coordinate $c = c_0+\alpha c_1+\ldots$. In addition, from Formulas (53)–(56), one can find the friction forces $F_{fr\pm}$, the energy loss E in the contact, the lubricant volume flux Q, and the additional pressure N_1.

Using integrals K_n, $\overline{K_n}$, L_n, and $\overline{L_n}$, the problem solution can be calculated analytically. However, due to the fact that the formulas are sufficiently complex for manual calculations, it is advisable to use some quadrature formulas (for example, the trapezoid rule) for calculation of some of the integrals, for example, the integrals for $F_{fr\pm}$, E, etc.

6. Examples of Some Specific Lubrication Problem Solutions and Discussion

Now, let us consider some results which can be extracted from the obtained approximate solutions. It will always be assumed that $\mu = 1$. The basic set the following values: $a = -10$, $R_e = 0.01$ m, $P = 2 \times 10^4$ N/m, $\mu_* = 6 \times 10^{-3}$ N·s/m^2, $\rho = 800$ kg/ m^3, $u_1 = 0$, $u_2 = 10$ m/s, $\epsilon = \gamma_0/\theta \approx 0.0008$, $\mu_p = 0.2$, $\lambda = 1$, $\alpha_0 = 1$, and $s_0 = 2$ will be taken. Below, all figures are presented in the dimensionless variables.

Figure 2 represents the plots of contact pressure in the lubricated contact obtained for different values of α_0 and λ. For $\alpha_0 = 1$ and $\lambda = 2$, the pressure distribution is very close to the one for the corresponding Newtonian fluid (see [13,29]). It is clear from these graphs that the maximum of the contact pressure tends to increase with the growth in α_0 and λ, while the size of the contact region tends to decrease due to decrease in c (see Figure 3). The variations in the contact pressure distribution compared to the corresponding Newtonian one remain insignificant for small values of the mobility factor α_0 even for a relatively large lubricant relaxation time λ. However, for relatively large values of the mobility factor α_0, the former variations in pressure $p(x)$ and exit coordinate c become significant. Figure 4 shows the dependence of the exit coordinate c on the lubricant additive viscosity μ_p, which qualitatively (being linear) resembles the dependence of c on λ from Figure 3 which is nonlinear.

For $\alpha_0 = 0$, the Giesekus model becomes the convected Maxwell model and the value of the exit coordinate c tends slightly to increase with increase in both λ and μ_p.

The dependence of the exit film thickness h_e on the lubrication relaxation time λ and lubricant additive viscosity μ_p is represented in Figures 5 and 6, which show beneficial (increasing) behavior of h_e with increase in α_0, λ, and μ_p. The only difference is in whether it happens linearly or nonlinearly. A similar behavior of the minimum lubrication film thickness h_{min} is shown in Figures 7 and 8.

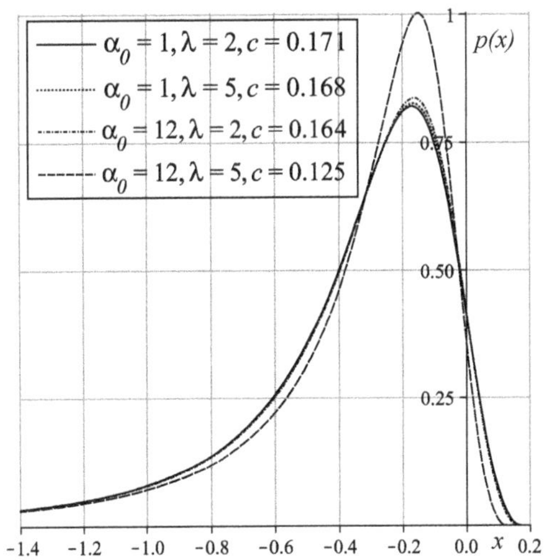

Figure 2. Plots of contact pressure distributions $p(x)$ for the basic set of input parameters and different values of α_0 and λ.

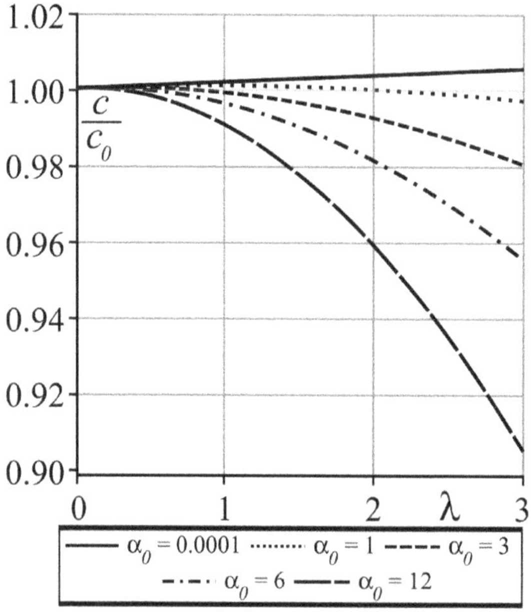

Figure 3. Plots of the dependence of the relative lubricated contact exit coordinate c/c_0 versus λ for different values of α_0.

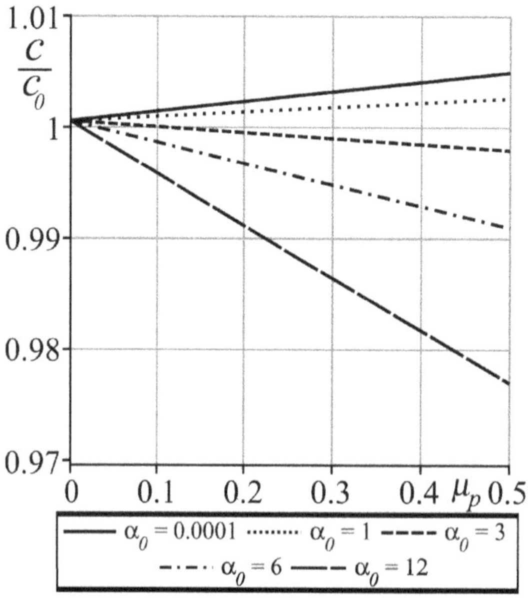

Figure 4. Plots of the dependence of the relative lubricated contact exit coordinate c/c_0 versus μ_p for different values of α_0.

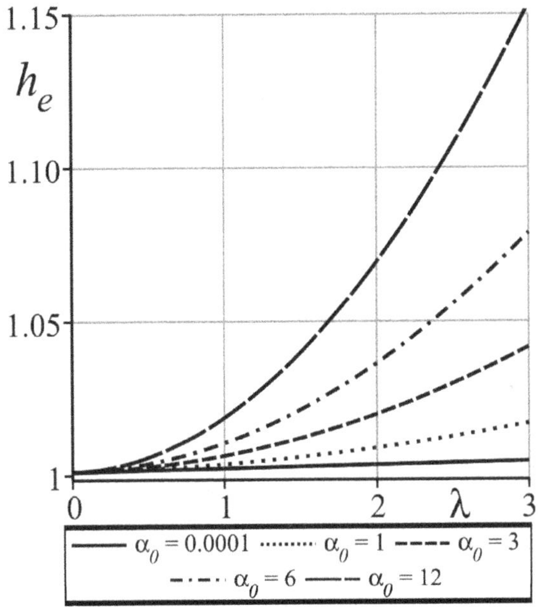

Figure 5. Plots of the dependence of the exit lubrication film thickness h_e versus λ for different values of α_0.

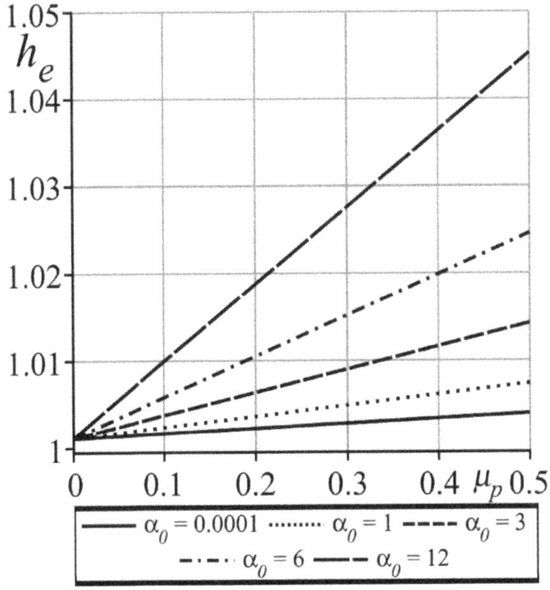

Figure 6. Plots of the dependence of the exit lubrication film thickness h_e versus μ_p for different values of α_0.

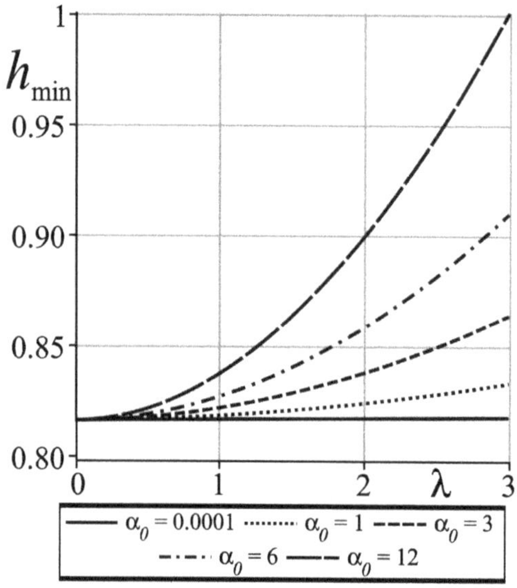

Figure 7. Plots of the dependence of the minimum lubrication film thickness h_{min} versus λ for different values of α_0.

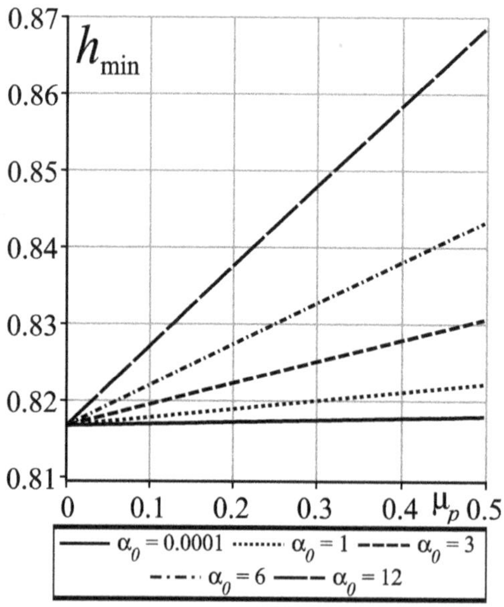

Figure 8. Plots of the dependence of the minimum lubrication film thickness h_{min} versus μ_p for different values of α_0.

As it is clear from Figures 9 and 10, depending on the value of the mobility factor α_0, the friction force F_+ may display an increasing or decreasing behavior with λ and μ_p. Specifically, for smaller values of α_0, the friction force F_+ decreases with λ and μ_p, while for larger values of α_0, the friction force F_+ monotonically increases with λ and μ_p. A similar behavior exhibits the energy loss in the contact E presented in Figure 11. The fact that the exit lubrication film thickness h_e always is greater than in the corresponding case of a Newtonian fluid and that one can pick the mobility factor α_0 value for which the friction force F_+ and energy loss E are just a bit higher than in the case of a Newtonian liquid provides an opportunity to optimize the lubricant design to increase lubrication film thickness and, at the same time, to practically not increase the frictional and energy losses. For example, for $\alpha_0 = 1$ and $\lambda = 3$, it is possible to achieve an increase in the minimum lubrication film thickness h_{min} by about 2% while the friction F_+ and energy E losses would increase by about only 1% compared to the lubricant with the corresponding Newtonian rheology.

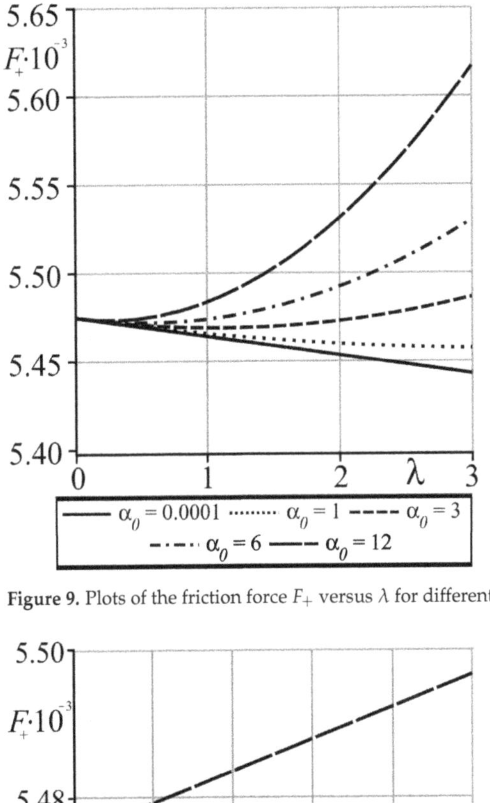

Figure 9. Plots of the friction force F_+ versus λ for different values of α_0.

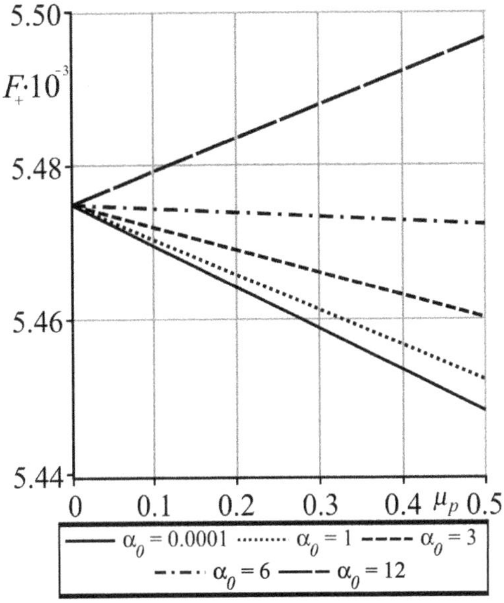

Figure 10. Plots of the friction force F_+ versus μ_p for different values of α_0.

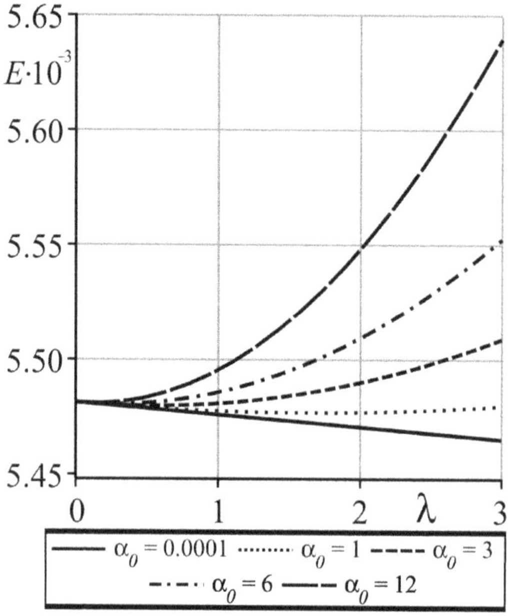

Figure 11. Plots of the contact energy loss E versus λ for different values of α_0.

The behavior of the lubricant flux Q as a function of the lubricant relaxation time λ and lubricant additive viscosity μ_p are depicted in Figures 12 and 13, and its behavior resembles the behavior of the exit film thickness h_e from Figures 5 and 6.

Some graphs of the additional pressure distribution N_1 versus x for $z = 0$ are presented in Figures 14 and 15. In a sense, the behavior of $N_1(x, 0)$ resembles the behavior of $p(x)$ from Figure 2.

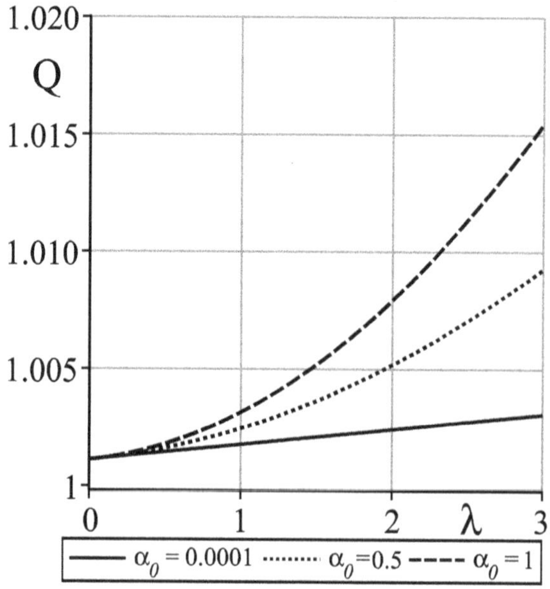

Figure 12. Plots of the lubricant volume flux Q versus λ for different values of α_0.

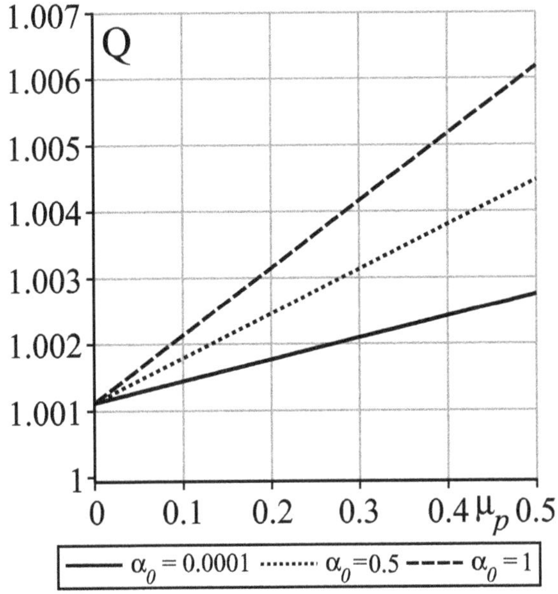

Figure 13. Plots of the lubricant volume flux Q versus μ_p for different values of α_0.

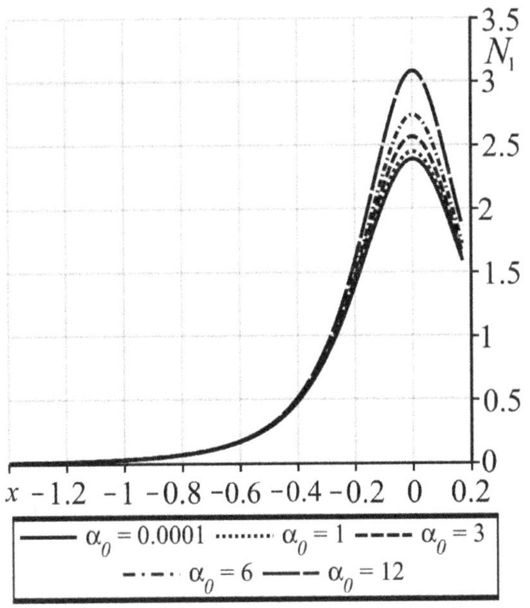

Figure 14. The plots N_1 is constructed at z = 0 and $\lambda = 1$ for different values of α_0.

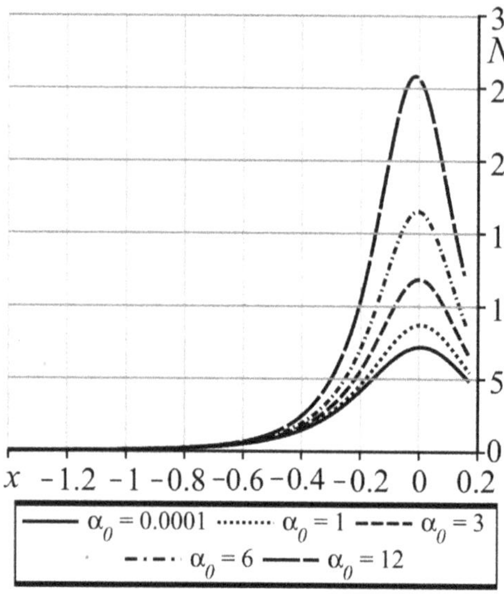

Figure 15. The plots N_1 is constructed at z = 0 and $\lambda = 3$ for different values of α_0.

7. One Problem Generalization

In the above analysis, a lubrication problem for two parallel cylinders of the same radius has been considered. Now, let us briefly consider the generalization of this problem on the case of two parallel cylinders with different radii R_1 and R_2. Then, in the same coordinate system as before, all equations of the problem formulation remain the same as before except for the boundary conditions for the lubricant velocity components, which now look as follows

$$u(x, -H_1(x)) = \overline{u_1},\ u(x, H_2(x)) = \overline{u_2}, \tag{66}$$

$$w(x, -H_1(x)) = -\overline{u_1}\frac{dH_1(x)}{dx},\ w(x, H_2(x)) = \overline{u_2}\frac{dH_2(x)}{dx}, \tag{67}$$

and the equilibrium condition

$$\int\limits_{x_i}^{x_e}[p(x) - \tau_{zz}(x, H_2(x))]dx = P, \tag{68}$$

where $z = -H_1(x)$ and $z = H_2(x)$ are the cylinder surfaces which are represented by equations

$$\begin{gathered} H_i(x) = \frac{h_e}{2} + \frac{x^2 - x_e^2}{2R_i}, i = 1,2,\ h(x) = H_1(x) + H_2(x), \\ H_i(x) = h_{i0}(x_0) + \alpha h_{i1}(x_0) + \ldots,\ i = 1,2,\ h_e = h_{e0} + \alpha h_{e1} + \ldots, \end{gathered} \tag{69}$$

$$\begin{gathered} h_0(x_0) = h_{10}(x_0) + h_{20}(x_0) = h_{e0} + \frac{x_0^2 - x_{e0}^2}{2R_e}, \\ h_1(x_0) = h_{11}(x_0) + h_{21}(x_0) = h_{e1} + \frac{x_{e1}}{R_e}\Big[\frac{x_0(x_0 - x_i)}{x_{e0} - x_i} - x_{e0}\Big], \end{gathered} \tag{70}$$

where R_e is the effective radius of the cylinders, $1/R_e = 1/R_1 + 1/R_2$.

The analysis of the system of these equations is conducted for $\alpha \ll 1$ in the manner similar to the one used above, but in (13), the variable z remained intact. The assumptions (11) and (23) regarding α and λ_1 remain the same. Then, using the same dimensionless variables as in (16) and asymptotic representations of solution components in which terms proportional to h_1 are omitted, one obtains Equations (29)–(32), in which terms proportional to h_1 are omitted. To derive the expressions for u_0, w_0, and u_1, one needs to use asymptotic boundary conditions (34) and (35) following from (66) and (67), in which $\frac{h_1(x_0)}{2}$ on the lower and upper surfaces should be replaced by $h_{11}(x_0)$ and $h_{21}(x_0)$, respectfully.

After that, the expressions for u_0, w_0, and u_1 have the form

$$u_0(x_0,z) = \frac{(\overline{u_2}h_{10}+\overline{u_1}h_{20})}{h_0} + \frac{(\overline{u_2}-\overline{u_1})}{h_0}z + (z-h_{20})(z+h_{10})\frac{1}{2\mu}\frac{dp_0}{dx_0}, \tag{71}$$

$$\begin{aligned} w_0(x_0,z) = & -u_1\frac{dh_{10}}{dx_0} + (\overline{u_1}-\overline{u_2})\frac{(z+h_{10})^2}{2h_0^2}\frac{dh_{20}}{dx_0} + (\overline{u_1}-\overline{u_2})\frac{(z+h_{10})(z-h_{10}-2h_{20})}{2h_0^2}\frac{dh_{10}}{dx_0} \\ & +((z+h_{10})(z-h_{10}-2h_{20})\frac{dh_{10}}{dx_0} - (z+h_{10})^2\frac{dh_{20}}{dx_0})\frac{1}{4\mu}\frac{dp_0}{dx_0} \\ & +(z+h_{10})^2(z-h_{10}-2h_{20})\frac{1}{6\mu}\frac{d^2p_0}{dx_0^2}, \end{aligned} \tag{72}$$

$$\begin{aligned} u_1(x_0,z) = & \ x_{e1}\frac{(x_0-x_i)}{(x_{e0}-x_i)}[(h_{20}-z)\frac{dh_{10}}{dx_0} + (h_{10}+z)\frac{dh_{20}}{dx_0}]\frac{1}{2\mu}\frac{dp_0}{dx_0} \\ & +[(z-h_{20})h_{11} - h_{21}(z+h_{10})]\frac{1}{2\mu}\frac{dp_0}{dx_0} + (\overline{u_2}-\overline{u_1})x_{e1}\frac{(x_0-x_i)}{(x_{e0}-x_i)}[(z-h_{20})\frac{dh_{10}}{dx_0} \\ & +(h_{10}+z)\frac{dh_{20}}{dx_0}]\frac{1}{h_0^2} + (z-h_{20})(z+h_{10})\frac{1}{2\mu}\frac{dp_1}{dx_0} - (\overline{u_2}-\overline{u_1})(zh_1+h_{12}h_{10}-h_{11}h_{20})\frac{1}{h_0^2} \\ & -\frac{3\lambda^2\mu_p}{\mu^2}\frac{dp_0}{dx_0}\{3(\overline{u_2}-\overline{u_1})^2(z-h_{21})(z+h_{11})\frac{1}{2h_0^2} + (\overline{u_2}-\overline{u_1})(z-h_{21})(z+h_{11})(z \\ & +\frac{h_{11}-h_{21}}{2})\frac{1}{h_0}\frac{1}{\mu}\frac{dp_0}{dx_0} + (z-h_{21})(z+h_{11})(z^2-h_{11}+h_{21}+\frac{h_{11}^2+h_{21}^2}{2})\frac{1}{4\mu^2}(\frac{dp_0}{dx_0})^2\} \\ & +\frac{\lambda\mu_p}{2\alpha_0\mu}(z-h_{20})(z+h_{10})\{(\overline{u_2}-\overline{u_1})^2\frac{1}{h_0^3}\frac{dh_0}{dx_0} + \frac{\overline{u_1}+\overline{u_2}}{2}\frac{1}{\mu}\frac{d^2p_0}{dx_0^2} - \frac{h_0^2}{4\mu^2}\frac{dp_0}{dx_0}\frac{d^2p_0}{dx_0^2} \\ & -\frac{h_0}{4\mu^2}\frac{dh_0}{dx_0}(\frac{dp_0}{dx_0})^2\} + \frac{Re_0}{\alpha_0\mu}(z-h_{20})(z+h_{10})\{-(\overline{u_2}-\overline{u_1})\frac{1}{24h_0^3}[(\overline{u_2}-\overline{u_1})[3h_{10}^2 \\ & +3(z+h_{20})h_{10}+z^2+h_{20}z+h_{20}^2] + 4u_1h_0(z+2h_{10}+h_{20})]\frac{dh_0}{dx_0} \\ & +[-(z+2h_{10}+h_{20})\frac{h_0\overline{u_1}}{3} + \frac{1}{4}(\overline{u_1}-\frac{\overline{u_2}}{3})[3h_{10}^2+3(z+h_{20})h_{10}+z^2+zh_{20}+h_{20}^2] \\ & -(\overline{u_1}-\overline{u_2})(2h_{10}^2+2(z+h_{20})h_{10}+z^2+h_{20}^2)(z+2h_{10}+h_{20})\frac{1}{15h_0}]\frac{1}{4\mu}\frac{d^2p_0}{dx_0^2} \\ & -\overline{u_1}(z+2h_{10}+h_{20})\frac{1}{12\mu}\frac{dh_0}{dx_0}\frac{dp_0}{dx_0} + [4z^4+4(5h_{10}+h_{20}-3h_0)z^3+(40h_{10}^2 \\ & +4(5h_{20}-12h_0)h_{10}+4h_{20}^2-12h_{20}h_0+15h_0^2)z^2 \\ & +(40h_{10}^3+8(5h_{20}-9h_0)h_{10}^2+(45h_0^2-48h_{20}h_0+20h_{20}^2)h_{10}+4h_{20}^3-12h_{20}^2h_0 \\ & +15h_0^2h_{20})z+20h_{10}^4+8(-6h_0+5h_{20})h_{10}^3+(45h_0^2-72h_{20}h_0+40h_{20}^2)h_{10}^2 \\ & +(45h_0^2h_{20}-48h_{20}^2h_0+20h_{20}^3)h_{10}+4h_{20}^4-12h_{20}^3h_0+15h_{20}^2h_0^2]\frac{1}{1440\mu^2}\frac{dp_0}{dx_0}\frac{d^2p_0}{dx_0^2} \\ & +[3h_{10}^2+3(z+h_{20})h_{10}+z^2+h_{20}z+h_{20}^2]\frac{h_0}{96\mu^2}\frac{dh_0}{dx_0}(\frac{dp_0}{dx_0})^2\}. \end{aligned} \tag{73}$$

The Reynolds equations of the first and second order are derived from the equation (for comparison, see (41))

$$\frac{d}{dx}\int\limits_{-H_1}^{H_2} u(x,z)dz = 0. \tag{74}$$

Using the obtained asymptotic solution representations, one derives the zero-order Reynolds equation

$$\frac{d}{dx_0}\int\limits_{-h_{10}}^{h_{20}} u_0(x_0,z)dz = 0, \tag{75}$$

and the first-order Reynolds equation

$$\frac{d}{dx_0}\{\int\limits_{-h_{10}}^{h_{20}} u_{1*}(x_0,z)dz + h_{11}\overline{u_1} + h_{21}\overline{u_2}\} = 0. \tag{76}$$

Substitution of the expressions for u_0 and u_1 from (71) and (73) into (75) and (76) leads to precisely the same Reynolds Equations (44) and (45) with boundaries (46) and (47) and additional conditions (49) and (50) derived for the case of cylinders of the same radius R; however, in this case, $1/R_e = 1/R_1 + 1/R_2$. The rest of the analysis is identical with the one conducted above for the case of cylinders of the same radius R.

Formulas for the friction forces $F_{fr\pm}$ and energy loss E obviously become transformed to the following ones

$$F_{fr-} = \int\limits_{x_i}^{x_e} \tau_{xz}(x, -H_1(x))dx,$$

$$F_{fr+} = \int\limits_{x_i}^{x_e} \tau_{xz}(x, H_2(x))dx, \tag{77}$$

$$E = \int\limits_{x_i}^{x_e} dx \int\limits_{-H_1}^{H_2} \tau_{xz}(x,z)\frac{\partial u(x,z)}{\partial z}dz.$$

8. Conclusions

This study is concerned with the new asymptotic solution of the steady hydrodynamic problem of lubrication of two rigid cylinders rolling and sliding over each other and then being separated from each other by a non-Newtonian fluid described by the Geisekus rheology. The problem is considered in the case when the lubricant mobility factor α, lubricant relaxation time λ_1, and characteristic dimensionless gap ϵ are of the same order of magnitude. Using the modification of the regular perturbation method, it was possible to simplify the original rheological equation of the Giesekus model and to obtain the first two Reynolds equations for determining the main and first terms of the asymptotic expansion of contact pressure with respect to small mobility factor α. The solutions of these equations are obtained analytically. Using these solutions, the coordinate of the exit point from the contact and the exit lubrication film thickness were determined. After that, it was numerically investigated the influence of the problem input parameters on contact pressure, lubrication film thickness, lubricant flux, friction force, contact energy loss, etc. For $\alpha = 0$, the Giesekus rheology turns into the convected Maxwell one. Graphs show that as the mobility factor α increases, the problem characteristics significantly deviate from the ones for the convected Maxwell case. Also, the solution demonstrates a significant dependence on the lubricant relaxation time. Therefore, the conclusion can be drawn that neglecting to take into account the lubricant mobility factor and lubricant relaxation time can lead to a significantly distorted picture of what is going on in a lubricated contact. The relatively simple solution obtained in the study would allow us to take into account the details of a complex rheology of lubricating oils.

Author Contributions: Conceptualization, I.I.K.; methodology, I.I.K. and S.S.V.; software, S.S.V.; validation, I.I.K. and S.S.V.; formal analysis, I.I.K. and S.S.V.; investigation, I.I.K. and S.S.V.; resources, I.I.K.; data curation, S.S.V.; writing—original draft preparation, I.I.K.; writing—review and editing,

I.I.K. and S.S.V.; visualization, S.S.V.; supervision, I.I.K. and S.S.V.; project administration, I.I.K.; funding acquisition, S.S.V. All authors have read and agreed to the published version of the manuscript.

Funding: The study was supported by the Russian Science Foundation grant № 19-19-00444.

Data Availability Statement: No new data were created or analyzed in this study. Data sharing is not applicable to this article.

Conflicts of Interest: The authors declare that the research was conducted in the absence of any commercial or financial relationships that could be construed as a potential conflict of interest.

References

1. Chapkov, A.D.; Bair, S.; Cann, P.; Lubrecht, A.A. Film Thickness in Point Contacts under Generalized Newtonian EHL Conditions: Numerical and Experimental Analysis. *Tribol. Int.* **2007**, *40*, 1474–1478. [CrossRef]
2. Yang, P.; Cui, J.; Jin, Z.M.; Dowson, D. Transient Elastohydrodynamic Analysis of Elliptical Contacts. Part 2: Thermal and Newtonian Lubricant Solution. *Proc. Inst. Mech. Eng. Part J J. Eng. Tribol.* **2005**, *219*, 187–200. [CrossRef]
3. Li, B.; Sun, J.; Zhu, S.; Fu, Y.; Zhao, X.; Wang, H.; Teng, Q.; Ren, Y.; Li, Y.; Zhu, G. Thermohydrodynamic Lubrication Analysis of Misaligned Journal Bearing Considering the Axial Movement of Journal. *Tribol. Int.* **2019**, *135*, 397–407. [CrossRef]
4. Wang, Z.; Hu, Y.; Wang, W.; Wang, H. Numerical Analysis of Point Contact EHL on Coated Substrates. In Proceedings of the ASME/STLE 2009 International Joint Tribology Conference, Memphis, TN, USA, 19–21 October 2009; ASMEDC: Memphis, TN, USA, 2009; pp. 471–473. [CrossRef]
5. Su, J.; Song, H.-X.; Ke, L.-L. Elastohydrodynamic Lubrication Line Contact Based on Surface Elasticity Theory. *J. Appl. Mech.* **2020**, *87*, 081004. [CrossRef]
6. Charles, P.; Elfassi, M.; Lubrecht, A.A. Double–Newtonian Rheology in a Model Piston-Ring Cylinder-Wall Contact. *Tribol. Int.* **2010**, *43*, 1902–1907. [CrossRef]
7. Katyal, P.; Kumar, P. On the Role of Second Newtonian Viscosity in EHL Point Contacts Using Double Newtonian Shear-Thinning Model. *Tribol. Int.* **2014**, *71*, 140–148. [CrossRef]
8. D'Agostino, V.; Petrone, V.; Senatore, A. Effects of the Lubricant Piezo-Viscous Properties on EHL Line and Point Contact Problems. *Tribol. Lett.* **2013**, *49*, 385–396. [CrossRef]
9. Kazama, T. A Comparative Newtonian and Thermal EHL Analysis Using Physical Lubricant Properties. In *Tribology Series*; Elsevier: Amsterdam, The Netherlands, 2002; Volume 40, pp. 435–446. [CrossRef]
10. Liu, Y.; Wang, Q.J.; Bair, S.; Vergne, P. A Quantitative Solution for the Full Shear-Thinning EHL Point Contact Problem Including Traction. In Proceedings of the ASME/STLE 2007 International Joint Tribology Conference, Parts A and B, San Diego, CA, USA, 22–24 October 2007; ASMEDC: San Diego, CA, USA, 2007; pp. 1031–1032. [CrossRef]
11. Cherizol, R.; Sain, M.; Tjong, J. Review of Non–Newtonian Mathematical Models for Rheological Characteristics of Viscoelastic Composites. *Green Sustain. Chem.* **2015**, *5*, 6–14. [CrossRef]
12. Bird, R.B.; Curtis, C.F.; Armstrong, R.C.; Hassager, O. Dynamics of Polymeric Liquids. In *Fluid Mechanics*, 2nd ed.; John Wiley & Sons: New York, NY, USA, 1987; Volume 1.
13. Kudish I.I.; Covitch M.J. *Introduction to Modeling and Analytical Methods in Tribology*; Chapman & Hall/CRC: Boca Raton, FL, USA, 2010.
14. Ree, T.; Eyring, H. Theory of Non–Newtonian Flow. *I. Solid Plast. System. J. Appl. Phys.* **1955**, *26*, 793–800. [CrossRef]
15. Cui, J.; Yang, P.; Jin, Z.M.; Dowson, D. Transient Elastohydrodynamic Analysis of Elliptical Contacts. Part 3: Non–Newtonian Lubricant Solution under Isothermal and Thermal Conditions. *Proc. Inst. Mech. Eng. Part J J. Eng. Tribol.* **2007**, *221*, 63–73. [CrossRef]
16. Su, J.; Song, H.; Ke, L.; Aizikovich, S.M. The Size-Dependent Elastohydrodynamic Lubrication Contact of a Coated Half-Plane with Non–Newtonian Fluid. *Appl. Math. Mech.* **2021**, *42*, 915–930. [CrossRef]
17. Tichy, J.A. Non–Newtonian Lubrication with the Convected Maxwell Model. *J. Tribol.* **1996**, *118*, 344–348. [CrossRef]
18. Huang, P.; Li, Z.; Meng, Y.; Wen, S. Study on Thin Film Lubrication with Second-Order Fluid. *J. Tribol.* **2002**, *124*, 547–552. [CrossRef]
19. Akyildiz, F.T.; Bellout, H. Viscoelastic Lubrication with Phan-Thein-Tanner Fluid (PTT). *J. Tribol.* **2004**, *126*, 288–291. [CrossRef]
20. Gwynllyw, D.R.; Phillips, T.N. The Influence of Oldroyd-B and PTT Lubricants on Moving Journal Bearing Systems. *J. Non–Newton. Fluid Mech.* **2008**, *150*, 196–210. [CrossRef]
21. Covitch, M.J.; Trickett, K.J. How Polymers Behave as Viscosity Index Improvers in Lubricating Oils. *Adv. Chem. Eng. Sci.* **2015**, *5*, 134–151. [CrossRef]
22. Sawyer, W.G.; Tichy, J.A. Non–Newtonian Lubrication with the Second-Order Fluid. *J. Tribol.* **1998**, *120*, 622–628. [CrossRef]
23. Taylor, R.; De Kraker, B. Shear Rates in Engines and Implications for Lubricant Design. *Proc. Inst. Mech. Eng. Part J J. Eng. Tribol.* **2017**, *231*, 1106–1116. [CrossRef]
24. Giesekus, H. A Simple Constitutive Equation for Polymer Fluids Based on the Concept of Deformation-Dependent Tensorial Mobility. *J. Non–Newton. Fluid Mech.* **1982**, *11*, 69–109. [CrossRef]

25. Giesekus, H. A Unified Approach to a Variety of Constitutive Models for Polymer Fluids Based on the Concept of Configuration-Dependent Molecular Mobility. *Rheol. Acta* **1982**, *21*, 366–375. [CrossRef]
26. Yoo, J.Y.; Choi, H.C. On the Steady Simple Shear Flows of the One-Mode Giesekus Fluid. *Rheol. Acta* **1989**, *28*, 13–24. [CrossRef]
27. Raisi, A.; Mirzazadeh, M.; Dehnavi, A.S.; Rashidi, F. An Approximate Solution for the Couette-Poiseuille Flow of the Giesekus Model between Parallel Plates. *Rheol. Acta* **2008**, *47*, 75–80. [CrossRef]
28. Abbaspur, A.; Norouzi, M.; Akbarzadeh, P.; Vaziri, S.A. Analysis of Nonlinear Viscoelastic Lubrication Using Giesekus Constitutive Equation. *Proc. Inst. Mech. Eng. Part J J. Eng. Tribol.* **2021**, *235*, 1124–1138. [CrossRef]
29. Kudish, I.I.; Volkov, S.S.; Aizikovich, S.M.; Ke, L. One simple case of lubricated line contact for double-layered elastic solids. *Probl. Strength Plast.* **2022**, *84*, 5–14. [CrossRef]
30. Kudish, I.I.; Pashkovski, E.; Volkov, S.S.; Vasiliev, A.S.; Aizikovich, S.M. Heavily Loaded Line EHL Contacts with Thin Adsorbed Soft Layers. *Math. Mech. Solids* **2020**, *25*, 1011–1037. [CrossRef]
31. Kudish, I.I.; Volkov, S.S.; Vasiliev, A.S. Lightly Loaded Hydrodynamic Thrust Bearing Lubricated by a Non–Newtonian Fluid. *Adv. Struct. Mater.* **2023**, *170*, 433–448. [CrossRef]
32. Kudish, I.; Pashkovski, E.; Patterson, R. Line Contact Lubricated by a Fluid Described by Non–Newtonian Giesekus Model. *IMA J. Appl. Math.* **2022**, *87*, 722–756. [CrossRef]
33. Van-Dyke, M. *Perturbation Methods in Fluid Mechanics;* Academic Press: New York, NY, USA; London, UK, 1964.
34. Kaplunov, J.D.; Nolde, E.V.; Shorr, B.F. A Perturbation Approach for Evaluating Natural Frequencies of Moderately Thick Elliptic Plates. *J. Sound Vib.* **2005**, *281*, 905–919. [CrossRef]

mathematics

MDPI

Article

Degenerated Boundary Layers and Long-Wave Low-Frequency Motion in High-Contrast Elastic Laminates

Lenser A. Aghalovyan [1,†], Lusine G. Ghulghazaryan [1,2,†], Julius Kaplunov [3,*,†] and Danila Prikazchikov [3,†]

1 Institute of Mechanics, National Academy of Sciences of the Republic of Armenia, Yerevan 0019, Armenia; lagal@sci.am (L.A.A.); lusina@mail.ru (L.G.G.)
2 Faculty of Mathematics, Physics and Informatics, Department of Mathematics and its Teaching Methods, Armenian State Pedagogical University after Khachatur Abovyan, Abovyan 0010, Armenia
3 School of Computer Science and Mathematics, Keele University, Keele, Staffordshire ST5 5BG, UK; d.prikazchikov@keele.ac.uk
* Correspondence: j.kaplunov@keele.ac.uk
† These authors contributed equally to this work.

Abstract: The effect of high contrast on the multiscale behaviour of elastic laminates is studied. Mathematical modelling in this area is of significant interest for a variety of modern applications, including but not limited to advanced sandwich structures and photovoltaic panels. As an example, the antiplane shear of a symmetric, three-layered plate is considered. The problem parameters expressing relative thickness, stiffness and density are assumed to be independent. The high contrast may generally support extra length and time scales corresponding to degenerated boundary layers and propagating long-wave low-frequency vibration modes. The main focus is on the relation between these two phenomena. The developed multiparametric approach demonstrates that those do not always appear simultaneously. The associated explicit estimates on contrast parameters are established. In addition, the recent asymptotic extension of the classical Saint-Venant's principle is adapted for calculating the contribution of the degenerate boundary layer or long-wave low-frequency propagation mode. The peculiarity of the limiting absorption principle in application to layered media is also addressed.

Keywords: laminate; high contrast; multiparametric; multiscale; asymptotic; Saint-Venant's principle; boundary layer; long-wave; low-frequency; radiation conditions

MSC: 74J05; 74B05; 74G50

1. Introduction

The mathematical modelling of layered elastic structures has been a subject of numerous publications; see, e.g., [1–4] and references therein. More recently, a growing interest in high-contrast laminates has been inspired by modern important applications, including laminated glass, photovoltaic panels, lightweight sandwich-type structures; see, e.g., [5–8]. In addition, we mention the related contributions [9–11].

Two of the most significant observations regarding high-contrast layered structures include a degeneration of the boundary layers, demonstrating slow decay into the interior, studied in [12–15], along with the emergence of a specific low-frequency shear vibration spectra, characterised by a small cut-off frequency with the value tending to zero as powers of small parameters standing for contrast (see [16]) treating three-layered symmetric papers for several mechanical and geometrical setups; see also [17]. The eigenvalue problem on the plate's transverse cross section also arises in the case of free longitudinal vibrations of a strongly inhomogeneous, three-component rod [18]. The presence of an extra low-frequency shear mode supports two-mode long-wave low-frequency asymptotic theories for asymmetric three-layered strips, subject to antiplane elastic deformation; see [19].

The cited paper also suggests a non-trivial asymptotic generalisation of Saint-Venant's principle for the considered high-contrast scenario, stating the consistent boundary conditions (see also [20]) for extension to the anti-plane shear of multi-layered structures. The extra approximate boundary condition in [19] ensures the equilibrium of stress couples applied along the contour of a soft layer. The number of extra asymptotic boundary conditions in [20] coincides with that of soft layers and may be interpreted in a similar manner.

The relation between the degenerated boundary layers and the extra low-frequency modes has not yet been studied systematically, with the exception of [19], addressing only one specific parametric setup. At the same time, the static analysis in [12–14] is mainly restricted to studying the roots of the related transcendental equation without a more general insight into the peculiarities of the stress and strain fields. The main objective of the present paper is elucidation of the link between these two phenomena. Enhanced understanding of this link is relevant for new technological advances over virtually all of the above-mentioned applications, including modern sandwich panels.

A basic example of the antiplane problem in dynamic elasticity for a three-layered high-contrast symmetric laminate with Neumann boundary conditions on the faces is considered. The results of the full multiparametric analysis are presented for the anti-symmetric low-frequency wave. In this case, the ratios of thicknesses, stiffnesses and densities of the layers are assumed to be independent of each other. The tackled example clearly illustrates a multiscale nature of the high contrast effect. Indeed, it brings an extra relatively large length scale (compared to thickness) corresponding to a degenerate boundary layer, as well an extra time scale (small compared to the time elastic waves take to propagate the distance of order thickness) associated with the lowest cut-off, in addition to laminate thickness. Obviously, the aforementioned length and time scales do not arise in non-contrast situations.

It is established that the extra low-frequency modes do not always accompany the degenerate boundary layers. It is shown that both uniform and non-uniform approximations of the associated dispersion curve are possible. The asymptotic bounds on the problem parameters, corresponding to various options are found. Numerical illustrations are presented.

Another emphasis of the paper is on the explicit solution of the forced problem for a semi-infinite symmetric three-layered strip with traction-free faces under a self-equilibrated edge stress. The decay condition in [19] generalising Saint-Venant's principle is adapted. Above the cut-off frequency, this transforms to a radiation condition aimed at excluding the long-wave propagating mode; see [21] for more detail. As a result, the amplitude of the mode of interest is found. In addition, the peculiarity of the implementation of the limiting absorption principle for layered structures is briefly addressed for the long propagating mode originating from the cut-off in question. Along with the analysis of the coexistence of slowly decaying boundary layers and low-frequency vibration modes, this also contributes to the novelty of the paper.

The paper is organised as follows. The problem is formulated in Section 2 with the dispersion relation and the associated eigensolutions derived in Section 3. Section 4 is concerned with multiparametric analysis of the dispersion relation. The main result of the paper addressing the general issue of the relation between the degenerated boundary layer and the low-frequency cut-off is also exposed in this section. The explicit solution to the forced time-harmonic problem for a semi-infinite three-layered strip is obtained in Section 5. Several concluding remarks are summarised in Section 6.

2. Problem Statement

Consider a semi-infinite, three-layered, symmetric, elastic laminate; see Figure 1. Let the horizontal axe pass through the middle of the inner layer. Then, the latter occupies the domain $\{0 \leq x_1 < \infty,\ -h_1 \leq x_2 \leq h_1,\ -\infty < x_3 < \infty\}$, whereas the outer skin layers are located at $\{0 \leq x_1 < \infty,\ h_1 \leq |x_2| \leq h_1 + h_2,\ -\infty < x_3 < \infty\}$. Below, we focus on antiplane shear motion, with the displacement field given by $\mathbf{u} = (0, 0, u_3(x_1, x_2, t))$, more specifically, on an antisymmetric problem, for which the out-of-plane displacement satisfies $u_3(x_1, -x_2, t) = -u_3(x_1, x_2, t)$.

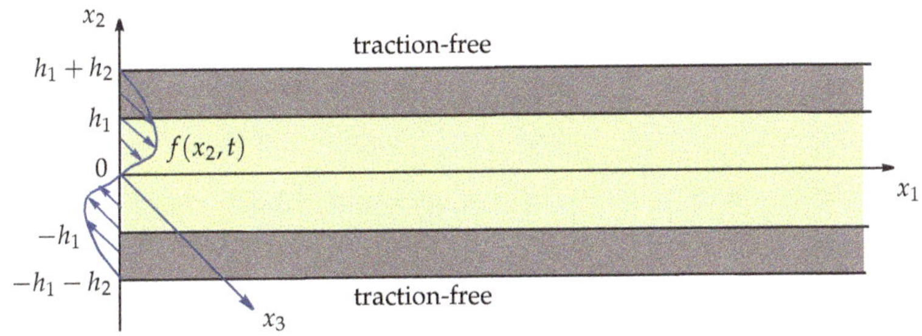

Figure 1. Schematic of a three-layered laminate.

The wave equations of motion are written conventionally as

$$\Delta u_3 - \frac{1}{c^2} \frac{\partial^2 u_3}{\partial t^2} = 0, \tag{1}$$

where $\Delta = \frac{\partial^2}{\partial x_1^2} + \frac{\partial^2}{\partial x_2^2}$ is the two-dimensional Laplacian in x_1 and x_2, and the transverse wave speed c is defined as

$$c = \begin{cases} \sqrt{\dfrac{\mu_1}{\rho_1}}, & |x_2| \leq h_1; \\[2ex] \sqrt{\dfrac{\mu_2}{\rho_2}}, & h_1 \leq |x_2| \leq h_1 + h_2. \end{cases} \tag{2}$$

Here, μ_i and ρ_i $(i = 1, 2)$ stand for the shear moduli and volume mass densities of the core and skin layers, respectively.

The traction-free boundary conditions at the faces $x_2 = \pm(h_1 + h_2)$ are assumed, i.e.,

$$\sigma_{32} = 0, \tag{3}$$

while the perfect contact conditions along the interfaces $x_2 = \pm h_1$ are formulated as

$$[u] = 0, \quad [\sigma_{32}] = 0, \tag{4}$$

where $[\,]$ indicate jumps of the associated quantities.

Here and below, the stresses σ_{3j} $(j = 1, 2)$ are given by

$$\sigma_{3j} = \begin{cases} \mu_1 \dfrac{\partial u_3}{\partial x_j}, & |x_2| \leq h_1; \\[2ex] \mu_2 \dfrac{\partial u_3}{\partial x_j}, & h_1 \leq |x_2| \leq h_1 + h_2. \end{cases} \tag{5}$$

In addition, let a shear edge loading be modelled by the boundary condition

$$\sigma_{31} = f(x_2, t) \tag{6}$$

at $x_1 = 0$, where $f(x_2, t)$ is a prescribed stress, being an odd function in x_2, for the sake of simplicity, in order to exclude the fundamental mode from the consideration. At infinity $x_1 \to \infty$, we impose decay or radiation conditions on the sought-for solution.

The formulated problem has a clear multiparametric nature, characterised by several dimensionless parameters, including the relative stiffness, density and thickness, given by

$$\mu = \frac{\mu_1}{\mu_2}, \quad \rho = \frac{\rho_1}{\rho_2}, \quad \text{and} \quad h = \frac{h_1}{h_2}. \tag{7}$$

Below all of them are assumed to be independent unless specified otherwise. The problem apparently provides the simplest step forward from 1D analysis in [18] to a 2D setup.

3. Dispersion Relation

First, we introduce the travelling wave ansatz for the displacement in the form

$$u_3 = A(x_2)e^{i(kx_1 - \omega t)}, \tag{8}$$

where k and ω conventionally denote the wave number and angular frequency, respectively, and $A(x_2)$ is a sought for odd function; hence, below, we may restrict ourselves to the interval $0 \leq x_2 \leq h_1 + h_2$.

Then, we infer from (1) that

$$A(x_2) = \begin{cases} A_1 \sin(\alpha_1 \eta), & 0 \leq \eta \leq \dfrac{h}{h+1}; \\ A_2 \sin(\alpha_2 \eta) + A_3 \cos(\alpha_2 \eta), & \dfrac{h}{h+1} \leq \eta \leq 1, \end{cases} \tag{9}$$

where

$$\eta = \frac{x_2}{h_1 + h_2} \tag{10}$$

is the dimensionless transverse variable; A_1, A_2 and A_3 are arbitrary real constants; and the parameters α_1 and α_2 are defined by

$$\alpha_1 = \sqrt{(1 + h^{-1})^2 \Omega_1^2 - K^2}, \quad \alpha_2 = \sqrt{(1 + h)^2 \Omega_2^2 - K^2}, \tag{11}$$

with

$$K = k(h_1 + h_2), \quad \text{and} \quad \Omega_j = \omega h_j \sqrt{\frac{\rho_j}{\mu_j}}, \qquad j = 1, 2, \tag{12}$$

standing for the dimensionless wave number and frequencies, respectively.

Next, inserting (8) and (9) into boundary and interfacial conditions (3) and (4), we arrive at the dispersion relation

$$\frac{\alpha_2}{\alpha_1} \tan\left(\frac{\alpha_1 h}{1+h}\right) \tan\left(\frac{\alpha_2}{1+h}\right) = \mu. \tag{13}$$

The eigenform amplitude $A(x_2)$ is then given by

$$A(x_2) = \begin{cases} C \cos\left(\dfrac{\alpha_2}{h+1}\right) \sec(\alpha_2) \csc\left(\dfrac{\alpha_1 h}{h+1}\right) \sin(\alpha_1 \eta), & 0 \leq \eta \leq \dfrac{h}{h+1}; \\ C[\tan(\alpha_2) \sin(\alpha_2 \eta) + \cos(\alpha_2 \eta)], & \dfrac{h}{h+1} \leq \eta \leq 1, \end{cases} \tag{14}$$

where C is an arbitrary constant.

In what follows, we consider the long-wave low-frequency behaviour of this dispersion relation, assuming that

$$|K| \ll 1 \quad \text{and} \quad \Omega_i \ll 1. \tag{15}$$

This setup has been earlier defined as a global low-frequency one (see [18]), since it is associated with polynomial variation of the sought-for displacement field across the thickness within each of the layers. The absolute value of K in (15) indicates that both propagating and evanescent modes are studied.

4. Multiparametric Analysis

Below, the dispersion relation (13) is subject to asymptotic analysis in three independent parameters (7) over the long-wave low-frequency domain (15). As mentioned above, the focus is on the relation between the two most importance consequences of high contrast, namely the degeneration of boundary layers and the shift of cut-off frequencies to the low-frequency domain. It is worth noting that the latter are related to specific extra scales, including a large length scale and slow time scale, respectively. For these scales, a typical decaying rate is much greater than the laminate thickness, and a characteristic time considerably exceeds that elastic waves take to propagate the distance between laminate faces.

Consider first the static limit at $\Omega_i = 0$. In this case, the transcendental equation has small imaginary roots

$$K = \pm i K_*,$$

where, at leading order,

$$K_* \approx \sqrt{\frac{\mu}{h}}\,(h+1) \ll 1, \tag{16}$$

provided that

$$\mu \ll \frac{h}{(h+1)^2}. \tag{17}$$

In contrast to homogeneous structures, the root K_* corresponds to the so-called degenerated boundary layer, which does not decay at the distances of order thickness away from the edge. The asymptotic behaviour (16) was earlier established in [12]. However, the two-parametric origin of the problem was not taken into consideration; namely, it was supposed that $h \sim 1$. As a result, the restriction (17) has not appeared in the cited paper.

Next, turn to the asymptotic analysis of the lowest cut-off frequencies, coming from the dispersion relation (13) at $K = 0$. In this case, the aforementioned global low-frequency regime ($\Omega_j \ll 1, j = 1,2$) takes place over the following parameter range:

$$\mu \ll h \ll \frac{1}{\rho}, \tag{18}$$

see [16,18] for more detail. Then, for the sought-for cut-off frequency

$$\Omega_2 = \Omega_* \approx \sqrt{\frac{\mu}{h}} \tag{19}$$

or

$$\Omega_1 = \sqrt{\frac{\rho}{\mu}}\,h\,\Omega_* \approx \sqrt{\rho h}. \tag{20}$$

It might be easily verified that a non-contrast setup, for which $\Omega_* \sim 1$, does not support low-frequency cut-offs.

Let us now investigate the inequalities (17) and (18) simultaneously. First of all, it is clear that for $h \lesssim 1$, i.e., excluding the laminates with thin skin, the condition (18) guarantees not only a low cut-off, but also a slowly decaying static boundary layer. More generally, both phenomena occur when

$$\mu \ll \frac{h}{(h+1)^2} \quad \text{and} \quad h \ll \frac{1}{\rho}. \tag{21}$$

In particular, for $h \gg 1$, i.e., for conventional sandwiches with thin skin, the low-frequency vibration mode appears along with the strongly localised static boundary layer (typical of homogeneous or weakly inhomogeneous structures), provided that

$$\frac{1}{h} \ll \mu \ll h \ll \frac{1}{\rho}. \tag{22}$$

Next, by taking leading-order Taylor expansions, we derive a shortened form of the dispersion relation (13). It is given by

$$\alpha_2^2 = \mu \frac{(1+h)^2}{h}, \tag{23}$$

or, equivalently,

$$\Omega_2^2 - \frac{K^2}{(1+h)^2} = \Omega_*^2. \tag{24}$$

As might be expected, the last equation implies the values of $K = K_*$ at $\Omega_2 = 0$, and also $\Omega_2 = \Omega_*$ at $K = 0$, which is in agreement with the estimates (16) and (19), respectively. Moreover, it is obvious that the approximation (24) of the full dispersion relation is uniformly valid over the domain

$$|K| \leq K_*, \quad 0 \leq \Omega_2 \leq \Omega_*, \tag{25}$$

provided that the conditions (17) and (18) are satisfied. The shortened dispersion relation (24) also shows that the quasi-static limit occurs at $\Omega_2 \ll \Omega_*$.

At the same time, above the cut-off frequency, the validity range of the approximation (24) is given by

$$K \ll \min\left\{1, \frac{1+h}{h}\sqrt{\frac{\mu}{\rho}}\right\} \tag{26}$$

and

$$\Omega_2^2 - \Omega_*^2 \ll \min\left\{\frac{1}{(1+h)^2}, \frac{\Omega_*^2}{\rho h}\right\}. \tag{27}$$

These strong inequalities generalise earlier considerations for one-parametric setups in [16,19].

Let us now present several graphical illustrations highlighting the observations above. First, we present a sketch characterising the domains corresponding to conditions (17) and (18) in parametric space. For the sake of clarity, we present a 2D graph in μ and h only, but bearing in the mind the restriction $h \ll \frac{1}{\rho}$, which is assumed to be valid. The highlighted domains in Figure 2 correspond to various ranges of the dimensionless parameters μ and h, defined by the conditions (17) and (18), for which the latter are either satisfied simultaneously (yellow), or the relation (18) holds but (17) is violated (green), or when both (17) and (18) fail (blue).

Now, let us illustrate the case of a small cut-off frequency along with a rapidly decaying boundary layer, i.e., when the the strong inequalities (18) are satisfied but condition (17) is violated. It is clearly seen from Figure 3, displaying the approximate curve (24) covering both domains of the real and imaginary wavenumber K. The approximate value in the static limit at $\Omega_2 = 0$ and $K \approx 3.5i$ on the approximate curve (24) is far off from the numerical solution of the exact dispersion relation (13) at $\Omega_2 = 0$, $K_* \approx 1.6i$. At the same time, the approximate curve is clearly passing very close to the exact point of the cut-off frequency $K = 0, \Omega_2 = \Omega_*$.

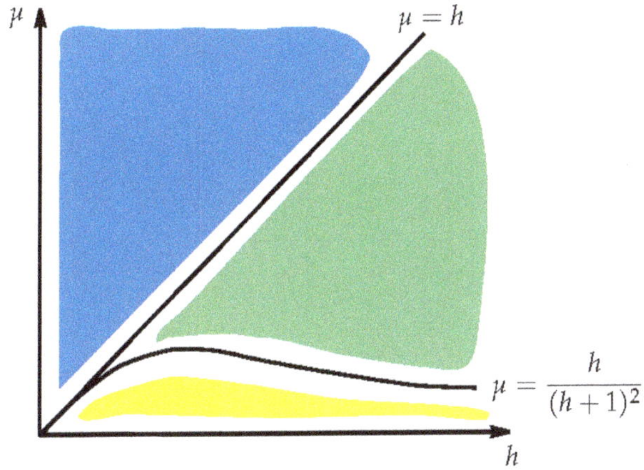

Figure 2. Validity of the strong inequalities (17) and (18) in μ-h plane.

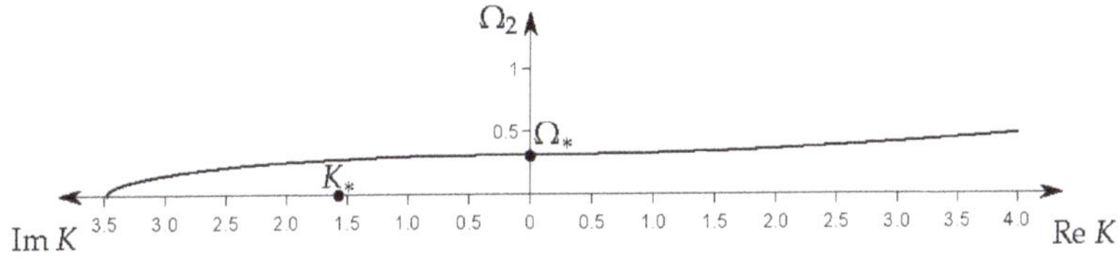

Figure 3. Approximation (24), the exact static limit K_*, and the lowest cut-off frequency Ω_* for the dimensionless parameters (7) chosen as $\mu = 1, h = 10, \rho = 0.01$.

An opposite scenario is presented in Figure 4, in which the parameters $\mu = 0.1, h = 1$ and $\rho = 10$ are chosen such that the condition (17) is satisfied, and the inequalities (18) do not hold. Consequently, it is observed that the approximate curve (24) passes close to the static limit K_*, but misses considerably on the cut-off frequency Ω_*.

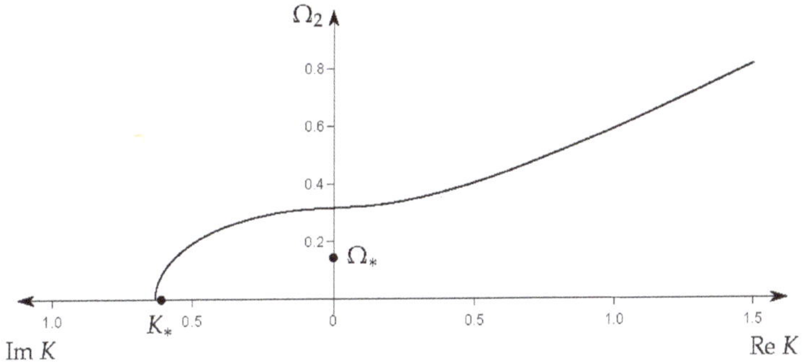

Figure 4. Approximation (24) and the exact static limit K_* and lowest cut-off frequency Ω_* for the dimensionless parameters (7) taken as $\mu = 0.1, h = 1, \rho = 10$.

Finally, we are displaying the setup when both conditions (17) and (18) are satisfied; see Figure 5. In this case, the approximate curve (24) is shown by a dashed line and is complemented by the exact dispersion (13), depicted by solid curve. It is also remarkable that approximation (24) provides a uniform approximation both for decaying and propagating modes.

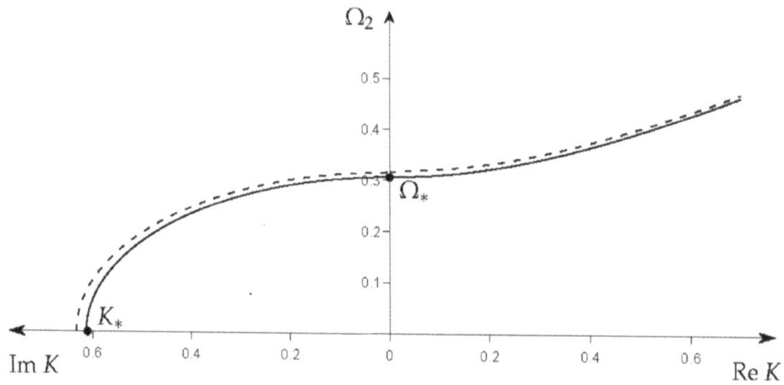

Figure 5. Dispersion curve (13) (solid line) and its approximation (24) (dashed line) for the dimensionless parameters (7) specified as $\mu = 0.1, h = 1$, and $\rho = 0.1$.

5. Forced Vibration

Consider now time-harmonic forced vibration of a semi-infinite laminate, starting from the formulation in Section 2. For the sake of simplicity, restrict ourselves to a one-parametric scenario in which $\mu \ll 1,\ h \sim 1$ and $\rho \sim \mu$. In this case, as follows from (14), the leading-order behaviour of the eigenform becomes

$$A(x_2) = \begin{cases} C\dfrac{h+1}{h}\eta, & 0 \le \eta \le \dfrac{h}{h+1}; \\ C, & \dfrac{h}{h+1} \le \eta \le 1. \end{cases} \tag{28}$$

To determine the constant C, we adapt the decay conditions in [19,20], generalising the classical Saint-Venant's principle for high-contrast laminates. For a three-layered symmetric structure excited by a prescribed (self-equilibrated) shear edge stress $P(x_2)$, an appropriate asymptotic condition becomes

$$\int_0^{h_1} x_2 P(x_2) dx_2 + h_1 \int_{h_1}^{h_1+h_2} P(x_2) dx_2 = 0. \tag{29}$$

This condition ensures the static equilibrium of the soft layer with respect to stress couples applied to both its edge and interfaces; in doing so, the first and second terms in (29) correspond to the contributions of edge $x_1 = 0$ and interfaces $x_2 = \pm h_1,\ x_1 > 0$, respectively. In fact, this condition is also applicable at leading-order to low-frequency vibration, see [22]. Below, the factor $e^{-i\omega t}$ is assumed to be omitted. Note that for the frequencies above the cut-off Ω_*, this condition becomes a radiation one, prohibiting long-wave propagating modes; see [21].

Let us now implement the substitution

$$P(x_2) = f(x_2) - \sigma_{31}(0, x_2), \tag{30}$$

in (29), where σ_{31} is calculated from (5) starting from the leading-order eigenform (28). This guarantees rapid decay (at the scale of the thickness) of the boundary layer corresponding to the discrepancy between the given edge load $f(x_2)$ and the stress $\sigma_{31}(0, x_2)$ associated with the low-frequency mode of interest. As a result, we arrive at

$$C \approx \frac{1}{ik\mu_2 h_1 h_2}\left[\int_0^{h_1} x_2 f(x_2)dx_2 + h_1 \int_{h_1}^{h_1+h_2} f(x_2)dx_2\right]. \tag{31}$$

The related wave number in the exponential term in (8) is approximated from (24) by

$$k \approx i\frac{1}{h_2}\sqrt{r}, \tag{32}$$

below the cut-off, and

$$k \approx \frac{1}{h_2}\sqrt{r}, \tag{33}$$

above the cut-off, where

$$r = \frac{\mu_1 h_2 - \rho_2\omega^2 h_1 h_2^2}{\mu_2 h_1}. \tag{34}$$

The last expression satisfies the Sommerfeld radiation principle, also giving room for the simple implementation of the alternative limiting absorption principle; see [23,24] and references therein. To this end, we assume in (34)

$$\mu_j = \mu_{j0}(1 - i\gamma_j), \quad j = 1, 2, \tag{35}$$

where $\gamma_j \ll 1$. As a result,

$$r = \frac{h_2\big(\mu_{10}(1 - i\gamma_1) - \rho_2\omega^2 h_1 h_2\big)(1 + i\gamma_2)}{h_1\mu_{20}(1 + \gamma_2^2)}. \tag{36}$$

It may be seen from the latter that when $\Omega_2 = \Omega_*$,

$$r = -\frac{i\gamma_1 h_2\mu_{10}(1 + i\gamma_2)}{h_1\mu_{20}(1 + \gamma_2^2)}. \tag{37}$$

hence, on the cut-off frequency only the absorption over the soft inner layer appears in the expression for $\mathrm{Im}(r)$. Clearly, for low-frequency propagating modes (with frequencies above the cut-off) the absorption in both soft and stiff layers needs to be taken into consideration. At the same time, as in most of the cases, the results coming from Sommerfeld and limiting absorption principles agree with each other.

6. Conclusions

The developed multiparametric framework for high-contrast elastic laminates demonstrates that degenerated boundary layers do not always coexist with specific low cut-off frequencies. Explicit conditions in the form of strong inequalities on the contrast problem parameters along with numerical illustrations are presented in Section 4.

For a semi-infinite strip, the amplitude of the mode varying slowly both in space and time (decaying or propagating) is determined in Section 5 using a generalisation of Saint-Venant's principle [19,20] oriented to strongly inhomogeneous laminates. For low-frequency propagating modes, it is shown that the implementation of the limiting absorption principle assumes taking into consideration small dissipation in both the core and skin layers, whereas on the cut-off frequency, only the absorption in the soft inner layer plays a role.

The basic example considered in the paper appears to be highly informative for elucidating a number of important multiscale phenomena characteristic of high-contrast laminates. In particular, it elucidates the general idea of the near-cut-off long-wave behaviour typical for elastic waveguides with strong transverse inhomogeneity, characterised by the transition from slowly decaying boundary layers to low-frequency propagating modes. The established framework allows for various extensions, including asymmetric multi-layered thinwalled structures of more general geometry; vector plane and 3D problems; and analysis of various boundary conditions along faces and interfaces, e.g., arising at the mathematical modelling of elastic coatings.

Author Contributions: Conceptualization, L.A.A. and J.K.; methodology, J.K. and D.P.; software, L.G.G.; validation, D.P. and L.G.G.; formal analysis, D.P.; writing—original draft preparation, J.K. and D.P.; writing—review and editing, L.A.A. and J.K.; visualization, L.G.G.; funding acquisition, L.A.A. and L.G.G. All authors have read and agreed to the published version of the manuscript.

Funding: This research was funded by the Science Committee of Republic of Armenia, within the framework of the research project № 21T-2C075.

Data Availability Statement: Data sharing not applicable.

Conflicts of Interest: The authors declare no conflict of interest.

References

1. Aghalovyan, L.A. *Asymptotic Theory of Anisotropic Plates and Shells*; World Scientific: Singapore, 2015.
2. Mikhasev, G.I.; Altenbach, H. *Thin-Walled Laminated Structures*; Springer: New York, NY, USA, 2019.
3. Reddy, J.N. *Mechanics of Laminated Composite Plates and Shells: Theory and Analysis*; CRC Press: Boca Raton, FL, USA, 2003.
4. Liew, K.; Pan, Z.; Zhang, L. An overview of layerwise theories for composite laminates and structures: Development, numerical implementation and application. *Compos. Struct.* **2019**, *216*, 240–259. [CrossRef]
5. Boutin, C.; Viverge, K. Generalized plate model for highly contrasted laminates. *Eur. J. Mech. A/Solids* **2016**, *55*, 149–166. [CrossRef]
6. Boutin, C.; Viverge, K.; Hans, S. Dynamics of contrasted stratified elastic and viscoelastic plates-application to laminated glass. *Compos. Part B Eng.* **2021**, *212*, 108551. [CrossRef]
7. Altenbach, H.; Eremeyev, V.A.; Naumenko, K. On the use of the first order shear deformation plate theory for the analysis of three-layer plates with thin soft core layer. *ZAMM-J. Appl. Math. Mech./Z. Für Angew. Math. Mech.* **2015**, *95*, 1004–1011. [CrossRef]
8. Aßmus, M.; Nordmann, J.; Naumenko, K.; Altenbach, H. A homogeneous substitute material for the core layer of photovoltaic composite structures. *Compos. Part B Eng.* **2017**, *112*, 353–372. [CrossRef]
9. Berdichevsky, V.L. An asymptotic theory of sandwich plates. *Int. J. Eng. Sci.* **2010**, *48*, 383–404. [CrossRef]
10. Boutin C. High-contrast multi-layered plates. Statics, dynamics and buckling. In *Mechanics of High-Contrast Elastic Solids: Contributions from Euromech Colloquium 626*; Altenbach, H., Prikazchikov, D., Nobili, A., Eds.; Springer International Publishing: Cham, Switzerland, 2023; pp. 39–63.
11. Kuznetsov, S.V. Detection of a hidden sandy layer in a stratified substrate by dispersion analysis. In *Mechanics of High-Contrast Elastic Solids: Contributions from Euromech Colloquium 626*; Altenbach, H., Prikazchikov, D., Nobili, A., Eds.; Springer International Publishing: Cham, Switzerland, 2023; pp. 119–132.
12. Baxter, S.C.; Horgan, C.O. End effects for anti-plane shear deformations of sandwich structures. *J. Elast.* **1995**, *40*, 123–164. [CrossRef]
13. Wijeyewickrema, A.C.; Horgan, C.O.; Dundurs, J. Further analysis of end effects for plane deformations of sandwich strips. *Int. J. Solids Struct.* **1996**, *33*, 4327–4336. [CrossRef]
14. Horgan, C. Saint-Venant end effects for sandwich structures. In Proceedings of the Fourth International Conference on Sandwich Construction, Stockholm, Sweden, 9–11 June 1998; Volume 1, pp. 191–200.
15. Avila-Pozos, O.; Movchan, A.B. Slow decay of end effects in layered structures with an imperfect interface. *J. Eng. Math.* **2003**, *45*, 155–168. [CrossRef]
16. Kaplunov, J.; Prikazchikov, D.A.; Prikazchikova, L.A. Dispersion of elastic waves in a strongly inhomogeneous three-layered plate. *Int. J. Solids Struct.* **2017**, *113*, 169–179. [CrossRef]
17. Ryazantseva, M.Y.; Antonov, F.K. Harmonic running waves in sandwich plates. *Int. J. Eng. Sci.* **2012**, *59*, 184–192. [CrossRef]
18. Kaplunov, J.; Prikazchikov, D.A.; Sergushova, O. Multi-parametric analysis of the lowest natural frequencies of strongly inhomogeneous elastic rods. *J. Sound Vib.* **2016**, *366*, 264–276. [CrossRef]
19. Kaplunov, J.; Prikazchikova, L.; Alkinidri, M. Antiplane shear of an asymmetric sandwich plate. *Contin. Mech. Thermodyn.* **2021**, *33*, 1247–1262. [CrossRef]
20. Prikazchikova, L. Decay conditions for antiplane shear of a high-contrast multi-layered semi-infinite elastic strip. *Symmetry* **2022** *14*, 1697. [CrossRef]
21. Babenkova, E.; Kaplunov, J. Radiation conditions for a semi-infinite elastic strip. *Proc. R. Soc. A Math. Phys. Eng. Sci.* **2005**, *461*, 1163–1179. [CrossRef]

22. Babenkova, E.; Kaplunov, J. Low–frequency decay conditions for a semi–infinite elastic strip. *Proc. R. Soc. A Math. Phys. Eng. Sci.* **2004**, *460*, 2153–2169. [CrossRef]
23. Nazarov, S.A. The Mandelstam energy radiation conditions and the Umov–Poynting vector in elastic waveguides. *J. Math. Sci.* **2013**, *195*, 676–729. [CrossRef]
24. Rodnianski, I.; Tao, T. Effective limiting absorption principles, and applications. *Commun. Math. Phys.* **2015**, *333*, 1–95. [CrossRef]

MDPI AG
Grosspeteranlage 5
4052 Basel
Switzerland
Tel.: +41 61 683 77 34

Mathematics Editorial Office
E-mail: mathematics@mdpi.com
www.mdpi.com/journal/mathematics

www.ingramcontent.com/pod-product-compliance
Lightning Source LLC
Chambersburg PA
CBHW040822250926
51850CB00007BA/1239
9783725872800